HOST-PLANTS OF CHRYSOMELIDAE OF THE WORLD

An Essay about the Relationships between the Leaf-beetles and their Food-plants

HOST-PLANTS OF CHRYSOMELIDAE OF THE WORLD

An Essay about the Relationships between the Leaf-beetles and their Food-plants

BY

PIERRE JOLIVET
AND TREVOR J. HAWKESWOOD

BACKHUYS PUBLISHERS
LEIDEN
1995

Front cover: *Leptinotarsa decemlineata* Say, the most famous chrysomelid beetle, which feeds on certain species of wild and cultivated *Solanum* (Solanaceae).

CIP-data Koninklijke Bibliotheek, Den Haag

Jolivet, Pierre

Host-plants of Chrysomelidae of the world: an essay about the relationships between the leaf-beetles and their food-plants / by Pierre Jolivet and Trevor J. Hawkeswood.
- Leiden : Backhuys Publishers. - Ill.
With index, ref.
ISBN 90-73348-42-0 bound
NUGI 821
Subject headings: chrysomelid beetles / host-plants.

Printed in The Netherlands.

Dedicated to the entomologists of the past and to those of the present who have patiently accumulated these precious data on the host-plants.

The authors

CONTENTS

PREFACE

This is the first essay of its kind which covers a major beetle family on a world-wide basis.

Of course, it has not been possible for the authors to quote all species from a family of which about 37.000 species have been described, and probably some 10.000 or more which are as yet undescribed.

The authors have listed the genera and their food preferences and, where necessary, have given descriptive tables for important genera. The bibliography covers most of the main references, since a complete bibliography would have been beyond the scope of the book. A detailed bibliography is available in the publications of the senior author.

Where possible, the biology of the genera and species has heen summarized, specifically for the leaf-miners, the root-feeders, the myrmecophilous species and the borers. The larval biology of three subfamilies, viz. the Aulacoscelinae, the Megascelinae and the Orsodacninae remains completely unknown, although it is strongly suspected that the larvae of the first group could be borers and those of the second and third subfamilies could be root-feeders. The classical division into 20 subfamilies has been adopted here because all other proposals, cladograms and family-trees are questionable.

The extensise observations by the senior author on the Chrysomelid fauna of the Tropics has been merged with the vast experiences of the Australian fauna of the junior author. Although the trophic relationships of the Chrysomelidae are globally known for 30% of the species, there are still important gaps in our knowledge of several parts of the world such as Madagasacar, tropical Africa and certain areas of South America. However, the food-selection of some species can be statistically deduced from examining the known spectrum of hosts.

This work is of major importance as it provides data on many genera which are on the verge of extinction at the eve of the 21st century . Despite extensive surveys in all parts of our planet, we are still practically ignorant of the food-plants of the Chrysomelinae of e.g. Tierra del Fuego, northern Australia and Africa. Even if it were only available in theory, such information would provide precious data for agricultural entomologists involved in biological control of weeds introduced to their various countries.

Professor Claude Caussanel
Director, Laboratory of Entomology,
Muséum National d'Histoire Naturelle, Paris

FOREWORD

There are probably in excess of 40,000 species of Chrysomelidae still extant in the world. Of these, probably no less than 37,000 have been described to date, although a few from the Holarctic region remain undescribed. In addition, sibling species probably still remain undescribed among the Palaearctic Alticinae and we cannot believe that all of the Chinese fauna, mostly in the northern mountains, the Sinkiang, Mongolia and Tibet has been totally circumscribed. Also, new species are discovered from time to time from Japan, Taiwan and Siberia. However, we can reasonably suppose that all of the Donaciinae, most of the Palaearctic Eumolpinae, Synetinae, Chrysomelinae and Galerucinae have been described. However, published information on their biology is another matter, but considerable efforts have heen made in the study of the larvae, distribution, host-plants, etc. Unfortunately, the life-history of the Asiatic Megalopodinae is still unknown with the exception of *Temnaspis* in China.

It is mainly in the New World tropics where new species are commonly found and described and many of these (i.e. Alticinae, Eumolpinae, Galerucinae, Megascelinae) live in the canopy of mesophilic and hydrophilic forests in lowlands and middle latitudes, and as such, are not accessible to the ordinary entomologist. In any case, a large number of species remain undescribed in both Asia and Africa, mostly from the canopy of what remains of the primary forests. The recent English expedition to Sulawesi has revealed the great diversity of these Indonesian forests and the abundance of undescribed Chrysomelidae and Curculionidae.

Spraying pyrethroids into the canopy provides a sampling method which reaches 95% of the existing fauna but this does not indicate the inter-relationships between the plants and insects. To fly a balloon with an attached raft over the top of the trees seems an interesting collecting method but is most likely to greatly disturb the normal activity of the insects. A large number of specialists are needed to study those millions of new insects both in vivo and in the laboratory. However, there are not enough taxonomists in the world and the contemporary trend of universities is not to train people in basic classification. Academics are most often preoccupied in statistical and cladistic studies of families and genera at the expense of specific taxonomic and ecological studies which are "kept for tomorrow". There will be plenty of research work remaining for the future generations, who, once the forests are logged and destroyed, will have to work mainly with the fossils and subfossils (i.e. dead insects) remaining in the museums, since only vestiges of the tropical forests will remain and most beetle species will be extinct. Some people say that the destruction of the large rain forests is inescapable, but nothing is really definite and responsibility for the protection of these forests rests with the various Governments who have the power of reversing the present destructive trend.

However, no matter what happens in the future, it appears wise to collect now as rapidly and extensively as possible, and perhaps the suggestion by a German entomologist of reuniting all the collections into an International Museum, for the disposal of future generations, is not such a bad one after all. Such a project should be sponsored by UNESCO (United Nations Educational Scientific Organization) or any similar organization such as the Environmental Protection Unit in Nairobi. The project should be financed by private funding since the simultaneous exploration of the forest canopies of America, Asia, Africa, Madagascar, Indonesia and Australia would require important human and financial resources. The first project of this type was formerly conceived by Julian Huxley in 1950, for the exploration of the Amazonian forest, but has lamentably failed because Brazil refused then to allow naturalists access to the forests, then relatively pristine. And what remains now? And what will still be left by the year 2000? Probably nothing or very little. Then what shall mankind do? Should we collect all that remains at the moment and restock it elsewhere, or save what remains of the forests in order to keep alive the invaluable biological material which is still extant? Will mankind be wise enough to save the forests before they are transformed into cattle ranches providing the meat for the hamburgers of the fast food restaurants? We doubt that man will make the wisest choice. Some philosophers may beg to disagree, but the most probable result is that everything will be destroyed and that the non-collected fauna will disappear before being studied and described.

The following work mainly results from the personal observations of the senior author in the Old and New World tropics, but also from the deciphering of approximately 20,000 index cards, containing all of the reliable literature. Any doubtful data have been omitted. The observations and records of the authors and those of many other entomologists, both early and recent, in the Holarctic and tropical zones have been assembled here. Very often the botanical relationships between families and genera of plants are confirmed by the selection of host-plants by closely related genera and species of leaf beetles.

We apologize for the fact that authority names are not always quoted after the plant names but this would have enlarged the book somewhat and the bibliography also is incomplete.

Reference is hereby made to the separate publications on the different subfamilies which the senior author has published between 1971 and 1992 and which contain all of the previous papers dealing with host-plants, biology etc.

In the last part of this book, not all plant families have been treated, but only examples from the most common host-plant families have been provided. A complete list of the families with their hosts would have been beyond the scope of the present book.

Pierre Jolivet, Paris, France
Trevor J. Hawkeswood, Brisbane, Queensland, Australia
September 1994

ACKNOWLEDGEMENTS

P. Jolivet: I would like to thank my colleague Niilo Virkki from Rio Piedras, Puerto Rico for providing a complete documentation of the biology of the Alticinae of his island. My special thanks are to him and to my colleagues from the Entomological Congresses with whom I have had so many valuable discussions. I am also indebted to Serge Doguet who accepted to critically read the section on Alticinae and to Eduard Petitpierre with whom I have published many papers on leaf-beetle host-plants. My thanks are also expressed to Mr. Morelet (INRA, Nancy, France) who provided the unpublished observations of *Chalcoides* feeding on *Venturia* on an aspen leaf. In the following bibliography, it has not been possible to list all of the thousands of papers and other publications dealing with the Chrysomelidae and their biology and host-plants. However, any reference not listed can be found by consulting my previous papers dealing with trophic selection of Chrysomelidae. I apologize to my colleagues if their papers have not been listed but a more complete list would have been an enormous bibliography. I would like to also thank Prof. Ross Arnett and Dr Wim Backhuys for editorial assistance. Finally I am much indebted to Ms Pamela Gilbert, librarian at the Natural History Museum London, who provided certain literature that I lacked (mainly of Indian and Chinese origin) and which were unobtainable in Paris; I thank her for her patience.

T.J. Hawkeswood: I would like to thank also the following people for correspondence over the years and for sending me copies of their chrysomelid papers and/or other literature: M. Peterson, I. Faithfull, P.I. Forster, R.W. Wells (Australia), Z.J. Buzzi (Brazil), L. LeSage (Canada), W.G. Eberhard (Costa Rica), K. Vig (Hungary), M. Schmitt (Germany), M.S. Mohamedsaid (Malaysia), L. Borowiec (Poland), H. Takizawa (Japan), D. Sassi (Italy), R.A. Crowson, B.O.C. Gardiner (United Kingdom), I.S. Askevold, D.M. Windsor, D.G. Furth, P. Bernhardt, G.A. Samuelson, C.E. Williams, F.G. Messina, T.N. Seeno, C.L. Staines, K.L. Olmstead, B.D. Farrell (United States of America) and E. Petitpierre (Spain).

1. GENERAL

1. Classification

There are several ways of dividing the Chrysomelidae into subfamilies. Since Chapuis (1874) with his system of 4 sections and 15 tribes, and Jacoby (1908) who adopted 5 sections and 16 tribes, there have been numerous classifications, e.g. Chen (1934, 1985), Gressitt (1942), Chûjô (1953), Monros (1959) and others. These are reviewed by Seeno and Wilcox (1982).

The Jacoby system was modified by Jolivet (1959), who added the Zeugophorinae, the Aulacoscelinae and the Synetinae, and removed the Megascelinae from the *Eupoda* and placed it in the section Cyclica together with the Lamprosomatinae and the Eumolpinae. This results in the following system with 19 subfamilies of very unequal composition (Table 1). The new subfamily Palophaginae has been included in the *Eupoda* near the Orsodacninae.

An adequate key to the known larvae (arranged in subfamilies) has been provided by Boving and Craighead (1953). Their key is more comprehensive than the earlier one by Chu (1949) and although both contain small mistakes, they are still useful for modern workers. Numerous keys to the adults have been published by various authors, including the ones quoted above. Our classification here is based mainly on male genitalia and wing venation and, until new data indicate otherwise, the Megascelinae have been placed together with the Eumolpinae, although their larvae, like the Aulacoscelinae, remain unknown. The larva of Orsodacninae, discovered by Van Emden during the 1940's and described later by Cox (1981) and Crowson (1981), appears to be a root feeder, although this is not absolutely certain since the larva was obtained from the egg only and was not observed feeding. Reviews of larval biology of the various subfamilies and genera are provided in later sections of this book.

The probable evolutionary relationships between the subfamilies are shown in the following table (Table 2), which has only been slightly modified since 1959. While there is widespread agreement about the cerambycidian origin of the Chrysomelidae and the Bruchidae, there are differing views concerning the phylogeny of the various chrysomelid subfamilies. There are at least five different lines which can be isolated but only three groups are really homogeneous, viz. the Cryptostoma (Hispinae and Cassidinae) (the most differentiated) the Trichostoma (Chrysomelinae, Galerucinae and Alticinae) anc the Camptosoma (Clytrinae, Cryptocephalinae and Chlamisinae).

Several subfamilies contain a large number of genera and species, e.g. the Alticinae which contain 514 genera and more than 7,000 described species. Others contain only a few genera and species, e.g. the Megascelinae which contain 2 genera and 120 species and the Synetinae which have 2 genera and 8 species.

2. Biology of larvae

All Chrysomelidae, both adults and larvae, feed on plants and the only exceptions (i.e. larval carnivory among the Clytrinae and the Chrysomelinae where ant eggs or their own eggs are devoured, and coprophagy among Lamprosomatinae adults)[1] are extremely rare, occasional and, with the exception of Clytrinae larvae, are not part of the normal life-cycle of the insects. Besides, it is probable that Clytrinae larvae are polyphagous in the ant nests where they consume excreta as well as ant eggs or the vegetal debris of the nest. The adults are polyphagous on leaves of trees or shrubs growing around the ant nests.

It is certainly in the grassy areas, on the young trees or shrubs, in the secondary forest clearings or forest regrowth that the Chrysomelidae are the most common or the most accessible to the collector. The Chrysomelidae of the forest canopy in the tropics are only starting to be collected and their relationships with their host-plants are known mostly through statistical studies. For most of these species, the real host-plant is only suspected.

The biology of the larvae varies considerably, ranging from those living below the leaves to the miners which dig between the palisade tissue, to those which hide between appressed leaves of monocots (Hispinae), to the stem-borers such as the Megalopodinae, to the gallicolous Sagrinae, to the aquatic Donaciinae, to the root-feeders and subterranean larvae (Eumolpinae, Synetinae, several Galerucinae and Alticinae) and to the myrmecophilous Clytrinae. Other larvae live openly above the leaves protected by their toxicity or by homochromy (crypsis).

The Camptosoma and the Lamprosomatinae are case-bearers and live inside a case composed of excreta (Clytrinae) or of fine particles of wood and other plant material glued together by excreta and resin (Lamprosomatinae). The abdomen of these larvae is inflated posteriorly and adapted to carry a case. Only Clytrinae larvae live normally inside or in the vicinity of ant nests. This behaviour also occurs among several Cryptocephalinae and a number of *Syagrus* (Eumolpinae) but this is very exceptional.

The Criocerinae have large larvae which usually feed externally on the host leaves. Many of them cover themselves with a viscous secretion and excreta, which are effective protections against their predators, mostly ants.

Very few species among the Criocerinae are case-bearers (e.g. *Plectonycha)* or gallicolous (e.g. *Ortholema*), or even miners in the larval stage (e.g. *Lema quadrivitta* Boheman). Nymphosis occurs within a cocoon, either glued to the host-plant, or buried in the ground.

Some larvae (i.e. Donaciinae) are aquatic and suck the juices of plants from roots or stems immersed in water. The abdomen of these larvae is terminated

---

[1] Recently Lazell et al. (1991) recorded the galerucine *Aplosonyx nigripennis* (Jacoby) as having been collected with ticks (Acarina) from an old wound site of the snake *Coluber dipsas* in Sulawesi. Normally *Aplosonyx* feeds on *Alocasia* (Araceae) and on *Cissus* (Vitaceae) depending on the species. *Aristobrotica angulicollis* (Erichson) in Brazil feeds on adult meloids (Mafra-Neto & Jolivet, 1994).

by a pair of spiny processes adapted to pump oxygen from the plant. One species, *Haemonia mutica* (Fabricius), lives in the Baltic Sea, a weakly saline environment, feeding on *Zostera*.

All Eumolpinae and Synetinae larvae, and several Alticinae and Galerucinae larvae, feed within the soil on the roots of various plants. These larvae are polyphagous and often feed on grass roots (Poaceae), either internally or externally, depending on the species. The larvae of Megascelinae and Aulacoscelinae are still undescribed and unknown, but it is possible that the larvae of Megascelinae are subterranean and feed on the rootlets of plants in a similar manner to that of the Eumolpinae. Many Alticinae larvae, several Galerucinae, Hispinae and Zeugophorinae larvae are leaf-miners. However, among Galerucinae and Alticinae, many species are external leaf feeders, while others feed on the plant roots. Among the Hispinae and several rare Cassidinae, the larvae are external feeders between the appressed leaves of monocots. Some Galerucinae larvae (e.g. *Exosoma*) feed inside bulbs of Liliaceae and Amaryllidaceae (Monocotyledonae) and are practically aquatic. Some larvae of *Longitarsus* (Alticinae) feed inside the berries of *Piper* (Piperaceae), but very few chrysomelid larvae live inside fruits. No chrysomelids feed inside dry seeds as do the Bruchidae. Most of the Chrysomelinae larvae and adults are external feeders of plant leaves.

The biology of several chrysomelid subfamilies is extremely complex, e.g. the Hispinae, where many larvae are miners, others live between the appressed leaves, others burrow into the stems, while others are external feeders. Such complexity is also paralleled among the Alticinae and the Galerucinae where several species eat flowers, buds, petioles or stems of plants. Root-feeders among the Alticinae larvae can consume only the cortex or the central area of the roots or rootlets. Nymphosis occurs on the leaves, or within the soil, the pupa being naked inside a cocoon, depending on the subfamily or the genus. Many American Cassidinae (Stolaini) metamorphosize in groups on a stem and remain attached by the end of the body. The mother remains with its offspring from the eggs to the pupae.

3. Biology of adults

The consumption of flowers, including stamens and ovaries is a rather rare phenomenon among the Chrysomelidae but it frequently occurs among the adults of *Diabrotica*, *Aulacophora* and other Galerucinae, among the Aulacoscelinae, Megascelinae, and among several Eumolpinae, such as *Cleptor inermis* Lefèvre in Australia (Hawkeswood & Jolivet, 1988). Pollen feeding is much more common and widespread among Criocerinae, Cryptocephalinae, Eumolpinae, Alticinae and several Chrysomelinae, such as *Hydrothassa* in Europe on *Ranunculus* flowers (Ranunculaceae). Eating pollen means absorption of additional complementary proteins to the diet. However, pollen consumption does not always correlate with pollination by chrysomelids, but pollination of the host plant sometimes does occur. For instance, Hawkeswood (1983) has quoted the small *Pyrgoides dryops* (Blackburn), a Chrysomelinae, as being an efficient pollinator of *Acacia leiocalyx* (Mimosaceae) in the Brisbane area, Queensland. Cycadales inflorescences and the beetle flowers of Magnoliaceae and many others attract leaf-beetles, either by the nectar, by the floral odors or by the pollen. It is known that floral nectar (but mainly extra-floral nectar) is rich in amino-acids, and several Chrysomelidae are known to drink these fluids.

For the Alticinae, Criocerinae and Cryptocephalinae, consumption of pollen represents a survival strategy and a food supplement in a similar way to that of certain butterflies and moths. Samuelson (1989) has recently summarized pollen feeding among the Alticinae.

Arnett (1969) first observed that pollen appeared to germinate in the gut of the Oedemeridae. Recently, Haslett (1983), observing what happens among the Syrphidae (Diptera), has discovered that there is no mechanical grinding of the pollen grains in the gut, but instead, exudation of the pollen contents occurs through one of the pores in the pollen exine. This strongly resembles an actively growing pollen tube and is certainly due to the insect digestive juices which produce the necessary stimulus. So digestion in many insects is undertaken by a process of pseudo-germination.

During periods of drought in the summer or in autumn, many Alticinae feed on leaves and pollen of plants not usually regarded as their natural larval host-plants. Frequently, Alticinae adults of different genera, will occur in Europe on *Quercus, Populus* and *Pistacia*, while in the tropics, other genera occur on guava, *Persea*, *Carica* and many other trees and shrubs. The larvae never feed on these so-called "secondary host-plants". Succulent leaves of the paw-paw (*Carica papaya*), for instance, are particularly attractive to *Phyllotreta* and others during summer.

The exotic or temperate leaf-beetles do not usually suck nectar, but there are sometimes extraordinary adaptations among myrmecophilous plants which act as alternative food for these chrysomelids. For example, *Coelomera*, a galerucine specialist feeder on *Cecropia* (Cecropiaceae) in South America, feeds on leaves, but also (as do the *Azteca* ants), upon the food-bodies (müllerian bodies), rich in proteins and glycogen, which grow on the plant trichilium at the base of the petioles. It is most probable that the adult beetle also drinks the nectar from the female inflorescences of *Cecropia*, but this has not been fully proven.

Only the good flyers such as the Alticinae, Galerucinae, Megascelinae, Hispinae and Cryptocephalinae, can be captured over the forest canopy. The Chrysomelinae, which are mostly heavy and often poor flyers (often brachy- or apterous), are extremely rare in the canopy of tropical forests (e.g. *Plagiodera*). In Europe and the USA, most of the Chrysomelinae and Cassidinae feed on herbs (e.g. Rubiaceae, Convolvulaceae, Asteraceae, Lamiaceae, Poaceae and Chenopodiaceae) or on young shrubs or new shoots of trees, e.g. *Chrysomela*, *Plagiodera* and *Calligrapha*. The partly solanophagous *Leptinotarsa,* although good flyers, prefer a sedentary existence on the tender low-growing grasses upon which they are host-specific, whereas *Platyphora* and some *Leptinotarsa* feed on arborescent *Solanum* and are not restricted to the most tender leaves. Recently, Medvedev and Pavlov (1987) have studied the mating behaviour of chrysomelids, a subject which had only been studied superficially during the last century. They found that mating appears to follow a complex and strictly determined ritual. Pheromones intervene in the selection of sexes as well as in host-plant selection.

4. Phytophagy, distribution, apterism and colour variation

Many brachypterous and apterous genera and species are found among the Chrysomelidae, from the Lamprosomatinae to the Cassidinae but not in the Eupoda and the Camptosoma. Only among the primitive subfamilies of Sagrinae, Orsodacninae, Zeugophorinae, Aulacoscelinae, Donaciinae, Synetinae, Megalopodinae, Criocerinae and the Camptosoma, are the species all macropterous. The wings are only somewhat reduced occasionally among the Cryptocephalinae and the Chlamisinae; in the last group because of the reduced elytra. The wings of the Megascelinae and the Hispinae are also all macropterous, but those of the Cassidinae can be brachy- or macropterous. The Cassidinae generally feed on low grasses, small shrubs and low-growing vines, (e.g. *Cordia, Ipomoea)*, while the Megascelinae are good flyers and also occur on low plants (e.g. Fabaceae) rather than in the forest canopy. Reduction of elytra is sometimes correlated with smaller wings among certain Galerucinae (e.g. *Arima)* which gives them mimetic resemblance to *Meloe* beetles (Meloidae). Apterous species are strongly associated with weeds, e.g. *Timarcha,* which feeds in the Old World on *Galium* (Rubiaceae), *Plantago* (Plantaginaceae), and several others which are slow-moving. Only the New World *Timarcha* feed nocturnally on shrubs, e.g. *Rubus* (Rosaceae) or on low-growing plants, e.g. *Fragaria* (Rosaceae). These beetles are also apterous.

Many species of *Chrysolina*, including those that are macropterous, are unable to fly because of wing muscle reduction and atrophy, so that flight is exceptional in this genus. However, *C. americana* (L.) and *C. aurichalcea* (Mannerheim) are two species which are still able to fly. The inability to fly or to fly well, appears to explain why all *Chrysolina*, like *Timarcha*, are closely associated with low-growing plant species and are not usually found on larger shrubs or small trees. Often the loss of wing muscles is compensated for by increased fecundity and larger eggs. Viviparous species are numerous among the Chrysomelinae, but viviparity is not closely linked with latitude or altitude; in other words it is not correlated with temperature. Viviparity occurs in numerous species but rarely through an entire genus, e.g. some *Chrysolina* and *Oreina* (*Chrysochloa*) are viviparity as well as many *Platyphora* (*Doryphora*). Of these, *Platyphora* fly rather well, at least well enough to enable them to reach the branches of small shrubs such as *Solanum* (Solanaceae). There does notappearto be any case of viviparity outside the Chrysomelinae, but this needs further verification. It is probable that viviparity is a survival strategy established in order to shorten the life-cycle. It is true that almost all the *Chrysolina* of the Arctic zone are viviparous. Parthenogenesis is exceptional among the Chrysomelidae, being rarely obligatory, and often accidental.

Apterism is a serious handicap in the modern world, since in cultivated areas, or even within the forests, apterous chrysomelids (e.g. *Timarcha*, *Chrysolina*, *Galeruca*) are often the victims of insecticides, fungicides and herbicides. It is a fact that these three genera and others are becoming rarer in western Europe and in California, USA. Another reason for the demise of *Timarcha* in Europe and California is the destruction of coastal areas containing their host-plants. *Elytrosphaera*, an apterous genus which is nearing extinction, barely survives in Brazil in remnants of coastal or montane (Atlantic) forests on the rare *Adenostemma brasilianum* (Asteraceae), which grows with extreme difficulty on cleared areas. In contrast, the closely related, macropterous *Platyphora*, which feeds on *Solanum* (Solanaceae), and *Coelomera* which feeds on *Cecropia* (Cecropiaceae), are able to survive in secondary growth forest and are thus not in any immediate danger of extinction.

Colonization of Europe by apterous and brachypterous species of Chrysomelidae has probably taken millions of years and has most likely arisen from an undetermined center of origin such as central Asia. On the other hand, the strong-flying *Leptinotarsa decemlineata* (Say) has invaded Europe very quickly and is a potential pest of potatoes where this plant is cultivated. The "doryphora" or Colorado potato beetle is probably invading Asia slowly from the Ural Mountains over which it has also already crossed. This beetle has all the necessary "preadaptations" (Vlasova, 1978). Its progression from the secondary focus in the Bordeaux area, France, is due to its adaptability to monoculture crops and its dispersal abilities. However, it does not survive the intense cold of northern Europe or Canada, and, like *Timarcha*, its distribution does not cross the limits of the Quaternary glaciers in the north. No other species of *Leptinotarsa* have shown any threat to agriculture, probably because they show stronger specificity towards their host-plants.

Several oceanic islands in the Pacific and Indian Oceans have been populated by "light-weight" chrysomelids such as small Alticinae or Eumolpinae, but not with large species since they are too heavy to be carried easily by the wind. Even in the case of continental islands, such as Juan Fernandez in the Pacific, well-known for its high plant endemism, the endemic chrysomelids are small and almost entirely represented by Alticinae. Another explanation for this phenomenon is that large species did not exist on the neighbouring continental landmass.

Adaptability of phytophagous species to a new land is generally linked with the acceptability of a new host-plant. Only potentially polyphagous genera such as those of the Eumolpinae are able to adapt easily to new plant groups (e.g. on the Mascarene Islands), although many Alticinae have opportunities to locate food-plants close to their usual hosts.

The colonization of oceanic islands is subject to these conditions and the

Cabo Verde islands are a remarkable example. These islands have been colonized only by "light-weight" species, i.e. 4 Criocerinae (*Lema* and *Crioceris*), one Galerucinae (*Luperus*), 12 Alticinae (4 *Phyllotreta*, 2 *Aphthona*, 3 *Longitarsus*, 2 *Argosomus*, 1 *Chaetocnema*) and one Cassidinae (*Aspidomorpha*). The last species is too heavy to be able to fly over 500 km of sea and has probably been imported from the continent with sweet potatoes (*Ipomoea batatas*, Convolvulaceae). For those Alticinae, the adaptation to new territory is not a real problem, since these beetles feed on Euphorbiaceae, Brassicaceae, Boraginaceae and Asteraceae, plants of which are very common and widespread. There are no Eumolpinae in Cabo Verde; none of the heavy continental species, such as *Euryope*, having crossed the sea. In any case, this genus appears to be host-specific on *Leptadenia* (Asclepiadaceae) which is absent from these islands.

Although oceanic islands such as the Mascarene Islands and Cabo Verde islands do not possess large chrysomelids (and no Chrysomelinae) as do continental islands such as the Canary Islands and Madeira, the situation is not the same with other beetle families such as the Tenebrionidae, Cerambycidae and Carabidae. One can always suspect that many organisms have been introduced accidentally by humans bringing in soil and plants from other places, and this has been confirmed often with Tenebrionidae (darkling beetles) and Diplopoda (millipedes), which are soil-dwelling invertebrates. In any case, the lack of Chrysomelinae is an indication of the "non-continentality" of those islands. However, there are always exceptions to any rule because islands like New Zealand, which lack Hispinae and Cassidinae, do in fact possess several small endemic Chrysomelinae. New Zealand is not an oceanic island but was part of Gondwanaland and as such has a very archaic fauna.

Among the genus *Chrysolina*, as stated previously, because of brachyptery or wing muscle reduction, flight is extremely rare. Despite this, the genus is established throughout the Holarctic, Africa, South-East Asia, Sri Lanka (Ceylon), Vietnam and Indonesia, although it never reached Australia. In Africa, the genus has reached South Africa and is distributed within mountain chains and the high plateaus of West and East Africa. Regions such as Hoggar, Sinai and the plateaus of Ethiopia and eastern Sudan, represent the relics of an older and larger distribution of *Chrysolina* which shrank with increasing desertification. It is strange that *Timarcha* have not left any relictual species in Sinai and south of Lybia. There are *Chrysolina* on the high plateaus of the Cameroons, east of Zaire (Kivu) and Angola which are always found amongst herbaceous and other vegetation. In Ethiopia, above 1500 m altitude, the species of *Chrysolina* aestivate during the dry season under the bark of dead trees.

The loss of wings is not closely linked with an insular or orophilous life style among chrysomelids, although statistically the cases are more common in these environments.

The brachypterous or apterous species are widespread and represent a high percentage of all chrysomelids. Several genera are always winged (e.g. *Lepti-*

notarsa, Platyphora, Aulacophora), but certain species such as *Chrysolina banksi* (Fabricius) and many Alticinae possess winged and brachypterous individuals within the species. These characters have typical Mendelian ratios and often brachyptery is recessive. Among *Chrysolina banksi* (Fabricius) (Jolivet, 1959), 75% of the examined individuals are brachypterous and 25% macropterous.

Most of the montane and subdesertic species show special adaptations, i.e. the former ones are generally diurnally active, while the latter ones feed nocturnally. In both cases, the resting period is often spent under stones which makes it difficult to determine their correct food-plants. For instance, *Timarcha metallica* (Laicharting) has been collected from areas containing *Vaccinium* (Ericaceae) so that this plant has always been suspected as the host for this beetle. However, recent observations have shown that, like other European species, it feeds on *Galium* (Rubiaceae). *Timarcha hummeli* Faldermann from the Caucasus, which lives in a mountainous area devoid of *Vaccinium*, undoubtedly feeds also on the Rubiaceae. *Timarcha intricata* Haldeman is a nocturnal feeder on *Rubus* (Rosaceae) in Oregon, United States of America, a middle altitude region, while *T. cerdo* Stål, a small, coastal and lowland species, feeds mostly on *Fragaria* (Rosaceae). As discussed later, certain larvae feed during the day, while other tropical species are nocturnal feeders—there is no apparent reason for this, except perhaps as a defense strategy.

There are usually no chrysomelids in the intensely hot, arid deserts living like those xerophilous Tenebrionidae under stones during the day. However, in some deserts, some Alticinae may be able to survive by feeding on certain scarce arid-adapted plants. In South Africa, some species, like the slightly flattened *Iscadida,* may take refuge under stones in dry areas, but are not strictly desert-inhabiting. In mountainous areas, *Metallotimarcha* are mainly nocturnal in habits and take refuge under stones during the day. There are no flattened Chrysomelidae which live exclusively under bark as do the strange African Tenebrionidae of the genus *Cossyphus*. However, like the tenebrionid *Gonocephalum,* certain species of *Chrysolina* often take refuge under the bark of trees during the summer in the Canary Islands, Ethiopia etc.

There are no altitudinal limits for Chrysomelidae and Alticinae can be found higher than 4000 m in the Andes. In the Ruwenzori, in Africa, Galerucinae are not found above 3000 m but Alticinae can be captured above that altitude.

Chen and Wang (1962) mentioned that in Altai and Tian-shan, the genus *Oreomela* is common in the alpine zone. *Xenomela* is also a montane genus. According to Chen (loc. cit.) chrysomelids from the Chinese deserts of Sinkiang, display adaptive features in colouration, structure and behaviour. The most significant adaptation concerns the response to subterranean concealment where the tarsal brushes are modified. The chrysomelids of the desert slopes appear to be highly modified, i.e. apterism is common and the tarsi are more specialized. Those species living under stones are flattened in shape, whereas those under the sand tend to assume a fusiform body-shape. The reduction of the tarsal pubescence, more pronounced in the female, is a character present in all desert Chrysomelidae of Sinkiang.

Many Cassidinae change colour either during their life-cycle (e.g. chemical pigments ingested by *Cassida* together with their food) or during a short time before they return to their former colouration (e.g. physical pigments in the bodies of *Metriona* and *Deloyala)*. These colour changes probably have mimetic, sexual and aposematic functions. The sequence of colour changes of certain American Cassidinae is from golden-yellow, to blue-green, to violet then to orange-brown and vice versa.

According to Knab (1909) certain adult chrysomelids change their colour upon the attainment of sexual maturity. Such nuptial colours appear to be well developed among some species of *Physonota, Chelymorpha, Zygogramma, Calligrapha, Disonycha* and *Omophoita*. Among certain *Cassida*, colour changes are dependant on age and chemistry of the host-plant and probably have a dual adaptive function, i.e. mimetic and sexual (from green to brown). Mimetism will be discussed in a later section.

5. *Egg-laying*

The eggs of chrysomelids are laid separately or are protected by excreta or by a hardened secretion forming a primitive (in *Timarcha*) or a more elaborate (in *Coelomera*) ootheca. In extreme cases, the eggs of certain species of *Coelomera* are laid inside the internodes of the stems of a myrmecophytic plant (i.e. *Cecropia*). The female beetle pierces a special non-vascular area on the *Cecropia* plant known as the prostoma which is normally penetrated by the queen *Azteca* ant (Andrade, 1981; Jolivet, 1987d). *Coelomera* and *Azteca* avoid each other in the internodes of the hollow stems but they can co-exist more or less peacefully elsewhere on the same tree (Jolivet, 1987d). Normally the ant has the tendency to reject the eggs of any intruder if they are accessible, however *Coelomera* protects its eggs either by laying them inside the stem, or as with other species of the genus, by laying them on the leaves beyond the reach of the ants (Jolivet, 1987d). As mentioned earlier, the larvae and adults of *Coelomera* feed on the leaves and the müllerian bodies which are the usual ant food. However, this chrysomelid is not the only beetle to feed on the müllerian bodies nor is it the only insect or arthropod to do so.

The female of the American Megalopodinae cuts the end of *Solanum* stems or those of another weed and deposits an egg inside. The Asian and African genera probably do the same with their host-plants. The female sucks the exudates oozing from the cut.

Several Hispinae insert an ootheca within the foliar tissues. Among oviparous leaf-beetles, the larvae are deposited in groups or separately on the leaves. Larval gregariousness occurs indifferently among viviparous or oviparous species.

Anbiru *et al.*(1987) have shown that females of *Pyrrhalta humeralis* (Chen) deposited more than half of their egg-masses on the one-year old branches of their host-tree, *Viburnum awabuki* (Caprifoliaceae) and that the percentage decreased when the branches were older. The majority of the egg-masses were deposited within a distance of 10 cm from the nearest buds on the terminal tips of the branches. Choice of young leaves is also undertaken by the female and not solely by the young larvae.

6. Means of defence of adults and larvae

Many adult chrysomelids such as *Timarcha* and *Coelomera* use reflex bleeding or prebuccal or femoro-tibial autohaemorrhage as protection against predators such as birds, ants and lizards. These systems are very efficient.

Most of the Chrysomelidae are toxic, so much so that the Kalahari tribes in south-west Africa utilize the poisons for their arrows. For instance, they squeeze the larvae of *Polyclada* or *Diamphidia* (Alticinae) and mix them with the larvae of *Lebistina* (Carabidae: Lebiinae) which are predators of the Chrysomelidae. It should be noted that the Indians or Mayans consume several types of beetles and their larvae but never feed on the extremely toxic chrysomelids. Toxicity by itself is excellent protection against birds, but in association with reflex bleeding, it is very efficient, especially against ants.

Other Chrysomelidae, when disturbed, drop to the ground (reflex immobilization) or jump (e.g. Alticinae) or fly away rapidly. Some like the Chlamisinae mimic death and resemble bird excreta or strongly adhere to the substratum by their tarsi which are provided with oily suckers (e.g. Cassidinae). The flat appearance of tortoise beetles provides them with further defence since potential aggressors are unable to hold onto their bodies easily. The extremely hard elytra of *Desmonota* are also extremely efficient against predators. They can be broken only with a hammer and the insects are often used as jewels by the elegant ladies. Toxicity, shape and adhesion to the substratum very efficiently protect the Cassidinae against all of their predators, especially spiders and ants.

Homochromy with the host-plant is also common among the Cassidinae. It is sometimes structural and sometimes linked with the host-plant chemistry as we have mentioned previously. When the colour changes rapidly, it is linked with the aqueous film of the cuticle. The reason for this adaptation is not clear but it can confuse a potential predator so it may have a cryptic function. Many Chrysomelidae have aposematic colouration of black and red or blue or metallic violet. Others have the elytra spotted against a yellow, orange or brown background. Several species of *Elytromena*, a subgenus of *Elytrosphaera* (Chrysomelinae), show pretty and aesthetic patterns.

Defence of the larvae varies widely among the various subfamilies. Those which are not protected by a larval case or which are not miners or borers, or not living between the appressed leaves of monocotyledons, can defend themselves with toxic secretions, enteric discharges or gregariousness.

Unless further research shows otherwise, parental care is restricted among the Chrysomelidae to *Stolaini* (Cassidinae) larvae in the Neotropics and to *Gonioctena* (*Phytodecta*) (Chrysomelinae) in the Holarctic (Jolivet, 1988a). Ring defence strategy or cycloalexy has been recently discovered among *Chelymorpha* (Cassidinae, Stolaini), *Platyphora* (Chrysomelinae) and all the *Coelomera* species (Galerucinae). Probably *Dircema* among Galerucinae exhibit the same phenomenon. It appears that among the Stolaini, this extraordinary behaviour is related to parental care of the larvae (*i.e.* in *Omoplata*). A similar phenomenon (i.e. parental care and ring defence behaviour) exists among the tenthredinid larvae, namely those of the Pergidae and Argidae (Hymenoptera). Among Chrysomelidae, resting occurs in a circle on the abaxial (lower) surface of the leaf of the host-plant. The larvae are arranged head against head, with the abdomens positioned outside of the ring (i.e. among Cassidinae and *Coelomera*) or tail against tail (extremity of the abdomen) with heads outside the ring (i.e. among *Platyphora*). These larvae also adopt menacing attitudes, which involve the raising of the abdomen, enteric discharges (i.e. in *Chelymorpha* and *Coelomera*), buccal discharges and biting attempts (i.e. in *Platyphora*). Among *Coelomera* and *Chelymorpha*, trophic activity is diurnal while resting is nocturnal, whereas among *Platyphora*, food is ingested during the night and resting is diurnal. It is most probable that Stolaini show the same phenomenon, although it may or may not be associated with parental care. All *Coelomera* and probably other *Platyphora* behave the same way, but most of the latter genus have solitary larvae. During cycloalexy, some larvae are usually situated inside the circle and an exchange occurs with the external larvae, the ones most exposed to aggression by predators. The system appears to be efficient because if the circle is broken, ants and Pentatomidae bugs immediately capture some of the larvae. The supra-anal shield of *Coelomera* larvae appear to also offer efficient protection.

Recent studies on the effect of co-ordinated groups of the subsocial species of *Plagiodera versicolor* (Laicharting) on feeding efficiency shows a positive correlation with group size (Breden & Wade, 1987).

7. Allotrophy or xenophagy

Allotrophy is defined as a drastic food change among animals in response to certain factors, but it has not really been well analyzed or understood (Jolivet, 1965, 1986b). Often allotrophy occurs under stress conditions and is a survival strategy, e.g. for the Alticinae which bore holes into *Quercus* (Fagaceae) leaves in Europe and into *Carica* (Caricaceae) and *Psidium* (Myrtaceae) in the tropics when the normal food-plant is absent. Many other cases exist among the Chrysomelidae and are mentioned by Jolivet (1986b) and elsewhere in this book.

Larvae of Clytrinae which are polyphagous and often carnivorous (they feed on ant eggs) produce strictly phytophagous adults. This case is not one of allotrophy but is part of their normal development like the Hydrophilidae which evolve from carnivory in the early larval stages to herbivory in the later life-stages, while the situation is reversed for tadpoles. However, it has been observed very often that adult chrysomelids are able to feed on their own eggs or on those of another individual. This is a method of regulating the number of larvae but probably also a way of increasing protein intake. An interesting case exists with *Oomorphus floridanus* Horn-normally a feeder on the Araliaceae, this species also consumes excreta of a rodent in Florida, USA. In Key Largo, *Hedera* (Araliaceae) is not common and possibly the beetle here is attempting to survive by feeding upon a replacement food like the Papilionidae and Nymphalidae (Lepidoptera) which willingly drink human excreta and urine in the tropics.

However, we define here "allotrophy" as mostly a drastic change in host-plant diet among phytophagous insects in order to replace a lack of normal host-plants. Very often this sudden change is due to a hidden oligophagy within the genus. The classical example is the case of *Agelastica alni* (L.) (Galerucinae), a European insect host-specific on *Alnus glutinosa* (Betulaceae) but which also feeds occasionally on *Populus* and *Salix* (Salicaceae) and *Betula* (Betulaceae) (all Amentiferae). From time to time, invasions of *A. alni* also occur on arborescent Rosaceae. This is not really surprising because oriental species of *Agelastica* feed on Rosaceae and in Europe the selection is latent among the genus, even if not normally expressed. Several other cases of allotrophy are known among Chrysomelidae and sometimes the change in food plant (biotype) preferences results in the formation of new races or even emerging species. In any case, allotrophy must not be confused with polyphagy. Among feeders on Polygonaceae, like *Gastrophy*sa (Chrysomelinae), no case of allotrophy has been shown by the insects. It is often the rule that food choice remains stable for most of the time. Imported plants like the soya bean (*Glycine max*, Fabaceae) can be adopted when introduced from outside the range of the normal host plants either by oligophagous feeders on Fabaceae (e.g. *Phaedonia inclu*sa (Stål) Chrysomelinae) or by polyphagous species. These cases do not involve allotrophy.

8. The host-plants

When one is searching for the host-plants, one must, if possible, first observe the larvae, because they feed more voraciously than the adults and they never feed on accessory or secondary, non-related plants. If the adult is observed, it must be observed feeding, because accidental findings usually do not represent true trophic relationships. One must be very careful when reading lists of host-plants which are often only a simple catalogue of captures with few real plant/insect relationships. Often lists written by applied entomologists are full of errors.

Our book is based on some 20,000 index cards (21 x 13 cm) filled with personal data or credible reports from the literature. Several "good" lists have been compiled and used but many "bad" lists which have been haphazardly compiled by agronomists have been systematically rejected.

It is sad to see that certain areas of the world are still virgin of observations. For instance, nothing is known about Madagascan host-plants and because of the markedly high endemism and the antiquity of the fauna and flora, it is an important gap to be filled. Few reliable data exist for the tropical African fauna in general. Otherwise, Palaearctic and Nearctic areas are very well studied (with the exception of *Chrysolina*) as well as South-East Asia (with the exception of India where most of the records are erroneous or misleading) and the Neotropical areas. Of course, there still remains much more to be discovered in tropical America and Australia, where the Paropsini, because of their pest status, have been well studied. Virtually nothing is known of the biology of Gondwanian Sagrinae from Australia to Madagascar and even Brazil, and of the biology of the Australian Alticinae and New Zealand Chrysomelinae. Many South American lists are full of mistakes but several excellent biological studies compensate for this.

Roughly, the host-plants of about 20% of the described leaf-beetles are known, but nothing is recorded on the biology of the Megascelinae, Aulacoscelinae and the African Megalopodinae.

Many new things remain to be discovered if one day, one is able to study biologically (and not only by chemical means) the tropical forest canopy. Many host-plants will thus be identified and linked with "their" chrysomelids. Attempts to study the forest canopy fauna, like those of Prof. Halle in Guyana and Africa using a raft and a lighter-than-air balloon, could result one day in a solution to this problem. It is hoped that in the near future specialists will be able to undertake real observations and to establish the trophic associations. The spraying of pyrethroids can assist in the discovery of many new species but is not helping in biological studies. Both methods, in association with the one utilizing bamboo walkways and cranes into the canopy, can be complimentary.

Very often the trees of the canopy belong to families which are represented among the herbaceous plants common in temperate climates. There could be relationships between the leaf-beetles and certain plant families whether the plants are grasses or trees. Anyhow, this question now remains hypothetical.

A pioneer work on the trophic relationships between plants and chrysomelids was presented by Maulik (1945) and drafted into his unpublished book. His adopted classification is mostly botanically-oriented with the leaf-beetle genera listed before the genus of plant. However, during Maulik's time, there were very few reliable data on host-plants, most of which were from the tropics. Much of the available data were erroneous or doubtful. The native place of Maulik, India, was then and still remains practically a "terra incognita" for the leaf-beetle/host-plant relationships. All of the recent Indian lists are at least 90% inaccurate. However, despite this problem, Maulik managed to complete an excellent series for the *Fauna of British India* which had been started by Jacoby in 1908. The three volumes written by Maulik provide very little information on the larvae, but he does quote a few larval hosts in his text.

The knowledge of trophic relationships of Chrysomelidae is important for the control of imported (introduced) weeds. We will describe later some examples of a few successful and other doubtful trials concerning biological control. In this endeavour, one must be very careful, because, for instance, the *Citrus*-feeding Chrysomelidae are not all restricted to the Rutaceae family. Some of these, like some Cassidinae and other Alticinae in response to sudden climatic changes, use only the young leaves as secondary food sources.

The solanaceous chrysomelids are numerous in the New World but are relatively scarce in the Old. This can be related to the center of origin of the family Solanaceae which is in Central and Southern America, and to their extensive diversity in the Neotropical region. In contrast, the attraction of Cassidinae to the Convolvulaceae is a world-wide phenomenon. Certain preferences of Chrysomelidae seem to reveal relationships between plant families that are widely separated in present classifications. For instance, the Rubiaceae and the Plantaginaceae, formerly placed together in the flora of Bonnier, but now widely separated, are the favoured food plants of *Timarcha* (and only of *Timarcha)*, together with the Dipsacaceae and the Scrophulariaceae. This indicates a close evolutionary relationship or at least a common chemistry among these plant families. There are several Galerucinae feeding on the Fagales in the Old World, but also three primitive Eumolpinae and one Megascelinae in Patagonia, South America, and one Chrysomelinae in Australia. It seems strange that those trees are so poorly colonized by leaf-beetles; but maybe this can be explained by the difficulty which leaf-beetles have in flying to reach the leaves of these tall rainforest trees. *Nothofagus* is an example of a Gondwanian tree which attracts Chrysomelidae in Patagonia and in Australia, and probably elsewhere since the fauna of these trees has been poorly studied to date. In Patagonia, *Mariamela wittmeri* Monros (Megascelinae) and *Hornius sulcifrons* (Fairmaire), *H. grandis* (Philippi) and *Psathyrocerus fulvipes* Blanchard (Eumolpinae), and in Australia, *Novacastria nothofagi* Selman (Chrysomelinae), are known to inhabit *Nothofagus*. In New Guinea, the *Nothofagus* niche must be more important but at present it is totally unexplored.

In the last section of this book, the leaf-beetles will be studied in detail in relation to some selected plant families.

Surprisingly, very few Chrysomelidae feed on the Rubiaceae, e.g. *Timarcha* (Chrysomelinae), *Sermylassa* (Alticinae) and *Neolochmaea* (Galerucinae), but plant families such as Asteraceae, Lamiaceae, Rosaceae and others attract a large number of species. The actual fauna is probably what remains of a large polyphagous, then oligophagous Tertiary fauna. Why some subfamilies are represented mostly by mono- or oligophagous species, such as Cassidinae, Hispinae, Criocerinae, Chrysomelinae, Megalopodinae and Alticinae, while others contain mostly polyphagous species, e.g. Galerucinae, Donaciinae, Clytrinae, Cryptocephalinae and Eumolpinae, remains a mystery. The Donaciinae, Eumolpinae, Clytrinae and Cryptocephalinae, being polyphagous in the larval stage (respectively aquatic, subterraneous, myrmecophilous and terrestrial near ground level), have probably retained this polyphagous tendency in the adult stage. The propensity of many Eumolpinae for selecting Asclepiadaceae/Apocynaceae is not accidental and appears confined to many tropical and temperate Eumolpinae. The biological study of radicolous Eumolpinae larvae has been undertaken several times but since it is well-known that it is very difficult to rear subterraneous larvae in the laboratory, such studies are not readily proceeding with good results. A behavioural study of the Asiatic subterranean genus *Clavicornaltica* (Alticinae) would be useful since up to the present time, only taxonomic information is available on this genus. It is likely that the adults themselves are polyphagous and feed on roots, but it is also very probable that certain species are more exclusively subterraneous than others. Closely related genera such as *Kamala* from India and *Mniophila* and *Mniophilosoma* from the Palaearctic region, probably exhibit similar behaviour, but *Clavicornaltica* possess wings which are not always reduced, so that some species of this genus must disseminate above ground.

Sometimes the age of the leaves can be more important than plant species, e.g. in the genus *Passiflora,* the condition will usually influence the feeding of *Monomacra* sp. and *Disonycha quinquelineata* Latreille (both Alticinae) (Thomas, 1987). Species of *Chrysomela* and *Paropsis* (both Chrysomelinae) usually select the young shoots of *Populus* (Salicaceae) and *Eucalyptus* (Myrtaceae) respectively.

Drift towards new plants, which are totally different from the usual (normal) host-plant(s), occurs among the Alticinae during stress and has probably no other origin.

One insect fed with senescent leaves can have its reproduction stopped and can be rapidly forced into diapause. The Colorado potato beetle, when feeding on old potato leaves, develops early diapause (de Wilde, 1969).

A recent paper by Larsson and Omart (1988), mentions that larvae of *Paropsis atomaria* Olivier (Chrysomelinae) survive and undergo nymphosis only if they had been raised to the third stage on young leaves, otherwise their development is disturbed. Only the new leaves, which have higher concentrations of oil and tannins, are favourable for normal development. Old leaves are also three times harder than the new ones. During periods of drought, the dynamics of *Paropsis* are strongly influenced, resulting in a slowing down of their development. Recently, during a trip to Papua New Guinea in February 1988, the senior author failed to find *Paropsis* larvae and adults on *Eucalyptus* saplings in Port Moresby, where they were abundant several years earlier during the same period. New leaves did not grow because of the drought which lasted for nearly two years before which, during the humid years, the *Paropsis* were abundant everywhere. In contrast, *Phyllocharis* were abundant in artificially watered areas.

While extra-floral nectaries mimicking butterfly eggs appears advantageous to *Passiflora* vines repelling *Heliconius* (Nymphalidae) females ready to lay eggs, the irregular leaf shape of some Moraceae plants seems to mimic feeding-damage and earlier colonization by herbivores (Niemelae & Tuomi, 1987). Such feeding-damage mimicry can function as a potential anti-herbivory adaptation if visually-searching herbivores avoid host-plants colonized earlier. This has not yet been recorded for leaf-beetles but it is a subject worth studying. No one knows, for instance, how *Coelomera* beetles choose their host-plants when they are already eaten by caterpillars or other insects. It is a fact that beetles, caterpillars and others exclude each other on young *Cecropia* trees (Cecropiaceae, formerly included in Moraceae).

Leaf aspect also appears to affect herbivory as demonstrated by Banerjee (1987). Leaves of tea cultivars differ in leaf geometry and leaf chewers feed mostly on horizontal leaves, leaf rollers and miners colonize semi-erect leaves and sap feeders inhabit the three types of leaves, with the largest accumulation on erect leaves. This is a new aspect of food selection by phytophagous insects which should be carefully investigated among the Chrysomelidae.

Rausher's work (1983-1984) on the host-plants of the American cassidine *Deloyala guttata* (Olivier), namely *Ipomoea pandurata* and *I. purpurea* (Convolvulaceae), are not conclusive. It seems, at least theoretically, that formation of races and sympatric speciation presuppose that one of the genotypes has a higher viability and fecundity on a plant than on another. Although the experiences described on the two *Ipomoea* species are not definite, we are already aware that populations of *Oreina cacaliae* (Schrank) (Chrysomelinae) vary in elytral length depending on the host plant, i.e. either *Senecio nemorensis* or *Petasites albus* (Asteraceae) (Kreslavsky *et al.*, 1976). This phenomenon could lead first to ecological races and later to true species.

It seems that botanical classification is not adequate in providing a clarified view of the trophic relationships of the leaf-beetles. At least, it is somewhat difficult with the present state of knowledge, to detect hidden relationships between the plants responsible for certain selections. Sometimes however, those relations are clear and well-defined and insects are often better botanists than the botanists themselves (e.g. the relationships between the Plantaginaceae and the Rubiaceae, see earlier comments on *Timarcha*). In the unpublished book by Maulik, it is impossible to distinguish oligophagous insect species from the polyphagous ones so that the use of these host-plant records in botanical or zoological classification is somewhat limited.

A synopsis of the chrysomelid subfamilies will be provided in the next chapter. This will be followed by a study of the plants selected by the leaf beetles. As it is impossible to quote all food-plant families, several important ones will be selected, of which a list of primary and secondary phytophagous beetles associated with those plant families and genera will be provided.

The classification of Takhtajan (1969) and Swain (1963) has been adopted for the plants. The authors have also been influenced by the splendid works of Cronquist (1981) and the classical books of Willis and Airy Shaw (1973) and Mabberley (1987). Within the beetle subfamilies, the selections are listed genus by genus or even species by species when it is really necessary, e.g. for *Timarcha* and *Chrysolina*. Since the publication of extensive lists of the host-plants of Chrysomelidae (Jolivet, 1966-1992), a large quantity of new data concerning certain families has been indexed. The new data do not modify the former conclusions although some of the more fanciful data have been rejected.

In the tables, the list of secondary host-plants (stress plants) are omitted. Only in the text itself are some additional (secondary) plants mentioned. It is evident that these plants, which have nothing to do with the real and normal botanical selections, have a survival value for the beetles, because they are succulent, non-toxic, or as in Poaceae, provide abundant and easily accessible pollen.

This book is dedicated to the entomologists of the past and to those of the present, who have patiently accumulated these precious data on the host-plants. Almost all of these people have been listed in the bibliography, but it is impossible to quote all of their references in the text after each plant genus. To offset this problem, at the end of each subfamily review, a list of references from the bibliography, pertaining to that subfamily, is provided.

2. SELECTION OF THE VARIOUS SUBFAMILIES OF CHRYSOMELIDAE

In the following pages the 20 subfamilies will be reviewed in taxonomical order from Sagrinae to Cassidinae. The data available are very unequal and actually the Criocerinae, Chrysomelinae, Hispinae and Cassidinae are the groups which have been the most studied.

SECTION I: EUPODA

2.1. Sagrinae

This subfamily, which has some affinities with the Cerambycidae but mostly with the Bruchidae, is chiefly Oriental in distribution but is also represented in Madagascar, Africa and tropical America. The genera *Rhaebus* and *Eubaptus* have been placed sometimes with the Chrysomelidae and sometimes with the Bruchidae but they are really intermediate between these two families. The Gondwanian genera of Sagrinae are distributed within Australia, Madagascar and Brazil. Of these regions, Australia possesses the largest number of genera and these are all primitive and diversified. The genus *Sagra* is distributed in Asia, Queensland, Papua New Guinea, tropical Africa and Madagascar. Their distribution begins at the northern soudanosahelian limit in the Burkina Faso (Upper Volta), not far from Bobo Dioulasso, Africa. The larvae of Sagrinae have been described by several authors, including Maulik (1941) for *Sagra* and Monros (1955) for *Atalasis*. Many other descriptions have been published after the initial works, but they do not add anything new.

The "sagriniform" bruchids live in the drupes or seeds of various plants, e.g. *Eubaptus* from Argentina in *Ruellia* (Acanthaceae) and *Rhaebus* from Oriental Asia in *Nitraria* (Zygophyllaceae). Such spermophagous behaviour and the morphology of these species strangely recalls the Sagrinae which are not really spermophagous but are gallicolous during the larval stages in the stems of their hosts. Unfortunately little or nothing is known about the various Gondwanian species living in Australia. They are from the genera *Coolgardica*, *Ametalla*, *Megamerus*, *Mecynodera*, *Neodiaphanops*, *Pseudotoxotus*, *Polyoptilus*, *Diaphanops*, *Duboulaya* and *Carpophagus*. *Megamerus* is also known from Madagascar where it is gallicolous and in the north-east of Brazil on the Natal coast. One *Mecynodera* has been also described from Papua New Guinea. *Mecynodera balyi* Clark from Australia apparently cuts the host stems like certain Cerambycidae and the Megalopodinae but Clark's observations were published in 1864 and the host-plant has never been identified. In New South Wales,

Australia, *Mecynodera coxalgica* (Boisduval) has been recorded recently (Webb, 1986) feeding from the flowers of *Kunzea* (Myrtaceae), presumably on the nectar and/or pollen. It is most probable that like *Sagra*, these Sagrinae are slightly polyphagous in the larval stage.

Carpophagus banksiae Macleay has been reported from the xerophytic shrub *Banksia* (Proteaceae), possibly from the flowers (the original record, i.e. Walker, 1906, is not specific) but it has also been recorded from *Macrozamia* (Zamiaceae, Cycadales) from temperate Australia. These relationships probably indicate an attractiveness to flowers common to many beetles and do not reflect true trophic relationships. All those genera are probably gallicolous but on which plants? We do not know!

Sagra is distributed in the tropical areas of the Old World. Its biology is well known and is certainly more related to cerambycid biology than to bruchid. Despite its enlarged hind femora, *Sagra* does not jump and this character appears to be hypertelic. The colours of the Asiatic species are more beautiful, more diverse and brighter than those of the African species, which are generally dull in colour.

All *Sagra* larvae are borers, i.e. they burrow into the stems of plants where they feed and grow. According to Maulik (1941), one to twenty larvae can be found within one gall formed around a parasitized host stem. The larva constructs a cocoon and nymphosis occurs in the middle of the swelling of the stem. The adult pierces a hole before crawling out. Hence the larva spends its entire life surrounded by edible tissue as in the Megalopodinae but unlike the Megalopodinae, gall production and nymphosis occurs above ground level and not below ground.

Maulik quoted the Asian *Sagra* feeding on Fabaceae (*Dolichos*, *Faba*, *Mucuna*, *Phaseolus*), Convolvulaceae (*Ipomoea)*, Verbenaceae (*Tectona*), Dioscoreaceae (*Dioscorea*), Rutaceae (*Citrus*), Rubiaceae (*Coffea*), Rhizophoraceae (*Rhizophora*) and Acanthaceae (*Thunbergia)*. This last genus is a vine which often surrounds the teak tree (*Tectona*). Since *Sagra* larvae are able to move easily from one plant to another during their life-cycle, it is possible that *Thunbergia* is not really a larval host-plant. *Sagra* larvae are very large in the last stage (c. 21 mm) and the adult can measure over 24 mm. *Sagra* are the largest chrysomelids along with several palm-feeding Hispinae. Many other plants, but mostly legumes, can harbor *Sagra* in the tropics: *Indigofera*, *Pueraria*, *Pachyrhizus*, *Vicia*, *Canavalia*, *Vigna* and *Eriosema* (Fabaceae) and *Cassia*, *Moghania* (Caesalpiniaceae). Apart from these, *Cucurbita* (Cucurbitaceae), *Cocculus* (Menispermaceae), *Terminalia* (Combretaceae) and *Durio* (Bombacaceae) are also recorded as larval host-plants. The galls are always produced on young plants or small shrubs. One of us (PJ) found many *Sagra* in Africa, near Bobo Dioulasso, on young *Eriosema* shrubs growing in a swampy area and at Hanoi, Vietnam, on the large, flexible legume *Pueraria*. Hatching occurs suddenly and en masse in the Sahel area, Africa, after the rainy season.

One *Sagra* has been found eating the leaves and young shoots of *Vanilla* (Orchidaceae) in Madagascar and *Solanum melongena* (Solanaceae) in Africa. These are adult food-plants but larval development in them has not been verified. However this indicates overall that the genus is polyphagous at least in the adult stage and that the hosts range from the Dioscoreaceae and Orchidaceae (Monocotyledonae) to families such as Rubiaceae and Cucurbitaceae (Dicotyledonae). In Africa, young imported *Eucalyptus* trees have been found to be infested with galleries. *Sagra* larvae are also polyphagous in the laboratory and will consume anything from carrots, lettuce and apples to various leaves. Experiments transferring larvae from one plant stem to another stem of a different species or genus have been successful.

Sagra eggs are deposited into the bark of shrubs and when *Rhizophora* are attacked in the Javan mangroves of Indonesia, the cocoons are often situated in a hollow root. *Coffea* roots (Rubiaceae) can also be attacked in Africa. Mostly the *Sagra* are polyphagous and feed on subligneous plants, mostly from the Fabaceae and Caesalpiniaceae.

Adults of *Sagra amethystina* (Guérin-Méneville) often extracted from galls on *Eriosema glomeratum* (Fabaceae)nearBoboDioulasso (Burkina Faso), often visit the Fire Tree, *Delonix regia* (Caesalpiniaceae) and are probably attracted by the secretions from the Bostrychidae tunnels. *Atalasis sagroides* Lacordaire, the only Argentinian species, lives on *Sphaeralcea bonariensis*, *Malvastrum coromandelianum*, *Gossypium* etc. and even occasionally on *Sida* and *Abutilon* (all Malvaceae). The larva has been described by Monros (1955-59) and the adults feed on the leaves, but mostly on the flowers, pollen and other floral organs of the Malvaceae. The larva produces a gall in the stems but the metamorphosis (nymphosis) apparently occurs in the ground(?). *Atalasis* live in the moderately dry areas of the north-east of Argentina, in contrast to *Sagra*, which always live in humid areas, even the Sahelian zone. The capture of *Atalasis* on the flowers of *Aenethum foeniculum* (Apiaceae) seems rather exceptional. Like *Sagra*, *Atalasis* larvae produce galls on the stems and each gall houses one larva. The stem becomes a brownish colour when parasitized and the bark cracks on the inflated parts, but this does not affect the leaves or the flowers. The galls are formed by dilation of the stem enclosing elongated chambers which increase in size with the growth of the larva which is situated in the pith region of the plant. Excreta accumulates at the extremities of the chambers. The larva eats the central pith areas of the stems and the surrounding fresh tissues and will starve to death if the stems dry up. When mature, the larva chews a circular hole in the wall of the gall and drops to the ground. In contrast to *Sagra*, nymphosis appears to occur in the ground but this has never really been observed. It is evident that the circular openings of the stem can also correspond to the emergence of the adults. The larva of *Atalasis sagroides* measures from 10 to 15 mm and has been described by Monros (1955); it is similar to those of *Sagra*.

Basically, much remains to be discovered about Sagrinae biology but mostly about the Gondwanian genera, of which only one genus is represented in northeastern Brazil. With much caution, one may deduce from observations on two genera, that the subfamily has cerambycid (*Megamerus*) or bruchid (*Carpophagus*) characters. However, the group also contains unique characters (e.g. in *Sagra* and *Atalasis*). It is possible that all species are gallicolous in subligneous plant stems, but this remains to be proved. We are still awaiting original observations from Madagascar, Australia and Brazil.

References listed in the bibliography pertaining to Sagrinae are as follows: Cantelo & Pholboon (1965), Clark (1864), Crowson (1946, 1991), Gressitt (1942, 1965), Jolivet (1977), Kalshoven (1951), Le Pelley (1959), Maulik (1941, 1945), Monros (1955, 1956, 1959), Tandon *et al.* (1975), Tayade (1978), Walker (1906), Webb (1986), Zaitsev (1982).

2.2. *Aulacoscelinae*

It was Maulik (1941) who first removed *Aulacoscelis* from the Sagrinae, followed by Crowson (1967). Monros created the Aulacoscelinae subfamily in 1953. Monros' revision dates from 1954, and this subfamily, though primitive (because of the the ring tegmen), appears to be at the origin of the Criocerinae, but it is also related to the Chrysomelinae because the upper Jurassic Protoscelinae are closely related to that subfamily (Medvedev, 1968). The subfamily is essentially Central American in distribution with an extension into Peru and Bolivia. Nothing has been recorded on their biology and host-plants and the larva remains undescribed. The adults have been captured on flowers of *Hechtia texensis* (Bromeliaceae), Asteraceae and *Cycas* and *Zamia* (Cycadales) but one usually cannot deduce any true relationships from these sporadic observations. However, recent observations by D. Windsor and H. Stockwell in Panama confirm the attractiveness of Cycadales flowers for *Aulacoscelis*.

References listed in the bibliography pertaining to Aulacoscelinae are as follows: Crowson (1967), Jolivet (1977), Maulik (1941), Medvedev (1968), Monros (1954b).

2.3. *Orsodacninae*

A little more is known about this subfamily which is mostly Holarctic in distribution. The larva remained unknown for a long time until it was finally described by Cox (1981) and Crowson (1981), partly from reared material and partly from the old collections of Van Emden in England. It was thought that the larva was a miner in the leaves or the buds of Rosaceae and related families, but this is not known for certain. The first stage larva has been described from eggs hatched in the laboratory. According to Cox (1981), the larva is closely related to the Galerucinae and Alticinae so that it would be expected to feed extensively or intensely on the roots of plants like *Syneta*. Only laboratory experiments using techniques adopted for rearing Eumolpinae on roots, could prove this assumption. The fact that *Orsodacne* larvae are common in tits' gizzards

(*Parus caerulens*) favours the theory that they are either stem or leaf-mining larvae.

The adults are floricolous like certain other chrysomelids and consume the pollen or anthers of mostly white flowers, e.g. Rosaceae (*Crataegus*, *Cerasus*, *Prunus*, *Pyrus*, *Malus*, *Mespilus*, *Rosa*, *Sorbus*, *Spiraea*, *Filipendula*), Caprifoliaceae (*Viburnum*), Orchidaceae (*Listera*), Celastraceae (*Evonymus*), Dioscoreaceae (*Tamus*), Oleaceae (*Ligustrum*), Aquifoliaceae (*Ilex*), Fagaceae (*Quercus*), Apiaceae (*Heracleum*, *Bupleurum*), Euphorbiaceae (*Mercurialis*) and Valerianaceae (*Valeriana*). It is evident that those host-plants of the adults do not necessarily mean they have a trophic relationship with the larvae. The larva is probably polyphagous if subterraneous, or associated with Rosaceae if a miner. However, we cannot really comment on the trophic relationships any further until more definite observations are made. In the United States, *Orsodacne atra* Ahrens has been captured during springtime on *Galium* (Rubiaceae), *Cercis* (Caesalpiniaceae), *Rhus* (Anacardiaceae), *Hamamelis* (Hamamelidaceae), *Hepatia* (Hepatiaceae), *Lindera* (Lauraceae), *Rubus*, *Spiraea*, *Prunus*, *Malus*, *Ame-lanchier* (Rosaceae), *Acer* (Aceraceae), *Cornus* (Cornaceae), *Carpinus* (Carpinaceae), *Betula* , *Alnus* (Betulaceae), *Corylus* (Corylaceae), *Salix* (Salicaceae), *Verbascum* (Scrophulariaceae), *Iris* (Iridaceae), etc. It can be seen that the adults, which are pollen feeders, are totally polyphagous. They never attack the leaves or petioles of the flowers. According to Cox, feeding on pollen which is rich in proteins, enables the female beetles to lay large numbers of eggs (20-35).

According to Kuschel and May (1990) and Crowson (pers. comm.), *Cucujopsis* develops in the male cones of *Araucaria* or *Agathis* (Araucariaceae) in Australia. The larva is very similar to those of the Cerambycidae and this is additional evidence for the Jurassic origin of the Chrysomelidae. According to Crowson (pers. comm.), the leaf-beetles were most probably developing as *Aulacoscelis* and *Carpophagus* in the reproductive organs or the "flowers" of the Mesozoic gymnosperms, such as the Bennettitales.

The references listed in the bibliography pertaining to this subfamily are as follows: Bedel (1891), Cox (1981), Crowson (1981), Jolivet (1948, 1957, 1977), Kuschel & May (1990), Wilcox (1979), Zaitsev (1988).

2.4. *Palophaginae*

This is a newly created subfamily (Kuschel & May, 1990), which is confined to Australia. The type genus, *Palophagus,* is represented by two known species, *P. bunyae* Kuschel from the Bunya Mountains, southern Queensland, and *P. australiensis* Kuschel, from north-eastern New South Wales (Kuschel & May, 1990). Adults and larvae of *P. bunyae* inhabit the male strobili of *Araucaria bidwillii* (Araucariaceae), where they feed on the pollen; the larvae usually inhabit the central pith of the strobili, chewing the stalks of the spent sporophylls. In the younger cones, the larvae, never more than one present at a time, feed on the pollen in the company of larvae of Nemonychidae weevils. The ambulatory ampullae on both dorsal and ventral surfaces enable the larvae to move with ease among the sporophylls. The larva pupates in the soil where it passes through an additional but inactive instar (hypermetamorphosis)(Kuschel & May, 1990). The fully grown larva, having produced a pupal cell, moults to an inactive pupal stage which is morphologically similar but lacks pigmentation and has simplified mouthparts. Moulting to the true pupal stage occurs after 5-6 months and the final ecdysis to adult after 30-50 days (Kuschel & May, 1990). The host-plant of the other species, *P. australiensis,* is not known (the type specimen was attracted to light) but it is suspected to be *A. cunninghamii* (Araucariaceae)(Kuschel & May, 1990) .

Kuschel and May (1990) also consider *Cucujopsis* as belonging to this subfamily, but we are following Crowson (1981) and Mann & Crowson (1981) in placing *Cucujopsis* in the Orsodacninae. Additionally, Kuschel & May (1990) place Palophaginae in the family Megalopodidae but we regard this in itself as a subfamily of Chrysomelidae.

The only reference on Palophaginae at present is that of Kuschel and May (1990).

2.5. *Zeugophorinae*

This subfamily is mostly Holarctic in distribution (*Zeugophora*) but it is also represented in the tropical regions of the Old World. Until now, it has never been found in the Neotropical area, but due to the very small size of the various species, the group may have been overlooked.

Zeugophora larvae are miners in the young leaves of trees or shrubs. There do not appear to be any exceptions. Larvae and adults occur on the same plants since the adults are phyllophagous. A few new observations have been undertaken on the host-plants of *Zeugophora*. In the Holarctic region, *Zeugophora* occur on the Salicaceae (*Salix*, *Populus,* mostly regrowth), Juglandaceae (*Juglans*), Corylaceae (*Corylus*) and Betulaceae (*Betula*). All these plant families are somewhat closely placed in the classification of Swain (1963), although Cronquist (1981) placed the Salicaceae some distance from the others.

The genus (or subgenus) *Pedrilla*, which is tropical in distribution, with incursions into Japan, has been recorded on *Santalum* (Santalaceae) in India, *Populus* (Salicaceae) in China and *Evonymus* and *Tripterygium* (Celastraceae) in Japan. The only common thing shared between all of these plants is that they are shrubs or trees since they are not closely related botanically. *Pedrilla* appears to be polyphagous, but may exhibit monophagy at the species level, since many other plants have been recorded from the Old World tropics. The presence of the adult of *Pedrilla annulata* Baly in China on *Isachne* (Poaceae) may be accidental or indicates pollen feeding.

The references listed in the bibliography pertaining to this subfamily are as follows: Gressitt (1965), Jolivet (1948, 1977), Kimoto (1964-66), Lee (1990), Reid (1989).

2.6. *Donaciinae*

This group is aquatic with polyphagous tendencies, since the larvae and adults devour aquatic or subaquatic plants. This selection is "ecological", i.e. it is linked with the biotope. The Nymphaeaceae appear to be the most common selection, both in the tropics and in the Holarctic zones. The Donaciinae are mainly distributed in the temperate Holarctic zone, but have made incursions into Central America, the West Indies, Indonesia, northern Australia and tropical and temperate Africa. They are only absent from South America, southern and central Australia, northern New Guinea and New Zealand. In New Guinea, they are extremely localized in the Fly River area in the south. It seems strange that they are absent elsewhere on this island, because the ecological conditions are favourable over the whole of the island.

Tropical Donaciinae are generally less brightly coloured (i.e. more brownish) than the temperate species. Perhaps this is linked to a more aquatic life, such as that of *Macroplea*. Although several *Donacia* have been recorded from Bangkok, Thailand and from Indonesia on *Nymphaea* (Nymphaeaceae) and *Nelumbo* (Nelumbaceae) and from India upon leaves and stems of *Nymphaea*, the senior author has captured them only with light traps in the Malay Peninsula. This suggests that the beetles have a hidden life in water but undergo temporary disseminating nocturnal flights. In Java and Sri Lanka, *Donacia aeraria* Baly has been captured on *Potamogeton* (Potamogetonaceae). *Donaciasta* have been captured at light in Rwanda, Africa, which indicates that the adults are not obligate aquatic beetles. In South Africa, *Donacia* live on *Eichornia* (Pontederiaceae), an American plant, and in Madagascar, *Donaciasta* larvae have been captured on *Potamogeton* (Potamogetonaceae) and *Myriophyllum* (Haloragidaceae).

The larvae of Donaciinae feed on roots and stems that are immersed in water and extract oxygen from the plant stems. The very closely related genera *Macroplea* and *Neohaemonia* are totally aquatic and move from the water for only a short time in order to disseminate the species. The adult and larva of *Macroplea* are chiefly found in small ponds and lakes on *Potamogeton* (Potamogetonaceae) and *Myriophyllum* (Haloragidaceae). Both the pupa and cocoon are found among the roots of these plants. The marine species, *Macroplea mutica* (Fabricius), lives in brackish water in the Baltic Sea (and perhaps also in the North Sea) on *Ruppia* (Ruppiaceae), *Zostera* (Zosteraceae), *Potamogeton* (Potamogetonaceae), *Myriophyllum* (Haloragidaceae) and *Zannichellia* (Zannichelliaceae). *Neohaemonia*, also entirely aquatic, lives on *Potamogeton* in the USA and on *Potamogeton* and *Myriophyllum* in Mongolia.

Freshwater *Macroplea* have been also recorded on *Typha* (Typhaceae) and *Carex*, *Scirpus* (Cyperaceae), but the references are not absolutely certain.

A number of *Donacia* host-plants are known from the northern temperate region. Some of these are larval host-plants, while others are exclusively adult food plants (i.e. terrestrial plants). Many of these plants are monocotyledons but others such as Nymphaeaceae and Nelumbaceae are dicotyledons. Some belong also to archaic plant families which possess "beetle flowers". The following have been recorded as host-plants in the literature: *Nymphaea, Nuphar* (Nymphaeaceae), *Nelumbo* (Nelumbaceae), *Alisma, Sagittaria* (Alismataceae), *Potamogeton* (Potamogetonaceae), *Sparganium* (Sparganiaceae), *Typha* (Typhaceae), *Phragmites, Baldingera, Glyceria, Arundo, Oryza, Phalaris, Scolochloa, Triticum* (Poaceae), *Carex, Scirpus, Cladium, Eriophorum, Eleocharis, Schoenoplectus* (Cyperaceae), *Brasenia* (Cabombaceae), *Nymphoides, Menyanthes* (Menyanthaceae), *Myriophyllum* (Haloragidaceae), *Eichornia, Pontederia* (Pontederiaceae), *Acorus, Peltandra, Symplocarpus, Orontium* (Araceae), *Butomus* (Butomaceae), *Ranunculus, Caltha* (Ranunculaceae), *Iris* (Iridaceae), *Zannichellia* (Zannichelliaceae), *Hydrocharis* (Hydrocharitaceae), *Polygonum* (Polygonaceae), *Eriocaulon* (Eriocaulaceae), *Juncus* (Juncaceae), *Tricyrtis* (Liliaceae) and *Equisetum* (Equisetaceae).

The adult eats the pollen from the *Nymphaea* flowers which appear to be the preferred food of this subfamily. All of the plant genera mentioned above are either real or accidental host-plants of the adults, but as previously mentioned, they are not necessarily larval food plants. For instance, *Iris* and *Equisetum* are evidently purely accidental plants upon which the adults may rest. Finally, from the above enumeration, only several plants can be considered as true hosts. No Poaceae or Cyperaceae can be included in the *Donacia* life-cycle because only the entirely aquatic plants are suitable for the larval and nymphal development. Both life-stages must be able to draw oxygen from the sap of the plant stems or roots.

Plateumaris has the same host-plants as *Donacia* but prefers mostly swampy areas and quiet waters, whereas *Donacia* live chiefly but not exclusively amongst running waters. Adult *Plateumaris* occur on *Arundo, Phragmites* (Poaceae), *Cyperus, Scirpus, Carex, Trichophorum, Cladium, Eriophorum* (Cyperaceae), *Caltha, Ranunculus* (Ranunculaceae), *Sparganium* (Sparganiaceae), *Lysichiton, Peltandra, Symphocarpus* (Araceae), *Sagittaria, Alisma* (Alismataceae) and *Iris* (Iridaceae). The larvae feed on roots where the pupae rest inside a cocoon or upon immersed stems.

Adults of several *Donacia* species appear to be, even in Europe, more aquatic than others. This is probably also the case with the African *Donaciasta* but the *Donaciacrioceris,* by their morphology, appear to be more terrestrial.

In general, the Donaciinae are polyphagous and show a selection of the ecological type, i.e. associated with the biotype: quiet waters, slow running streams or swamps.

Macroplea and *Neohaemonia* appear to be slightly more selective, with one species being marine. *Donacia* adults frequent and feed upon a number of plants, but they prefer the Nymphaeaceae and related families. Many plants recorded as hosts for the adults are only accidental because they grow close to the true host-plants in or near the water. *Plateumaris* occur on Poaceae and Cyperaceae where they probably feed on pollen only, like *Donacia.*

Monocotyledons appear to dominate the selection but numerous dicotyledons feed larvae and adults. This is only dependent on the relative frequency of the two plant groups in aquatic or subaquatic environments.

References listed in the bibliography pertaining to Donaciinae are as follows: Askevold (1987, 1990), Bertrand (1965), Borowiec (1983), Chen (1966), Gressitt & Kimoto (1961), Grillas (1988), Jolivet (1968a,b, 1970, 1972c, 1977), Otto (1985), Paik (1967), Pavlov (1984), Szel & Adam (1989).

2.7. *Megalopodinae*

This is mainly a pantropical family which has incurred into the Palaearctic zone in the Far East. Apart from the American species, their biology is virtually unknown but some host-plants and behavioural observations have been recorded in Asia. As in the Clytrinae, males of Megalopodinae are larger than the females. As early as 1845, Lacordaire recorded femoro-tibial haemorrhage in this group, but Monros and other biologists (including P. Jolivet) have not observed this in any species. Lacordaire also mentioned that adults display a strong, grinding of the prothorax against the mesothorax which has been confirmed in the field and is comparable to the sounds produced by the Criocerinae or by other means by the Cassidinae. On the other hand, the biology of Megalopodinae is very different from that originally conceived by Lacordaire. The larvae are not case-bearers like the Clytrinae but are borers like the Cerambycidae. The group is related to the Sagrinae but nymphosis takes place in the ground and the larvae are not gallicolous. Their flight is very fast when the sun is present unlike the Sagrinae which do not fly readily.

The first detailed observations on Megalopodinae biology have been provided by Tella (1952) in Brazil and by Monros (1954a) in Argentina. *Megalopus* and *Agathomerus* cut the apex of the host-plants for several centimetres, especially those of *Lycopersicon* and several *Solanum*, including *S. tuberosum*, *Capsicum* and *Nicotiana* (Solanaceae). At about 10 cm below the cutting, the female produces a circular incision around the stem. This incision, which can be spiral, is not deep and does not induce the fall of the stem. The cutting of the apex of the stem and the subapical incision are both made with the mandibles. The adults also lick the juices (sap) which exudes from the cuts. The free top of the stem is smooth and the cut is very neat. The lower incision is accomplished by successive biting. At 2-3 mm from the top of the cylinder delimited by the female, a small opening leads to an elliptical chamber where an egg is deposited. Egg-laying occurring inside the stem is a very rare occurrence among leaf-beetles.

The larvae of *Agathomerus* and *Megalopus*, described by Monros, live inside stems and feed on plant tissues. They are yellowish and covered with bristles which probably allow them movement inside the galleries. Larvae of both genera have a maximum length of 26 mm.

The remaining aspects of their development have been observed by Brazilian entomologists (e.g. Tella, 1952; Santos, 1980, 1981a,b). There are four larval instars and a prenymphal stage. Nymphosis in *Agathomerus* occurs in the ground. Tella (1952) recorded *Agathomerus sellatus* Germar as a pest of Solanaceae. He described the stem incision which reaches the cambium and has the purpose of diminishing the sap circulation to allow egg development. The female introduces her ovipositor inside the stem in order to deposit the egg. The elliptical egg is yellowish in colour and measures 2.8 × 0.87 mm. Only one egg is deposited per day in each cavity and 44 eggs is the maximum laid by one female during her life. The adults of Megalopodinae feed on the tender parts of the Solanaceae host-plants. The duration of the development is as follows: egg incubation: 5 days; 1st stage (larva): 7 days; 2nd stage: 7 days; 3rd stage: 12 days; 4th stage: 10 days; prenymphosis: 258 days; pupation: 50 days; longevity of the adult: 60 days.

The senior author has personally observed a Brazilian species of *Agathomerus* cutting the stems of *Adenostemma* (Asteraceae). This genus also feeds on *Alternanthera* (Amaranthaceae). *Mastostethus* lives also on *Ricinus* (Euphorbiaceae) and *Jacaranda* (Bignoniaceae). *Plesioagathomerus* feeds on Verbenaceae and *Bothromegalopus* on Solanaceae etc.

In the Orient, *Temnaspis* feeds upon *Fraxinus* (Oleaceae), *Rosa* (Rosaceae) and *Trichosanthes* (Cucurbitaceae), while *Clytraxeloma* feeds on *Sophora* (Fabaceae) and *Poecilomorpha* on *Arachis* (Fabaceae). In Africa, *Colo-baspis* has been recorded from *Solanum* (Solanaceae) and *Colobaspis fraxini* Komiya has been recorded on *Fraxinus* (Oleaceae) in Taiwan by Komiya (1986).

We are ignorant of the biology of Asian and African species. Curiously, no observations on Megalopodinae behaviour have been published in Japan, Korea or Vietnam. However, *Temnaspis* is relatively abundant on *Fraxinus* in Korea, where adults cut the stems like the American species (Seung Lak An, personal communication). This behaviour appears to be general among the subfamily, but has to be restudied in America and reviewed in Asia and Africa. A recent paper by Yu (1977) described the biology of *Temnaspis nankinea* (Pic) in China. It is similar to that of the American species. Recently, *Sphondylia tomentosa* Lacordaire has been recorded feeding on *Rhus zeiheri* (Anacardiaceae) in South Africa (L. Schultze, pers. comm.). Its behaviour is similar to the American and Asian species. Finally, all those genera appear to be polyphagous like the Sagrinae but they show a tendency towards Solanaceae in America and towards Oleaceae in Asia. All of the known host-plants of Megalopodinae are dicotyledons.

References listed in the bibliography pertaining to Megalopodinae are as follows: Jolivet (1977), Komiya (1986), Monros (1954a), Santos (1980, 1981a, b), Tella (1952), Yu (1977).

2.8. *Criocerinae*

The Criocerinae differ from Sagrinae in having a narrow prosternum, thinner femora, smaller size etc., but they also differ in having the tegmen of the aedeagus in a V and not in a ring. The Donaciinae are more closely related to the Sagrinae. The Criocerinae are a homogeneous group, spread all over the world, even on small oceanic islands such as the Mascarene Islands, but they are absent from New Zealand. Most of the Australian and Papua New Guinean species belong to the genus *Stethopachys* which favour various Orchidaceae. The subfamily is of great economic importance and lives on both mono- and dicotyledonous plants, although the first group are more common as food plants. Since the study by the senior author on the host-plants of the family (Jolivet, 1977), a remarkable review has been published by M. Schmitt (in Jolivet *et al*, 1988). The present knowledge on host-plants and biology is reliable and adequate to provide us with a clear view of the trophic preferences of the group except for data on species from Madagascar which have not been studied well. One can only hazard a guess that the Criocerinae of that large island display the same preferences. One Madagascan *Petauristes* (*P. grandis* Klug) feeds on *Vanilla* (Orchidaceae).

From 933 described species in 1913, the Criocerinae are presently represented by a little less than 1400 known species. It is very probable that new species will be added in the near future, because many cryptic species exist on the islands where they are rare and localized. The Criocerinae are not a normal part of the forest canopy fauna and it does not seem probable that many species will be collected from that niche. The host-plants of about 300 species of Criocerinae, i.e. 21.4%, are known. Although this figure may seem low, the sampling has been adequate to establish the selection trends of the group with sufficient accuracy.

Although they differ from the Sagrinae in many morphological characters, the Criocerinae have certain transitional genera, e.g. *Ortholema* (*O. abnormis* Heinze), a gallicolous Chinese species. The American *Neolema* display the beginnings of spermophagy, i.e. they feed on the ovules of the Commelinaceae, but it is not very marked. The Criocerinae are a transitional group between the *Eupoda* and the Chrysomelinae. Their wing venation belongs to the classic "chrysomelid" type.

The biology of the larvae may vary remarkably from one genus to another or from one species to another. Several larvae are leaf-miners (*Neolema*) on Commelinaceae, but other larvae of *Neolema* are free-living. Some are covered with excreta, such as *Lilioceris* and *Plectonycha*. The nymphosis takes place inside a cocoon, either in the ground at the base of a plant (e.g. *Oulema, Crioceris, Lilioceris*) or on the leaves (*Plectonycha*). The cocoon can be simple or elaborate in design. The larvae are often solitary but can be gregarious in *Lema* (*Quasilema*) *latipennis* Pic, in Argentina. They live together in a leaf-mine and move together to the next leaf (Monros, 1959). It should be noted that *Neolema* are not the only ones to devour fruits, since *Crioceris* also often consume *Asparagus* fruits mostly during the second generation.

The species (both larvae and adults) that attack monocotyledons have a tendency to consume the host leaf tissues in a straight line between the parallel veins, leaving one of the two epidermides uneaten. The species feeding on dicotyledons appear to be more voracious and destroy the whole foliar lamina and the young stems with the exception of the largest veins (Monros, 1959).

The classification of the Criocerinae is relatively straightforward but many species remain misplaced in their proposed subgenera. For instance, *Bradylema* previously considered as a separate genus, is probably a *Petauristes*, a subgenus of *Lema*. As with many species of the subfamily, *Petauristes* feed on orchids but not exclusively. It is very difficult to clearly divide *Lema* into subgenera and many species have been grouped under *Lema* sensu lato.

In the paper by the senior author (Jolivet, 1977), the affinities of different genera were discussed as well as the adaptations of certain species for feeding on Solanaceae and the occasional specificity of the beetles to certain genera or species of that plant family. Experiments have shown that species feeding on Solanaceae reject absolutely the Liliaceae, but that certain Liliaceae-feeding species accept accidentally the Solanaceae (allotrophy). The sudden host-plant change from *Lilium* to *Salix,* quoted several times in the literature by Maulik (1947), Saulnier (1947) and more recently by Berti and Rapilly (1976) for *Lilioceris merdigera* (L.) and *L. chodjanii* Berti and Rapilly, should be reverified. No one has really observed the beetle feeding on *Salix*, but only resting on the tree. This habit may be related to a climatic factor (stress) and could be similar to the secondary host-plant choice of the Alticinae. Perhaps a survival factor? In any case, this has nothing in common with Tempere's observations (1946, 1967) of *Lilioceris lilii* (Scopoli) shifting from *Lilium* to *Solanum dulcamara, S. laciniatum* and *Atropa belladonna* (Solanaceae) in Gironde, south-west France. This phenomenon can possibly be linked with the selective potential of the subfamily as a whole.

The following is a summary of the food selections arranged by genus or subgenus:

1- *Lilioceris* Reitter, 1912.

World-wide genus. On Liliaceae (sens. lat.) mostly, i.e. *Lilium*, *Scilla*, *Fritillaria*, *Tricyrtis*, *Nomocharis*, *Polygonatum*, *Convallaria*, *Asparagus*, *Asphodelus*, *Muscari*, *Anthericum*, *Majanthemum.* Also on *Allium* (Alliaceae), *Smilax* (Smilacaceae), *Pandanus* (Pandanaceae), *Dioscorea* (Dioscoreaceae), *Xanthorrhoea* (Xanthorrhoeaceae) and Orchidaceae (various genera). All of these plants belong to the Monocotyledonae. Two species have been captured on *Cycas* (Cycadaceae) in New Guinea and Vietnam. The adults and larvae feed on the unrolled fronds. More recently, one species, *L. nigripes* (Fabricius), has been recorded feeding on the young fronds of *Bowenia spectabilis* (Zamiaceae) in north-eastern Queensland (Hawkeswood, 1992). Of all the cases of allotrophy, the only certain one is the shift to Solanaceae by *Lilioceris lilii* in the southwest of France. The *Salix* case, previously mentioned, is more doubtful. Other reports are only of accidental cases of adult perching on non-food-plants, viz. *Triticum* (Poaceae), *Cassia* (Caesalpiniaceae), *Fagus* (Fagaceae), *Schefflera* (Araliaceae) and *Corylus* (Corylaceae). The beetles do not feed on these plants.

2- *Crioceris* Müller, 1764.

The entire genus appears to be strongly associated with Liliaceae (sens. lat.), mostly Asparaginae (*Asparagus*). The larvae and adults feed on phyllocladia and stems. Some species of *Crioceris* are restricted to certain species of *Asparagus* and reject others (even under experimental conditions) (Schmitt, 1988). Some species are more polyphagous (on *Asparagus*) than others.

3- *Elisabethana* Heinze, 1928.

In Africa, *E. militaris* has been recorded on *Convallaria majalis* (Liliaceae), which appears impossible because this plant does not occur in Africa. However, its food-plants are most likely to be other Liliaceae of a similar growth habit.

4- *Sigrisma* Fairmaire, 1888.

On *Asparagus* (Liliaceae) in East and South Africa (Schmitt, 1988).

5- *Lema* Fabricius, 1798.

The large genus *Lema* has been divided into at least 5 subgenera. It is very difficult to correctly place certain species. Theoretically, *Neolema* feeds on Commelinaceae, *Quasilema* on Solanaceae, *Petauristes* on Orchidaceae and *Lema* (sensu stricto) on Poaceae, but there are so many exceptions that it is very difficult to generalize. However, it seems that many of these subgenera appear totally artificial when the food selections within them are analyzed. Many records concerning *Lema* are doubtful and arise from captures by sweeping nets, i.e. without any reference to a real host-plant. Such examples include Rubiaceae (Kenya), Brassicaceae (Switzerland), Theaceae (*Thea*) and Polygonaceae (China), Geraniaceae (Venezuela), Apiaceae, Convolvulaceae, Verbenaceae (Brazil), Fabaceae (Madagascar and Mexico), Malvaceae (Ghana) and Pedaliaceae (*Sesamum*)(Nicaragua). It is difficult to determine the role of those plants in secondary trophisms by these observations. A strong-flying insect such as *Lema* may land anywhere and it is evident that very often those supposed host-plants are only observations of accidents. The record of *Lema* (*Quasilema*) *concolor* Leconte by Pallister (1953) "feeding on ferns at high altitudes in Mexico" is surprising and probably represents a secondary food-plant. According to Hendrix (1980), as quoted by Schmitt (1988), only six Chrysomelidae are known to feed on ferns (Pteridophyta). Actually there are many more, mostly Alticinae, which mainly frequent the tree ferns of tropical mountains (see Chapter 3). The observations of *Lema* on Poaceae in most cases concern pollen feeding but the normal food-plants are mostly *Dactylis*, *Avena*, *Zea*, *Hordeum*, *Triticum* and *Secale*.

The following list, arranged in subgenera, is far from being complete:

5a- Subgenus *Lema* Fabricius, 1798 (sensu stricto).

This cosmopolitan subgenus occurs on more than 20 species of Poaceae, Cyperaceae, Commelinaceae, Zingiberaceae, Liliaceae and Dioscoreaceae (Monocotyledonae) and Asteraceae, the most evolved family of the Dicotyledonae. The taxonomy of the subgenus needs to be revised.

5b- Subgenus *Petauristes* Latreille, 1829.

As with *Stethopachys*, this subgenus is specialized on Orchidaceae. Host-plants for about 15 genera have been recorded. Some species may also be captured on other monocots, e.g. Musaceae, Dioscoreaceae, Iridaceae, Zingiberaceae, Costaceae, Pandanaceae, Commelinaceae and Arecaceae. Also captured on *Lycium* and *Physoclaina* (Solanaceae) in Africa.

5c- Subgenus *Quasilema* Monros, 1951.

This Neotropical subgenus is specialized on Solanaceae (at least 14 genera) but the various species have restricted preferences. Numerous North American studies (e.g. Force, 1966; Kogan & Goeden, 1970 a,b,c) have recorded the reaction of *Lema trilineata* White to different Solanaceae. Some are preferred over others while some such as *Lycopersicon*, *Solanum melongena* and *Browallia speciosa*, are totally rejected. *Lema bilineata* Germar has been introduced from Argentina to South Africa (Omer-Cooper & Miles, 1951). This species is also very selective. Some records mention several *Quasilema* on Commelinaceae, Dioscoreaceae and Smilacaceae, but these could be misidentifications. Accidental captures, without determined trophic relationships, have been reported from Convolvulaceae, Saxifragaceae, Fabaceae, Grossulariaceae and Pedaliaceae.

5d- Subgenus *Neolema* Monros, 1951.

All species of *Neolema* feed on *Commelina* and *Tradescantia* (Commelinaceae) in Brazil. Certain species are leaf-miners in Argentina (e.g. *L. quadrivittata* Boheman), while others are not (e.g. *L. dorsalis* Olivier). The latter species feeds on leaves and ovules of Commelinaceae (Monros, 1959).

5e- Subgenus *Pachylema* Monros, 1951.

On Commelinaceae and Poaceae in the Neotropics.

5f- Subgenus *Oulema* Gozis, 1886.

All *Oulema* species, whether they are Holarctic, Asiatic or African, feed on Poaceae: *Imperata*, *Phragmites*, *Leersia*, *Isachne*, *Digitaria*, *Eleusine*, *Setaria*, *Oryza*, *Pennisetum*, *Zizania*, *Glyceria*, *Agrostis*, *Echinochloa*, *Holcus*, *Sorghum*, *Panicum*, *Triticum*, *Dactylis*, *Hordeum*, *Avena*, *Secale*, *Zea*, *Festuca*, *Bromus*, *Agropyrum*, *Brachypodium*, *Lolium*, *Poa*, *Arundo*, *Arrhenatherum*, *Calamagrostis* etc. Several *Oulema* have been wrongly recorded from *Nasturtium* and *Rorippa* (Brassicaceae) in Poland, Switzerland, etc. In addition, the following purported hosts are incorrect: *Rumex* and *Emex* (Polygonaceae), *Salix* (Salicaceae), *Angelica* (Apiaceae), *Filipendula* (Rosaceae), *Curcuma* (Zingiberaceae), *Galium* (Rubiaceae), *Lythrum* (Lythraceae), *Metasequoia* (Taxodiaceae), *Dioscorea* (Dioscoreaceae) and *Cirsium* (Asteraceae). Feeding trials have been conducted on Commelinaceae and Polygonaceae without success using a rice-feeding *Oulema* species (Kuwayama, 1954). *Oulema* larvae are free-living on leaves and are covered with mucous and excreta.

5g- Subgenus *Hapsidolemoides* Monros, 1951.

Artificial subgenus (Schmitt, 1988). Several species on *Cirsium* (Asteraceae) in the New World.

6- *Ortholema* Heinze, 1943.

On *Paspalum* (Poaceae) in Asia where the larva is probably gallicolous.

7- *Plectonycha* Lacordaire, 1845.

On *Boussingaultia* and other Basellaceae (Dicotyledonae) in tropical America.

8- *Stethopachys* Baly, 1861.

On various Orchidaceae (e.g. *Dendrobium*, *Cymbidium*) in Asia and Australia.

Concluding comments on the Criocerinae: Still nothing is known of the biology of 11 tropical genera. There is no reason to believe that their biology and host-plants could be significantly different from the others so that we could easily extrapolate from the information already known about the other closely related species or genera to guess the hosts of these unknowns. It can be seen that the Criocerinae are not polyphagous but are very specific towards their host-plants. The placement of species among the subgenera of *Lema* is not entirely satisfactory. This can sometimes explain the incoherence of certain selections. Many species display a choice for Zingiberaceae, Poaceae, Orchidaceae, Dioscoreaceae etc. Several genera appear to be rather selective, i.e. *Neolema* on Commelinaceae, *Lema* (sens. str.) on Asteraceae and Poaceae, *Oulema* on Poaceae, *Quasilema* on Solanaceae and *Petauristes* and *Stethopachys* on Orchidaceae, etc. The experiments undertaken on solanophagous species such as *Lema* (*Quasilema*) *trilineata* White, show that only certain species or genera of Solanaceae are consumed but still with the same relish according to the taste. *Datura* and *Physalis* are among the plants most favoured (Force, 1966). The larvae sometimes nibble Convolvulaceae, Polygonaceae, Liliaceae, Chenopodiaceae and Brassicaceae, but these plants are finally refused by the adults, and the larvae cannot develop on them. In South Africa, *Lema bilineata* Germar, a species imported from Brazil and Argentina where it feeds on *Salpichroa rhomboidea* or *S. oreganifolia*, has adapted to *Physalis, Datura, Nicotiana* and *Nicandra*. All of these plants are American Solanaceae introduced to Africa. Several cases of so-called allotrophy of *Lilioceris* from Liliaceae to Solanaceae, in France and elsewhere, outline very well the hidden tendency of many species towards the monocotyledons and Solanaceae. However, the mechanism of this selection is not well understood. Also, the fact that two species of *Lilioceris* from South-East Asia eat the leaves of Cycadaceae and one from Australia feeds on Zamiaceae, can be interpreted as a primitive characteristic of this group of leaf-beetles.

References listed in the bibliography pertaining to Criocerinae are as follows: -Battenfield *et al.* (1982), Berti & Rapilly (1976), Borisenko & Burdun (1985), Cantelo & Pholboon (1965), Force (1966), Gressitt (1965), Gressitt & Kimoto (1961-63), Hawkeswood (1985, 1987, 1992), Hendrix (1980), Hoxie & Wellso (1974), Jolivet (1953, 1971b, 1977), Kogan & Goeden (1970a,b,c), Kostromitin & Muraveva (1982), Kuwayama (1954), Maulik (1947), Monros (1959), Mummery & Valadon (1974), Omer-Cooper & Miles (1951), Pallister (1953), Samedov & Mirzoeva (1982), Saulnier (1947), Schmitt (1988), Tempere (1927, 1946, 1967), Wellso & Hoxie (1988).

2.9. Synetinae

There is not much information to be added to the previous review paper on this subfamily by Jolivet (1977), because nothing really new has been published on Synetinae recently. The work of Wilcox (1979) on "Leaf beetle host-plants in northeastern North America" merely confirms the host selections for the group. Despite the fact that the larvae are subterraneous and root-feeding, the "eumolpine" relationships seem highly unlikely.

The group is really enigmatic systematically. Several authors (Crowson, 1953; Kurcheva, 1967) attempted to place the Synetinae among the Eumolpinae, but even Kurcheva who studied the larvae, now admits that this relationship is very doubtful. The subfamily has been very well studied by Edwards (1953) and its systematic position discussed by Jolivet (1957).

For *Syneta* and *Tricholema*, Edwards has proposed the creation of the family Synetidae, which has been reduced here to the position of subfamily. As has been mentioned previously (Jolivet, 1957), the wing venation of the genera *Syneta* and *Tricholema* is as aberrant as their external morphology. However, this venation is of the eupodic type and lacks eumolpine characters, although it possesses an unique medio-cubital system. The medio-cubital spot of the Eumolpinae is absent. Jolivet concluded that *Syneta* has unquestionable orsodacnine affinities (i.e. morphology, cubito-anal system), but differs from that group by its unique larval and wing characteristics. The creation of a special subfamily is therefore necessary at the end of the *Eupoda*. The remote affinities with the Galerucinae cannot be fully rejected. The morphology of the *Syneta* aedeagus is also very peculiar. It is in no way eumolpine and its tegmen is V-shaped.

The *Syneta* larva had been described by Kurcheva (1967) as new, but the larva of *S. albida* LeConte had been described previously in 1913 from the USA by Wilson and Moznette. The larva was obtained by digging around rootlets of fruit trees. The *Syneta* female drops her eggs haphazardly on the ground where they hatch after two or three weeks. There is no protection for the eggs.

Like the Donaciinae, the Synetinae are polyphagous, but whereas Donaciinae larvae feed underwater on roots and stems and Zeugophorinae larvae are leaf-miners, *Syneta* larvae live in the ground like Eumolpinae and nibble on roots and rootlets. Food selection is rather broad and not restricted. Extensive sampling of the dominant species of Synetinae in the northern cold forests has indicated their selection is one of an ecological type upon Gymnospermae and Dicotyledonae. The *Syneta* species live either at high altitudes in the USA (including Alaska) and Canada, or in the European taiga. *Syneta adamsi* Baly has a larger distribution in the Far East (i.e. from Oriental Siberia, Manchuria, Sakhalin, Kuriles and Japan from Hokkaido to Kyushu and Tsuhima). The general tendency, despite several exceptions, e.g. *Syneta seriata* LeConte, is to occupy the high altitude forests or the taiga. Several species, e.g. *Syneta albida* LeConte, are pests of fruit trees in the USA. Plants attacked include *Prunus* (plum, peach), *Cydonia* (quince), *Pyrus* (pear), *Ribes* (currant), cherry etc. Basically, the genus *Syneta* is purely Holarctic in the cold zone, but with several incursions into the south of the USA, on the Pacific coast and in the east up to Georgia. The trophic selections of Synetinae are as follows:- Gymnospermae-*Abies, Tsuga, Picea, Larix, Pinus* (Pinaceae); Angiospermae (Dicotyledonae)-*Betula, Alnus* (Betulaceae), *Salix* (Salicaceae), *Carpinus* (Carpinaceae), *Fagus, Quercus* (Fagaceae), *Corylus* (Corylaceae), *Prunus, Pyrus, Cydonia, Rubus* (Rosaceae), *Ribes* (Grossulariaceae), *Acer* (Aceraceae), *Aesculus* (Hippocastaneaceae) and *Cornus* (Cornaceae). Overall, the selection is rather broad but many species show the dual choice of Gymnospermae and Betulaceae.

References listed in the bibliography pertaining to Synetinae are as follows: Crowson (1953), Edwards (1953), Jolivet (1957, 1977), Kurcheva (1967), Lee (1990), Wilcox (1979), Yu (1988).

Concluding comments on Eupoda:

The section Eupoda contains 9 somewhat heterogeneous subfamilies some of which exhibit primitive characteristics. The most primitive groups are the Sagrinae (gallicolous and polyphagous), Orsodacninae (root-feeders? and polyphagous) and Zeugophorinae (miners and oligophagous). Two groups, the Donaciinae and Synetinae, show a loose selection, which is more ecological than botanical. The Aulacoscelinae, which form a bridge between the Sagrinae and the Trichostoma, have a food selection which is poorly known. The Megalopodinae and Criocerinae are the only evolved groups with a restricted food choice. The Criocerinae and Synetinae are the only ones of the section to possess a V-shaped tegmen of the aedeagus which indicates a certain evolution from the ancestral archetype which possessed a ring tegmen like all of the other Eupoda. It should be noted that apart from the Eupoda, only the genus *Timarcha*

(Chrysomelinae) has a ring tegmen with a ciliated cap, a fact that remains an inexplicable anachronism.

SECTION II: CAMPTOSOMA

2.10. Clytrinae

The general tendency of this subfamily is towards myrmecophily or submyrmecophily in the larval stage and exceptionally in the adult stage among several species frequenting the stipular thorns of the East-African *Acacia* (Mimosaceae)(Selman, 1962). The larvae of these strange phytophagous insects feed in and around ant nests on branches and refuse, but equally upon ant eggs, exuviae, dead ant bodies and excreta. The adults are polyphagous on young leaves of shrubs (Jolivet, 1952 b,c, 1978; Erber, 1988). Reference is made here to the papers of P. Jolivet concerning detailed study of the selections of this subfamily.

In the work on the food preferences of the Chrysomelidae (Jolivet, Petitpierre & Hsiao, 1988), emphasis was made that the Clytrinae feed upon an enormous quantity of plants (37 families), both mono- and dicotyledons. However, the general tendency, perhaps strongly influenced by the local dominants of the forest flora, appear to be Mimosaceae in America, Fagaceae and Betulaceae in Europe and Malvaceae and Fabaceae in Africa. The choices depend mostly on the availability of the plants, on leaf succulence and upon leaf age. Adults feed on flowers, buds, very young leaves and pollen, without any discrimination. Normally the males, after hatching at spring time, congregate on the shrubs and after several days, the females arrive. After mating, those females become reclusive. They either drop their eggs, which may be strangely decorated, onto the ground *(Clytra)*, or stick them to a leaf with long setiform pedicels of excremental origin (*Labidostomis*). The eggs, decorated with excreta and secretions from the female, resemble seeds and are often carried away by the ants into their nests. In other cases, the larva clings to the ants in order to penetrate the nest (Samsinak, 1956). Not all Clytrinae larvae live inside the nests, but some live under stones in the vicinity of ant nests *(Labidostomis)* like several Tenebrionidae, Cetoniinae (Scarabaeidae), etc. Some rare species do not appear to be myrmecophilous, but myrmecophily is the rule in the temperate and tropical zones of the world.

The Clytrinae are extremely rare in the islands, even the large ones such as New Guinea, Australia and Madagascar. They are absent from the Mascarene Islands, most of the Pacific archipelagoes and all of the oceanic islands of the world. It appears that the complexity of their biology is the main reason for that absence, although most of the required ants exist on those islands. Also, the weight of the adult can be an obstacle in wind transportation.

Although there is an attraction of certain species of Clytrinae towards certain genera of ants, there is not a strong specificity. Submyrmecophilous species live as detritus-feeders near the ant nests. Many genera of ant hosts have been recorded, i.e. *Formica*, *Solenopsis*, *Camponotus*, *Atta*, *Megaponera*, *Cataglyphis*, *Lasius*, *Tetramorium*, *Messor*, *Plagiolepis*, *Aphaenogaster*, *Tapinoma*, *Crematogaster*, etc. The phenomenon of myrmecophily has been recorded everywhere in America, Africa, Europe, Asia and Australia.

Generally, the males of Clytrinae are more visible than the females and are associated with flowers, whereas the females remain within the bushes or disperse in order to stick their eggs on stems or leaves, or they may simply drop them onto the ground below.

An interesting case of Batesian mimicry between the male of the buprestid, *Chrysobothris humilis* Horn, and a clytrine, *Saxinis deserticola* Moldenke, can be explained by the fact that the males of the buprestid remain on the stems on the outside of the shrubs (Hespenheide, 1976). The buprestid mimics the males of *S. deserticola* which also remain on the outside of the stems and are also toxic and similar to their females (in regard to reflex bleeding of the toxic blood) which remain hidden. For the males of the *Chrysobothris* and *Saxinis*, the fact that they remain on the outside of the Fabaceae shrubs is part of a reproductive strategy common to both beetles. The female buprestids require more protection, but the predation on the males is reduced by the mimetism. Buprestid females are very different from the males, being metallic green in colour.

A study of the food preferences has been provided in a previous paper (Jolivet, 1978). A summary of the data presently available on this subfamily, listed genus by genus, is provided below:

2.10.1. Tribe Clytrini (Mostly found in the Old World)

1- *Labidostomis* Germar, 1817.

Labidostomis feed in the adult stage on the leaves of *Salix*, *Populus* (Salicaceae), *Quercus* (Fagaceae), *Corylus* (Corylaceae), *Ulmus* (Ulmaceae), *Betula* (Betulaceae), *Prunus*, *Cerasus*, *Rosa*, *Spiraea*, *Crataegus*, *Amygdalus* (Rosaceae), *Hordeum*, *Triticum*, *Phalaris*, *Secale* (Poaceae), *Gossypium* (Malvaceae), *Beta*, *Chenopodium* (Chenopodiaceae), *Vitis* (Vitaceae), *Lotus*, *Trifolium*, *Anthyllis*, *Astralagus*, *Lespedeza*, *Caragana*, *Dorycnium*, *Medicago* (Fabaceae), *Rhus* (Anacardiaceae), *Cichorium*, *Chrysanthemum*, *Artemisia* (Asteraceae), *Citrus* (Rutaceae), *Rumex*, *Atraphaxis*, *Polygonum* (Polygonaceae), *Nicotiana* (Solanaceae), *Tamarix* (Tamaricaceae), *Pistacia* (Pistaciaceae), *Cistus* (Cistaceae), *Angelica* (Apiaceae), *Elaeagnus* (Elaeagnaceae), *Euclidium*, *Descurainia* (Brassicaceae) and *Salvia* (Lamiaceae). *Labidostomis* species are really polyphagous and often attack the young buds of vine trees, willow shrubs, grass inflorescences, fruit trees and young leaves of oak. Several low-growing plants are equally fed upon in poorly or non-wooded subdesertic or mountainous regions. They feed on Poaceae in order to obtain pollen.

2- *Miopristis* Lacordaire, 1848.
On Pistaciaceae and Anacardiaceae in North Africa according to Furth (1985b).

3- *Macrolenes* Chevrolat, 1837.
These species live in Europe on Fagaceae and in North Africa on fruit trees and various mediterranean shrubs, e.g. *Rhus* (Anacardiaceae), *Pistacia* (Pistaciaceae), *Zizyphus* (Rhamnaceae) and *Tamarix* (Tamaricaceae).

4- *Antipus* Degeer, 1778.
Few data exist on this genus. On cotton flowers (*Gossypium,* Malvaceae) in Mozambique (Hargreaves, 1948).

5- *Tituboea* Lacordaire, 1848.
According to Medvedev (1962), several species of this genus are associated with ants of the genera *Messor*, *Plagiolepis* and *Aphaenogaster* in Europe and Asia. On various plants: Fagaceae, Pistaciaceae, Anacardiaceae, Fabaceae, Vitaceae, Rosaceae, Tamaricaceae, etc. Also feeds on pollen of Poaceae, catkins of *Quercus*, capitula of Asteraceae, corymbs of *Sambucus*, etc.

6- *Anomoea* Agassiz, 1846.
According to Moldenke (1970), the larva of one species consumes detritus on the ground without being associated with ants. The general tendency of the American species is towards Fabaceae, Caesalpiniaceae and Mimosaceae, although other host-plants have been recorded.

7- *Diapromorpha* Lacordaire, 1848.
In Africa, on *Acacia* (Mimosaceae) and in Asia on *Derris* (Fabaceae), *Thea* (Theaceae), *Cannabis* (Cannabaceae), etc.

8- *Peploptera* Lacordaire, 1848.
On *Hibiscus* (Malvaceae) in Ghana and on *Acacia* (Mimosaceae) in Kenya.

9- *Aspidolopha* Lacordaire, 1848.
On *Derris* (Fabaceae) in Malaysia.

10- *Aetheomorpha* Lacordaire, 1848.
Recorded from Cambodia on *Zea mays* (Poaceae) feeding on pollen.

11- *Lachnaia* Chevrolat, 1837.
Lachnaia larvae live inside *Cataglyphis* ant nests and the adults consume leaves, flowers and pollen of various plants, mostly Asteraceae, Fagaceae, Anacardiaceae, Rhamnaceae, Rosaceae, Vitaceae, Fabaceae and Pistaciaceae as well as several other plants.

12- *Clytra* Laicharting, 1781.

Clytra larvae are common within the large nests of *Formica rufa* as well as in the nests of *Lasius, Messor, Camponotus* and *Cataglyphis*. Adults are common during spring on fruit trees. Adults have been found on 42 genera belonging to 25 plant families, which highlights the degree of polyphagy exhibited by this genus. They are also found in Europe on introduced plants.

13- *Clytrasoma* Jacoby, 1908.

Clytrasoma have been captured in Senegal on *Pennisetum* (Poaceae) feeding on pollen.

14- *Herma* Weise, 1900.

On *Brachiaria* flowers (Poaceae) in Kenya. Also polyphagous.

15- *Melitonoma* Chevrolat, 1837.

On rice (*Oryza*, Poaceae) in Kenya and cotton (*Gossypium,* Malvaceae) in Nigeria.

16- *Smaragdina* Chevrolat, 1837.

The larvae are mainly associated with *Tetramorium* ants. In the New World, *Smaragdina* are mostly associated with Mimosaceae, but sometimes with *Quercus* (Fagaceae) as in Europe. The Palaearctic species feed on young shrubs of *Quercus* and *Castanea* (Fagaceae), *Corylus* (Corylaceae), *Alnus* and *Betula* (Betulaceae), *Crataegus, Sorbus, Prunus* (Rosaceae) and also on low herbs such as *Rumex* and *Polygonum* (Polygonaceae). Also recorded from Salicaceae, numerous Fabaceae, Asteraceae, Solanaceae, Convolvulaceae, etc. Various Poaceae provide pollen. The genus is cosmopolitan.

17- *Chilotomina* Reitter, 1912.

Mostly recorded from arborescent Rosaceae in Europe.

18- *Cheilotoma* Chevrolat, 1837.

On various low herbs and shrubs in Europe and Asia Minor. The selections are of the classic type: Salicaceae, Ulmaceae, Rosaceae, Corylaceae, Fabaceae and Polygonaceae.

19- *Coptocephala* Chevrolat, 1837.

Adults of this genus are polyphagous and common on the flowers of Apiaceae. They also occur on bushes of *Rhus* (Anacardiaceae), *Quercus* (Fagaceae) and other low plants such as Lamiaceae, Asteraceae and Dipsacaceae.

20- *Hockingia* Selman, 1962.

The adult lives as a symphile within the stipular thorns of *Acacia drepanolobium* (Mimosaceae) in East Africa, being associated with the ants *Crematogaster sjostedti* and *Camponotus rufoglaucus*. In association with

Hockingia, one cryptocephaline and several eumolpines are commonly found. The adult is probably detriticolous and carnivorous upon ant eggs like the typical clytrine larva.

2.10.2. Tribe Megalostomini (New World)

21- *Megalostomis* Chevrolat, 1837.
This American genus appears to be strongly associated with Mimosaceae (*Acacia, Prosopis*) and Caesalpiniaceae (*Cassia*). Larvae inhabit ant nests.

22- *Coscinoptera* Lacordaire, 1848.
The larvae live in ant nests (Cockerell, 1891). The adults prefer Mimosaceae, but accept a great variety of plants such as Rosaceae, Fagaceae, Polygonaceae, Anacardiaceae and several Fabaceae.

23- *Euryscopa* Lacordaire, 1848.
Many *Euryscopa* frequent mostly shrubby species of Fabaceae. Occasionally *Euryscopa* feed on Polygonaceae, *Quercus* (Fagaceae) and various Asteraceae.

2.10.3. Tribe Babiini (New World)

24- *Helioscopa* Gistel, 1848.
On *Acacia* (Mimosaceae) in Brazil. All Babiini larvae as far as known live in ant nests.

25- *Dachrys* Erichson, 1847.
On Rhamnaceae in Patagonia.

26- *Temnodachrys* Monros, 1951.
Only on shrubs of *Sesbania* (Fabaceae), *Piptadenia*, *Acacia*, *Prosopis* (Mimosaceae) and *Rosa* (Rosaceae) in Argentina and Brazil.

27- *Babia* Chevrolat, 1837.
Adults of this genus appear to be associated everywhere in Latin America with Mimosaceae (*Prosopis*, *Acacia)* and Fabaceae (*Robinia, Desmodium*) but certain species have been collected from *Hydrangea, Carya, Ceanothus* and various Lamiaceae.

28- *Stereoma* Lacordaire, 1848.
On *Acacia* (Mimosaceae) in Brazil. Also captured on leaves of *Cocos* (Arecaceae) and *Lantana* (Verbenaceae).

29- *Urodera* Lacordaire, 1848.
Restricted to the legume families Mimosaceae (*Prosopis, Acacia*), Caesalpiniaceae (*Cassia*) and Fabaceae (*Sesbania*).

30- *Saxinis* Lacordaire, 1848.

The species of this genus are closely associated in America with Fabaceae, Mimosaceae and Caesalpiniaceae. However, there are some records from *Quercus* (Fagaceae), Rosaceae, Polygonaceae, Rhamnaceae, Hydrangeaceae and Asteraceae. Larvae are myrmecophilous.

2.10.4. Tribe Ischiopachini (America)

31- *Ischiopachys* Chevrolat, 1837.

On *Prosopis, Mimosa* and *Piptadenia* (Mimosaceae) and *Caesalpinia* (Caesalpiniaceae) in Argentina. The genus has also been collected from *Borreria* (Rubiaceae), *Carica* (Caricaceae) and *Citrus* (Rutaceae).

Concluding remarks on Clytrinae:

The food-plant preferences of 31 clytrine genera out of a total of 63 known genera, have been recorded (i.e. 49.2%). However, this number does not correspond to the number of species observed, because many genera are rare and composed of only a few species. Several African genera, such as *Protoclytra*, have never been observed feeding, but they are frequently found on low herbs, as are other species. It seems peculiar that the greatest number of species are found in Africa (31.1%) whereas there are none known from New Zealand and only a few occur in the Australo-Papuan region (0.1%). It is in the Old World (i.e. Asia, Europe, continental Africa) and the Neotropical regions that they are the most common. Few species have been recorded from Madagascar.

Generally, among the Clytrinae there exist feeding preferences, but not trophic selections. The group is essentially polyphagous and adapted to the availability of bushes and low herbs growing in lowlands and at high altitudes. If the Clytrinae appear to show a predilection for young trees and bushes (i.e. Mimosaceae, Fabaceae, Caesalpiniaceae, *Quercus*, etc.), then it is directly related to the hidden life of the females and the succulence of the young leaves of the host-plants. The geographical selections are only determined by the frequency of the botanical associations according to the regions and the availability of ant nests. The larvae penetrate the nests actively (phoresy) or passively (in the form of eggs) and this behaviour is valid for almost all of the species. However, there are several species which are not completely myrmecophilous and prefer to remain in the vicinity of the ant nests. These feed on detritus, ant excreta and/or dead ant bodies. Feeding on pollen (mostly that of the Poaceae) is probably related to the requirement of proteins and amino-acids in the diet.

Polyphagous adult Clytrinae do not however feed on every plant group. For instance, Rubiaceae, Euphorbiaceae and Asclepiadaceae are avoided most of the time. Many of these are latex-bearing plants which are very toxic to most insects. However, there are very rare reports in the literature of Clytrinae from the toxic *Glochidion* (Euphorbiaceae) and *Borreria* (Rubiaceae) which need to be verified since they may be accidental. In addition, several genera of Clytrinae have species which feed on *Derris* (Fabaceae) in Asia (*Aspidolopha*) and Africa (*Diapromorpha*). These species must possess the ability to detoxify this highly poisonous plant which is used by humans as a potent insecticide or as a fish killer. Likewise for *Lonchocarpus* (Fabaceae) and *Azadirachta indica* (Meliaceae). There are also many other insects feeding on these "insecticidal" plants.

References listed in the bibliography pertaining to Clytrinae are as follows: Cockerell (1891), Erber (1969, 1988), Furth (1985b), Hargreaves (1948), Hespenheide (1976), Jolivet (1952b, c, 1978), Jolivet *et al.*(1988), Medvedev (1962, 1970), Moldenke (1970), Samsinak (1956), Selman (1962).

2.11. *Cryptocephalinae*

Very few Cryptocephalinae are myrmecophilous or submyrmecophilous like the clytrine *Labidostomis*. However, at least one species, *Isnus petasus* Selman, is known where the adult lives as a symphile inside the stipular thorns of *Acacia drepanolobium* (Mimosaceae) in East Africa. Generally, the case-bearing larvae of Cryptocephalinae are free-living like Chlamisinae and Lamprosomatinae.

The Holarctic Cryptocephalinae are not very specific in their food choice. Several species feed on Gymnospermae (Pinaceae), but this is exceptional and the great majority feed on Betulaceae, Fagaceae and related plant families and on Asteraceae. The choice of young shrubs or regrowth on older trees is rather general. There are very few differences between the food preferences of the Clytrinae and those of the Cryptocephalinae, although the beginning of specialization appears within the latter subfamily. Among the Monocotyledonae, only the Poaceae are accepted by the adults which consume the pollen and the flowers but not the leaves. Cryptocephaline adults usually eat very little but sometimes adults and larvae damage young leaves.

The larvae of Cryptocephalinae are detriticolous, but sometimes phytophagous like the adult. Details on larval biology are not well described. Very exceptionally some species appear to live symbiotically with *Lasius* ants (Jolivet, 1978). Others feed on plant leaves (Le Sage, 1984a; Erber, 1988), or have a hidden life at the base of the plants amongst roots.

2.11.1. Tribe Stylosomini

1- *Stylosomus* Suffrian, 1848.

The species of *Stylosomus* are recorded from *Tamarix*, *Myricaria* (Tamaricaceae), *Spiraea* (Rosaceae) and several other plants such as Betulaceae, Fagaceae and Ericaceae. Polyphagous but with tendency towards monophagy on *Tamarix* in subdesertic areas.

2.11.2. Tribe Pachybrachini

2- *Griburius* Haldeman, 1849.
The tendency of this American genus is towards the Mimosaceae (*Vachellia*, *Prosopis*) and Fabaceae (*Desmodium*) but the species also feed on Asteraceae, Ericaceae and Fagaceae.

3- *Metallactus* Suffrian, 1866.
A Brazilian species has been recorded from local *Psidium* and imported *Eucalyptus* (Myrtaceae).

4- *Acolastus* Gerstaecker, 1855.
In Senegal on *Pennisetum* (Poaceae).

5- *Pachybrachis* Chevrolat, 1837.
Genus distributed in the Palaearctic region and the New World. Depending on the area, the trophic preferences are usually difficult to determine (Jolivet, 1978). The Holarctic species apparently feed on *Populus, Salix* (Salicaceae), *Quercus* (Fagaceae), *Alnus*, *Betula* (Betulaceae), *Ulmus* (Ulmaceae) and *Corylus* (Corylaceae), but the genus as a whole is polyphagous and not really selective. *Pachybrachis* has also been found on Chenopodiaceae, Thymelaeaceae, Cornaceae, Polygonaceae, Asteraceae, Myricaceae, Oleaceae, Pinaceae, Elaeagnaceae, Rosaceae, Fabaceae, Rhamnaceae, Lamiaceae, Hydrangeaceae and Anacardiaceae. In the Neotropical realm, the choices are also eclectic. In Mexico, several Mimosaceae (*Poponax, Vachellia, Senegalia)*, Asteraceae (*Cirsium, Eupatorium)*, Rhamnaceae (*Ceanothus*), Onagraceae (*Oenothera*), Lamiaceae (*Salvia*) and Convolvulaceae (*Ipomoea*), have been recorded (Moldenke, 1971). One species, *P. contortus* Baly, has been recorded from Brazil on *Psidium guajava* (Myrtaceae), while another, *P. gayi* Blanchard occurs in Argentina on Fagaceae (*Nothofagus* and *Castanea*).

6- *Thelyterotarsus* Weise, 1882.
Recorded from Central Asian steppes on the available plants (Lopatin, 1984): Polygonaceae, Pistaciaceae, Apiaceae, Lamiaceae, Rosaceae, Fabaceae, Tamaricaceae, Chenopodiaceae and Asteraceae. Polyphagous.

2.11.3. Tribe Achenopini

7- *Achenops* Suffrian, 1857.
On *Mimosa* (Mimosaceae) in Natal.

2.11.4. Tribe Monachini

8- *Lexiphanes* Gistel, 1848.

The adults of certain species such as *L. saponatus* (Fabricius) are totally polyphagous. LeSage (1984b) has reared the larva on *Cassandra* (Ericaceae). Recorded from Asteraceae and Myrtaceae in tropical America and from Onagraceae, Sambu-caceae, Naucleaceae and Asteraceae in temperate America.

9- *Adiscus* Gistel, 1857.

On *Carpinus* (Carpinaceae) in Taiwan (Takizawa, 1978a), *Aparosa* (Euphorbiaceae) in Vietnam (Medvedev & Dap, 1983) and *Mytilaria* (Hamamelidaceae) in China (Juanjie *et al.*, 1980).

10- *Elaphodes* Suffrian, 1859.

In Australia (New South Wales) on *Acacia* (Mimosaceae).

11- *Ditropidus* Erichson, 1842.

Recorded from *Grevillea* (Proteaceae), *Eucalyptus* (Myrtaceae) and *Acacia* (Mimosaceae) in Australia. Hawkeswood (1983) recorded *D. phalacroides* Baly feeding on pollen and flowers of *Acacia leiocalyx* (Mimosaceae).

12- *Isnus* Weise, 1898.

At least one species is myrmecophilous living within the stipular thorns of the East African *Acacia drepanolobium* (Mimosaceae). The adult lives symbiotically with the ant *Crematogaster nigriceps* Emery (Selman, 1962) and probably feeds on debris, ant eggs and bodies of dead ants. The larval biology is unknown but probably takes place inside the thorns.

13- *Coenobius* Suffrian, 1857.

In the Far East, South-East Asia and the Pacific regions, the species of *Coenobius* feed on a great variety of non-related plants. Polyphagous. One species, *C. ensimacarangae* Gressitt, feeds on *Macaranga carolinensis* (Euphorbiaceae) in Palau. This plant is a myrmecophyte in association with the ant *Crematogaster*.

2.11.5. Tribe Cryptocephalini

14- *Cryptocephalus* Geoffroy, 1762.

Enormous, cosmopolitan genus, polyphagous. The trophic preferences (tender leaves, young shrubs) are reminiscent of the Clytrinae. A total of 372 species have been examined for host-plants among the total of 1300 described species (28%). Ten species feed exclusively or partially on Coniferae (Pinaceae, Taxodiaceae and Cupressaceae). The remainder have a very broad host-plant selection, ranging from the Poaceae to the Asteraceae, although families such as Arecaceae and most of the Monocotyledonae are rejected as food. The Poaceae are mostly utilized as a source of pollen. The main preferences in the Holarctic zone are the Mimosaceae, Rosaceae (*Rosa*, *Crataegus*), Fagaceae, Asteraceae and Betulaceae. In Australia, *Eucalyptus* (Myrtaceae) and *Acacia* (Mimosaceae) are often eaten and *Hibiscus* (Malvaceae) in Africa together with many Fabaceae. In Argentina, Proteaceae (*Lomatia*), Ericaceae (*Pernettya*), Rhamnaceae (*Discaria*) and Elaeocarpaceae (*Aristotelia*), are common host-plants. It is impossible here to quote all of the host-plants recorded for the 372 species.

15- *Melixanthus* Suffrian, 1854.
On Poaceae, but also on Salicaceae, Euphorbiaceae, Myrsinaceae and Melastomataceae in the Far East.

16- *Jaxartiolus* Jacobson, 1922.
In the Asian steppes, this genus occurs on Tamaricaceae and Chenopodiaceae, which are practically the only available plants (Lopatin, 1975, 1984).

17- *Diachus* LeConte, 1880.
Polyphagous. On Salicaceae, Rosaceae, Fabaceae, Rutaceae, Anacardiaceae, Lauraceae, Myrtaceae, Malpighiaceae, Polygonaceae and various Asteraceae in North America. The case-bearing larva is free and lives on clover flowers (*Trifolium*, Fabaceae)(Hatch, 1971). The adults feed mostly on buds and tender leaves or even on fruits such as strawberries (*Fragaria*, Rosaceae).

18- *Bassareus* Haldeman, 1849.
Bassareus species are recorded in the USA on various plants including *Alnus* (Betulaceae), *Cephalanthus* (Naucleaceae), *Diospyros* (Ebenaceae) and *Rubus* (Rosaceae).

19- *Triachus* LeConte, 1880.
Polyphagous. On Ericaceae, Myricaceae, Fagaceae and Anacardiaceae. On *Randia* flowers (Rubiaceae) in Puerto Rico, South America.

20- *Chariderma* Baly, 1877.
On *Eucalyptus* (Myrtaceae) in Australia.

21- *Cadmus* Erichson, 1842.
Australo-Papuan genus, found on *Glochidion*, *Acalypha*, *Macaranga* and *Homalanthus* (Euphorbiaceae), and *Eucalyptus* (Myrtaceae). *Cadmus macarangae* Gressitt feeds on *Macaranga* sp. in New Guinea, like *Coenobius macarangae* Gressitt. This plant, a true myrmecophyte, is associated with the ant *Cremato-gaster*.

22- *Brachycaulus* Fairmaire, 1843.
On *Eucalyptus* (Myrtaceae) in Australia. Also on *Xanthorrhoea* (Xanthorrhoeaceae) in Queensland (Hawkeswood, 1985) but this record needs verification.

23- *Loxopleurus* Suffrian, 1859.
Recorded on *Acacia* (Mimosaceae) flowers or leaves in Australia.

Concluding remarks on Cryptocephalinae:
The Cryptocephalinae were recorded from 34 plant families by Maulik (1945) but 70 are listed here. Although they are essentially polyphagous, the Cryptocephalinae have certain preferences, as previously mentioned, and are closely related to the Clytrinae. The adults visit the flowers of Asteraceae, Poaceae and Cyperaceae, mostly for pollen. *Casuarina* leaves (Casuarinaceae) are also occasionally eaten. Several Cryptocephalinae appear to be more or less specialized on Coniferae. Of 49 known genera, 23 have been listed here. There is no reason why other genera do not show similar selections.

Very few Cryptocephalinae are myrmecophilous either in the larval stage (exceptional) with *Lasius* ants or in the adult stage (*Isnus*) with the *Crematogaster* ants. It is possible that the genera feeding on *Macaranga* in New Guinea and the Pacific (*Coenobius*, *Cadmus*), have a special adaptation enabling them to be tolerated by the *Crematogaster* ants inhabiting these trees.

Some larvae feed on vegetable matter and other decaying material, or upon cryptogamic plants growing on stones (Erber, 1988). Other larvae feed on green leaves (i.e. *Lexiphanes*, *Cryptocephalus*) or dry leaves.

References listed in the bibliography pertaining to Cryptocephalinae are as follows:- Burlini (1968), Erber (1969, 1988), Hatch (1971), Hawkeswood (1983, 1985), Jolivet (1978), Juanjie *et al.* (1980), LeSage (1984b, 1985, 1986), Lopatin (1975, 1984), Maulik (1945), Medvedev & Dap (1983), Moldenke (1971), Reid (1988), Selman (1962), Takizawa (1978a), Yaroshenko (1985).

2.12. *Chlamisinae*

Maulik (1945) recorded only the following plants as hosts for this subfamily: Fagaceae, Lauraceae, Rosaceae, Aceraceae, Rhamnaceae and Ericaceae. His conclusion was that the few known data did not suggest any food specialization for the group. Since Maulik's work appeared, new observations have been published which show that these beetles are remarkably specific at the plant species level, but that at the subfamily level, the choices are extremely diversified. Tree leaves, shrubs or herbs are chosen according to the species. Adults are winged, but rarely fly. When closed, their elytra are connected along the suture. The mimicry exhibited by the adults (which resemble seeds) as well as that by the larva, provides them with excellent camouflage. The larval case resembles a caterpillar or the excreta of a large insect and well protects the larva against ants (Wallace, 1970). The adults are often provided with a bright or metallic colour on the body and elytra. They are relatively protected by the retraction of the head into the prothorax, the retraction of the legs into latero-ventral grooves, the hardness of the integument and perhaps by a certain toxicity.

Normally the Chlamisinae feed on plant leaves even though they have been purported to feed on moss and lichen (Erber, 1988). Wood (1966) recorded larvae of *Chlamisus cribripennis* (LeConte) feeding on leaves, stems and fruits of blueberries (*Vaccinium*, Ericaceae) and the adults feeding on stem bark. Theoretically, the Chlamisinae appear to be very specific, but the choices sometimes seem very diversified within a given species while sibling species can have very different selections. This is the case with the genus *Chlamisus* as described by Brown (1943, 1952, 1959). Brown recorded populations which, according to their geographical distribution, live in Canada exclusively on either *Salix* (Salicaceae), *Corylus* (Corylaceae) or *Alnus* (Betulaceae).

We are also aware of the case of *Chlamisus plicatus* Fabricius in the USA, normally on *Rubus* (blackberry)(Rosaceae) but which can also feed on *Vaccinium* (Ericaceae). In Argentina, Monros (1952a) mentioned that adult Chlamisinae feed on the compound leaves/phyllodes of *Acacia* (Mimosaceae) and on the leaves of Sapindaceae, Malvaceae, Sterculiaceae and Euphorbiaceae. Others nibble on tender buds on the stems of Asteraceae, while others live inside Malvaceae flowers. *Exema* larvae have been found mostly on Asteraceae (*Hymenoclea*, *Artemisia*, *Aster*, *Solidago*) in North America. The Canadian species of *Chlamisus*, with the exception of the *C. eubati* Brown group, appear to be strictly monophagous. *Chlamisus eubati* Brown feeds almost exclusively on *Rubus* (subgenus *Eubatus*), which is a type of second degree monophagy. However, several species appear to be oligophagous, but two factors must be taken into consideration, i.e. the earlier confusion between the different taxa and a possible allotrophy.

Outside their mimetic, toxic and cryptic protection, the Chlamisinae possess other means of defence, i.e. rapid flight and reflex dropping.

The following is a summary of the known host-plants arranged genus by genus and sometimes species by species. It should be noted that often the genus *Chlamisus* is treated as sensu lato.

1- *Pseudochlamys* Lacordaire, 1848.

One Neotropical species, *P. megalostomoides* Lacordaire, is recorded by Moldenke (1971) on *Sida* (Malvaceae) and on *Ipomoea* leaves (Convolvulaceae) in Mexico, but also in Brazil on *Waltheria* (Sterculiaceae). (It should be noted that the Malvaceae and Sterculiaceae are closely related and are placed within the Malvales).

2- *Diplacaspis* Jacobson, 1924.

Three unidentified species have been recorded on *Neltuma*, *Acacia* and *Vachellia* (Mimosaceae) and another one on *Larrea* (Zygophyllaceae) in Mexico.

3- *Chlamisus* Rafinesque, 1815.

The trophic selections of *Chlamisus* are extremely complex. Certain published observations are merely spatial coincidences, but most of them are valid and represent real choices. Several species are polyphagous as indicated by the diversity of the recorded host-plants. There are often "sibling" species or biotypes which are strictly specialized and often almost indistinguishable morphologically. As Brown (1952) states, the study of the genus is complicated because several species can change abruptly from their "normal" host-plant. For instance, *Chlamisus eubati* Brown was originally thought to be host-specific on *Rubus* (subgenus *Eubatus*)(second degree monophagy) but was later also found on *Potentilla fruticosa* (Rosaceae)(oligophagy). Other examples are known and the sibling species, sympatric or allopatric, can be several (up to four) in a given complex. Several unidentified *Chlamisus* species have been found in the Neotropical region on the following host-plants:- *Ocotea* (Lauraceae), *Melampodium*, *Baccharis* (Asteraceae), *Larrea* (Zygophyllaceae), *Sida* (Malvaceae), *Salvia* (Lamiaceae), *Mimosopsis* (Mimosaceae), etc. (Moldenke, 1971; Maulik, 1945).

Another comment needs to be made about the preferred host-plants of *Chlamisus*. The genus is absent from Europe, like the whole subfamily, but *Rubus* (Rosaceae) is selected as a food-plant in North America and Japan. Betulaceae, Fagaceae, Corylaceae, Ericaceae and Theaceae are the normal host-plants in the cold Holarctic zone. The range of selections change in the tropics and mostly trees belonging to the following families are selected:- Sterculiaceae, Mimosaceae, Caesalpiniaceae, Malpighiaceae, Myrtaceae and Myricaceae. In some cases, mixed selections of Malvaceae and Sterculiaceae occur in North America and confirm the relationships of the two plant families. A superficial view of the generic and specific choices of around fifty species of *Chlamisus* from the Nearctic, Neotropical, Indo-Australian and Sino-Japanese regions, are outlined in Table 3.

4- *Neochlamisus* Karren, 1972.

This genus has rather broad selections but is essentially associated with the same plant families as *Chlamisus*. i.e. Rosaceae, Betulaceae, Ericaceae, Salicaceae, Fagaceae, Myricaceae, Corylaceae, Ulmaceae, Platanaceae, Mimosaceae and Polygonaceae.

5- *Exema* Lacordaire, 1848.

Unidentified species of *Exema* have been recorded from Mexico by Moldenke (1971) on the following plants: *Ceanothus buxifolia* (Rhamnaceae), *Senegalia greggii* (Mimosaceae), and on the following Asteraceae: *Hymenoclea salsola*, *Viguiera dentata*, *V. pauciflora*, *Brickellia vernicosa*, *Verbesina sphaerocephala*, *V. greenmanii*, *Wedelia filipes*, *Eupatorium havanense*, *Baccharis heterophylla*, *B. ramulosa*, *Encelia halmifolia* and *Franseria ambrosioides*. The host-plants of the various *Exema* species are outlined in Table 4. Since the papers by Brown (1943, 1952, 1959) appeared, sibling complexes have been separated and several species, such as *E. canadensis* Pierce, formerly considered polyphagous, are actually now known to be strictly monophagous. The non-Asteraceae host-plants recorded previously for *E. canadensis*, i.e. *Sambucus*, *Corylus*, *Fragaria*, *Prunus*, *Rubus*, *Ulmus*, *Salix*, *Betula* and *Cornus*, are very doubtful. Brown (loc. cit.) recorded only *Solidago* spp. in Canada but other Asteraceae quoted in the literature are also mentioned in Table 4. A total of 34 genera of Asteraceae have been recorded as host-plants for *Exema*.

6- *Fulcidax* Voet, 1806.

The Neotropical species are mainly known to feed on Malpighiaceae. *Fulcidax bacca* (Kirby) is recorded from *Banisteria argyrophylla*, *B. stellaris* and *Mascagnia cordifolia* (Malpighiaceae) in Brazil. Also on *Acacia* sp. (Mimosaceae) and *Psidium guajava* (Myrtaceae) in Argentina. *Fulcidax cuprea* Klug is also known from *Banisteria stellaris* and *Mascagnia cordifolia* (Malpighiaceae) in Brazil.

7- *Aulacochlamys* Monros, 1952.

Adults on various *Solanum* (Solanaceae), including a shrub, *S. robustum*, in Brazil and Argentina.

8- *Melittochlamys* Monros, 1948.

On *Psidium* (Myrtaceae) in Brazil and Paraguay (Fiebrig, 1910; Monros, 1952a).

Concluding remarks on Chlamisinae:

It is difficult to generalize about the food preferences of the Chlamisinae. Adults and larvae are mostly phyllophagous, but can easily feed on bark or fruits. The larva, a case-bearer, is free-living like those of the Lamprosomatinae and most of the Cryptocephalinae. Rosaceae, Fagaceae, Betulaceae, Ericaceae and Asteraceae are the favoured food-plants in the temperate zone. Malpighiaceae, Myrtaceae, Sapindaceae, Melastomataceae and Myricaceae in the neotropics. It is rather surprising that the Polygonaceae, a common food-plant family for the Clytrinae and Cryptocephalinae, have not been recorded for the Chlamisinae. The trophic selection of 8 from 12 described genera are known.

References listed in the bibliography pertaining to Chlamisinae are as follows:- Bokermann (1964), Briggs (1905), Brown (1943, 1952, 1959), Erber (1988), Fiebrig (1910), Jolivet (1978), Karren (1989), LeSage (1982, 1984a), Maulik (1945), Moldenke (1971), Monros (1952a), Neal (1989), Root & Messina (1983), Wallace (1970), Wood (1966, 1978).

SECTION III: CYCLICA

2.13. Lamprosomatinae

The Lamprosomatinae are a small subfamily with a free-living case-bearing larva. The subfamily has been divided into two tribes. Their biology is not well known apart from that of *Oomorphus concolor* Sturm described in a recent paper by Kasap & Crowson (1976). Nothing is known of the larval and adult biology of the Sphaerocharini. Despite Monros (1949) and Kasap and Crowson (1976), we have separated the Lamprosomatinae from the Camptosoma, because of their external morphology and wing venation which are typically eumolpine (Jolivet, 1959).

The early observations indicated that Lamprosomatinae larvae feed on leaves and stems, but also on plant detritus. However, like other case-bearers, the larva of the Lamprosomatinae is most likely to be predominantly detriticolous, although certain exotic species may also feed on bark.

Despite the apparent relative polyphagy of the larva, adults and larvae of Lamprosomatinae are theoretically monophagous and the tendency of the Holarctic species of the genus *Oomorphus* appear to be towards the climbing, arborescent or herbaceous Araliaceae. The host-plant specificity of exotic species appears also to be relatively strong. Unfortunately, the biology of certain American species, even in the USA, is very poorly known. The hidden life of those insects and the nocturnal activity of the larvae are an obstacle to serious observations. The Lamprosomatinae appear to hibernate in the larval stage, which is rather rare among Coleoptera as a whole, but not uncommon among case-bearing larvae. Nymphosis occurs at spring time.

The following summarizes the known data on host-plants. No biological data are known on Sphaerocharini and Neochlamysini, Ethiopian or Neotropical species. Also nothing has been reported on *Oomorphus* from tropical areas. Recent data have been provided by Jolivet (1974), Kasap and Crowson (1976) and Erber (1988).

2.13.1. Tribe Lamprosomini

1- *Oomorphus* Curtis, 1831.

This genus is Holarctic and Neotropical. The Palaearctic species, *Oomorphus* (sensu stricto) *concolor* Sturm, feeds as both larva and adult on *Hedera helix* (Araliaceae), as recently confirmed by Kasap and Crowson (1976). However, old reports of this species from *Astrantia major* and *Aegopodium podagraria* (Apiaceae) also seem very plausible, because the Araliaceae and Apiaceae are closely related botanically and are grouped together with the Cornaceae into the Umbelliflorae. Another Japanese species, *Oomorphus japanus* Jacoby may feed on *Artemisia vulgaris* (Asteraceae) (Kimoto, 1964) but this seems very doubtful and disputable. Probably, the record represents a fortuitous association. A Puerto-Rican species, *Oomorphus* (*Histerogaster*) *longifrons* Lacordaire, feeds on *Eugenia jambos* (Myrtaceae) and one South African species feeds on the leaves of *Protea repens* (Proteaceae)(Coetzee & Latsky, 1986). The North American species, *Oomorphus floridanus* Horn has been recorded as feeding on rodent excreta (*Neotoma floridana*) by Woodruff (1985, pers. comm.). The observation was undertaken in Key Largo, Florida.

2- *Oomorphoides* Monros, 1956.

The Far Eastern *Oomorphoides* all feed on Araliaceae. These are *O. alienus* Bates from Taiwan on *Tetrapanax papyriferum; O. cupreatus* (Baly) from Korea and Japan on *Aralia elata, Kalopanax septemlobus*, *K. pictus, Hedera rhombea*; *O. loochooensis* Chûjô from the Ryukyus on *Acanthopanax* sp., *Schefflera octophylla* and *Aralia elata*; *O. nigrocoeruleus* (Baly) from Japan on *Hedera rhombea, Kalopanax septemlobus, K. sciadophylloides, K. innovans, K. pictus, Acanthopanax spinosum* and *Aralia elata; O. okinawensis* (Chûjô) from Okinawa on *Aralia elata*; *O. formosanus* Chaio from Taiwan on *Hedera*; *O. tonkinensis* Chûjô from Vietnam on *Aralia armata* and *A. dasyphylla*; *O. violaceonigera* Chûjô from Taiwan on *Aralia* sp.; *O. goosanicus* (Chen) from China on *Aralia sinensis* and *O. sakishimanus* Kimoto & Gressitt from the Ryukyus on *Dendropanax trifidum* (Takizawa, 1978a; Juanjie *et al.* 1980; Dap, 1981). All Vietnamese *Oomorphoides* feed on Araliaceae (Medvedev & Roginskaya, 1988).

3- *Lamprosoma* Kirby, 1818.

Various unidentified species have been collected in Brazil on imported *Eucalyptus* (Myrtaceae)(feeding on bark), on *Chorisia speciosa* (Bombacaceae) and on *Eugenia uniflora* (Myrtaceae). In Mexico, the finding of *Lamprosoma* sp. on *Vachellia farnesiana* (Mimosaceae) is probably accidental. *Lamprosoma acaciae* Monros from Argentina feeds on the bark of *Acacia* (Mimosaceae), *L. azurea* Germar from Brazil feeds on the trunk of *Psidium guajava* and *Eugenia uniflora* (Myrtaceae), *L. bicolor* Kirby in Brazil feeds on *Terminalia catappa* (Combretaceae), *L. chorisiae* Monros from Argentina feeds as larvae and adults together on the bark of *Chorisia speciosa* and *C. insignis* (Bombacaceae), *L. minima* Monros from Argentina is common on the trunks of *Acacia caven* (Mimosaceae) and *L. splendidum* Lacordaire feeds on *Psidium guajava* (Myrtaceae).

Concluding remarks on Lamprosomatinae:

All the Holarctic species of Lamprosomatinae appear to feed on Araliaceae, despite several old observations which record different plants as hosts. Neotropical species appear to feed on Myrtaceae, Combretaceae, Bombacaceae and Mimosaceae. The Holarctic species feed on vines or shrubs while the exotic species attack trees, on which larvae and adults nibble the bark. The case-bearing larvae of Lamprosomatinae lead a very cryptic nocturnal life in the temperate countries or are homochromically hidden upon tree trunks in the tropics. Nothing has been reported on the life histories of the African species. Feeding on excreta by an American species of *Oomorphus* is a very rare occurrence for a group normally associated with Araliaceae, but is understandable for a detriticolous species.

References listed in the bibliography pertaining to Lamprosomatinae are as follows:- Coetzee & Latsky (1983), Dap (1981), Erber (1988), Jolivet (1959, 1974, 1978), Juanjie *et al.* (1980), Kasap & Crowson (1976), Kimoto (1964), Medvedev & Roginskaya (1988), Monros (1949), Takizawa (1978a).

2.14. Megascelinae

It may be surprising to see Megascelinae placed among the Cyclica, but the real systematic position of this subfamily remains to be determined. Only the larvae, when found, can assist in the correct placement of the Megascelinae. Those larvae are probably subterraneous and feeding on plant rootlets like those of the Eumolpinae.

In reference to their wing venation, the Megascelinae, although devoid of a medio-cubital patch or "rubbing" spot (Jolivet, 1957), have a wing venation which is typically eumolpine. Their wing structure is probably due to the folding system of their wings. Primitive eumolpine species and lamprosomatines do not possess the medio-cubital patch or sub-cubital binding patch. This spot has been interpreted in different ways but is actually part of a pushing system for the wings by the abdomen.

Seeno and Wilcox (1982) placed the Megascelinae between the Criocerinae and Clytrinae while Mann and Crowson (1981) placed it between the Megalopodinae and the Lamprosomatinae. Bechyne and Bechyne (1969) emphasized the eumolpine characters of the Megascelinae and brought the subfamily near the Madagascan *Hemidacne* and the Neotropical *Psathyrocerus* (Eumolpinae). Their thesis was based on wing venation, the shape of the aedeagus and the male tarsi.

Actually, the general habitus of the Megascelinae and the Eumolpinae is sufficiently different to keep them apart, but they are without doubt very closely related.

The food preferences of the Megascelinae are outlined as follows:

1- *Megascelis* Sturm, 1826.

This genus is essentially American in the Neotropical region. Some *Megascelis* species have been recorded from bushes of *Desmodium*, nibbling on the lamina, in Argentina, on *Phaseolus* sp. in Honduras, on *Dolichos lablab*, *Glycine max*, *Phaseolus lunatus*, *P. vulgaris* and *Vigna sinensis* in Venezuela, and on *Cajanus bicolor, Canavalia ensiformis, Desmodium purpureum, Glycine max, Medicago sativa, Melilotus indica, Phaseolus lunatus, P. vulgaris, Robinia pseudoacaciae*, *Samanea saman*, *Trifolium repens*, *Vigna sinensis* and others elsewhere. All the above plants are Fabaceae. The selection has been confirmed by the recent discovery of *Megascelis frenata* Lacordaire on *Phaseolus* sp. (Fabaceae) in Venezuela (Bechyne, 1980) and *M. puella* Lacordaire on *Phaseolus* sp. in Honduras (Passoa, 1983).

Despite their relationships with legumes, the genus *Megascelis* has also been collected occasionally in Venezuela on *Annona cherimola*, *A. muricata* (Annonaceae), *Avena sativa*, *Zea mays* (Poaceae), *Cyclanthera pedata* (Cucurbitaceae), *Gladiolus hybridus* (Iridaceae), *Gossypium hirsutum* (Malvaceae), *Helianthus annuus* (Asteraceae) and *Porana paniculata* (Convolvulaceae). These latter records are probably accidental and do not represent true trophic relationships.

2- *Mariamela* Monros, 1951.

Genus from Patagonia. *Mariamela wittmeri* Monros appears to be exclusively associated with *Nothofagus* (Fagaceae) trees from the patagonian steppes. As will be seen later, at least two Eumolpinae (viz. *Hornibius sulcifrons* Fairmaire and *H. grandis* Philippi) feed on *Nothofagus* in Chile (in the extreme south at Magellanes) and one Chrysomelinae, *Novocastria nothofagi* Selman, also feeds on *Nothofagus* in the Australian, sub-humid, temperate rain forests. The primitive genus *Psathyrocerus* from Chile is also recorded from *Nothofagus*. Concluding remarks on Megascelinae:

It appears that the genus *Megascelis* feeds mostly on the Fabaceae and that the only described species of *Mariamela* has been captured only on *Nothofagus*. The two plant families are not related. The remaining undescribed larvae probably feed on rootlets of plants like Eumolpinae. Megascelinae adults are active in sunlight and fly rapidly. They often reach the forest canopy. They feed on the margins of leaves and do not appear to show any stridulation, jumping or reflex immobilization.

References listed in the bibliography pertaining to Megascelinae are as follows:- Bechyne (1980), Bechyne & Bechyne (1969), Jolivet (1957, 1987b), Mann & Crowson (1981), Monros (1951), Passoa (1983), Seeno & Wilcox (1982).

2.15. Eumolpinae

The Eumolpinae are remarkable in their general aspect, being compact and rounded, and for many species, in their metallic colouration. The tarsi have the third segment deeply divided. The cotyloid cavities are rounded. The subfamily is relatively important with a certain economic impact, mostly in regard to the polyphagous or frugivorous species. Entire genera, such as *Colaspis*, are considered garden pests in the tropics. It is in tropical America that the greatest number of species occur.

The Eumolpinae feed on 111 mono- and dicotyledonous families and also upon several Gymnospermae, probably a secondary adaptation. Several sections such as Euryopites and Corynodites are specific on Gentianales (Asclepiadaceae and Apocynaceae). These appear to be the ancient selections for the subfamily.

The polyphagous species can be frugivorous, granivorous or florivorous. Other species such as *Syagrus ortebiensis* Selman, are probably myrmecophilous in the adult stage. They live like the clytrine *Hockingia curiosa* Selman or the cryptocephaline *Isnus petasus* Selman inside the stipular thorns of East African *Acacia* (Mimosaceae). These chrysomelids must feed, both as adults and larvae, on detritus, ant eggs and excreta. *Syagrus, Isnus, Hockingia* appear to live as symphiles with *Crematogaster* ants. At least, *Isnus* and *Hockingia* have trichomes which are probably licked by the ants (Selman, 1962; Hocking, 1970). The *Syagrus* larvae are probably root-feeders but nothing is certain. It is possible that the ants transport the eggs and larvae into the thorns but this is hypothetical in the light of present knowledge. In the adult stage, the beetles are unable to escape through the stipular holes. These chrysomelids are protected from the ants by secretions, retractile appendices, some degree of mimetism with their hosts, etc. Many Eumolpinae are polyphagous. It is very difficult to determine a primitive type of food selection and then to trace a phylogenetic tree based on that tropism.

Kurcheva (1958, 1967, 1975) described the larvae of *Chloropterus versicolor* (Morawitz), *Bromius obscurus* (L.), *Chrysochus asclepiadeus* (Pallas), *Pachnephorus tessalatus* Daftschmid, *Bedelia kokandica* (Solsky), *Chrysochares asiaticus* (Pallas), *C. auratus* (Fabricius), *Adoxinia spinipes* Reitter, *Basilepta fulvipes* Motschulsky, and *Colasposoma dauricum* Mannerheim. The larvae of *Paria canella* (Fabricius) and *Chrysochares auratus* (Fabricius) have been described by Boving and Craighead (1931) and those of *Notodonta puncticellis* (Say), *Chrysochares auratus* (Fabricius), *Fidia viticida* Walsh, and *Paria canella* (Fabricius) by Peterson (1960) in the USA. Paterson (1931, 1932, 1936) had previously studied the biology of *Euryope terminalis* Baly and *Corynodes pusis* Marsh in South Africa. Other authors have either described the larva or the biology of other species, e.g. *Colaspis hypochlora* Lefèvre in Central America (Salt, 1928), *Myochrous denticollis* (Say) in the USA (Kelly, 1915; Blake, 1950), *Syagrus rugifrons* (Say) in South Africa (Broodryk, 1965) and *Syagrus calcaratus* (Fabricius) in Uganda (Gardner, 1934). More recently, Stehr (1991) has described the larvae of *Paria fragariae* Wilcox and *Fidia viticida* Walsh.

All Eumolpinae larvae are endogenous and feed on roots of the adult host-plant or any other non-related plants. The larvae are totally polyphagous. Many species favour the roots of the Poaceae such as *Paspalum* in the USA, but they can also attack the roots of the grapevine (*Vitis*, Vitaceae) and those of fruit trees etc. The larvae live generally in groups, chew on the roots and discard the bark. The feeding incisions are sometimes very deep, enough to shelter the larvae themselves. The damage is very similar to that caused by the larvae of Scarabaeidae or Curculionidae. Superficially, all eumolpine larvae resemble each other. Several species (e.g. *Typophorus viridicyaneus* (Crotch)) migrate from roots to tubers and even attack potatoes (*Solanum tuberosum*, Solanaceae) or sweet potatoes (*Ipomoea batatas*, Convolvulaceae). The larvae of Eumolpinae lack ocelli and are generally whitish or brownish in colour. There exist some morphological differences between the tribes (i.e. setae, spinulae, etc.) but generally no differences between the larvae within the tribes. The prominent sclerotization and the thickening of the anterior parts of the frons is certainly an adaptation for digging, in addition to the strong mandibles, the maxillae and the posterior part of the body.

The adults lay their eggs upon leaves in rearing cages but normally the eggs are laid at ground level into soil crevices and the larva digs immediately after hatching.

The biology of several species is well-known and it appears that there is a common pattern except for the myrmecophilous species. In the case of *Myochrous denticollis* (Say), the larva lives within the soil in the corn fields inside small earthen cells provided with a small tunnel which are attached to the corn roots (Kelly, 1915). These cells are 10-12 cm deep and the attacks occur during spring. The female lays her eggs in masses of 10 to 50 into soil crevices or into empty corn stems. The eggs hatch in 6-10 days. This scheme is the same for *Colaspis hypochlora* Lefèvre, a pest of bananas (*Musa* spp., Musaceae) in Central America (Salt, 1928). The female digs into the soil with her ovipositor and lays her eggs into crevices at a depth of 5 to 10 mm below the ground surface. A female can lay up to 118 eggs which hatch after 6 days. The larvae feed on Poaceae roots at a depth of 5 to 8 cm. They burrow deeper (up to 25 cm) during drought conditions when they construct small spherical chambers where nymphosis occurs. The adult chews on the banana fruits, but also feeds on leaves of a great number of plants belonging to ten non-related families including the Poaceae. The adaptation to banana is a secondary one (*Musa* is introduced to America) from a normal phyllophagous diet.

2.15.1. Tribe Hornibiini

1- *Hornibius* Fairmaire, 1888.

Hornibius grandis Philippi & Philippi and *H. sulcifrons* (Fairmaire) live in Patagonia like *Mariamela* on various *Nothofagus* such as *N. antarctica*, *N. pumilio*, *N. betuloides* and *N. obliqua* (Fagaceae). Jerez *et al.* (1992) have described the biology and larvae of *H. grandis* in detail. The larvae are borers in buds (first stage) and later feed on leaves; pupation occurs in soil.

2.15.2. Tribe Stenomelini

2- *Stenomela* Erichson, 1847.

Two genera are known in this tribe. The genus *Stenomela* from Chile comprises the sole species *S. pallida* Erichson. It feeds on *Gunnera chilensis* (Gunneraceae), a herbaceous plant with enormous leaves. *Stenomela* has a Gondwanian distribution but is absent from New Caledonia. The genus

Bohumiljania Monros has been described from New Caledonia, but nothing has been recorded on its biology and host-plants. It was described by Jolivet (1957) under the name of *Stenomela caledonica*.

2.15.3. Tribe Nodinini

3- *Basilepta* Baly, 1860.

This Asiatic genus is essentially polyphagous. Various species have been recorded from *Rosa, Malus, Pyrus, Prunus, Fragaria* and *Crataegus* (Rosaceae) and on many other plants, i.e. *Alnus* (Betulaceae), *Alangium* (Alangiaceae), *Populus, Salix* (Salicaceae), *Carya, Pterocarya* (Juglandaceae), *Maesa* (Myrsinaceae), *Litchi* (Sapindaceae), *Artemisia, Helianthus, Chrysanthemum, Inula, Cirsium* (Asteraceae), *Ficus* (Moraceae), *Glycine, Lespedeza, Glycyrrhiza* (Fabaceae), *Rumex, Polygonum* (Polygonaceae), *Melia* (Meliaceae), *Diospyros* (Ebenaceae), *Ampelopsis, Vitis* (Vitaceae), *Gossypium, Hibiscus* (Malvaceae), *Viburnum* (Caprifoliaceae), *Ilex* (Aquifoliaceae), *Castanopsis, Quercus* (Fagaceae), *Phyllanthus, Mallotus* (Euphorbiaceae), *Chamaecyparis* (Cupressaceae), *Cryptomeria* (Taxodiaceae), *Pinus* (Pinaceae), *Cocos* (Arecaceae), *Styrax* (Styracaceae), *Symplocos* (Symplocaceae), *Melastoma, Clidemia* (Melastomataceae), *Ochroma* (Bombacaceae), *Rhodomyrtus* (Myrtaceae), *Theobroma* (Sterculiaceae), *Musa* (Musaceae) and *Canna* (Cannaceae). This great variety in selections indicates a very large polyphagy, although several species feed on *Rubus* in Vietnam, China, Korea and Japan (i.e. *Basilepta distincta* Jacoby, *B. ruficolle* Jacoby, *B. leechi* Jacoby, *B. melanopus* Lefèvre, *B. lameyi* Lefèvre, *B. fulvipes* (Motschulsky), *B. davidi* Lefèvre and *B. bicolor* Lefèvre) and are thus monophagous in the adult stage. One of us (P. Jolivet) has very often observed the mating of two species of *Basilepta* on *Rubus* (Rosaceae) in Vietnam. These taxa are of different colours, i.e. *B. melanopus* Lefèvre is brown and *B. lameyi* Lefèvre is blue. Some polyphagous species such as *B. subcostatum* Jacoby and *B. viridipennis* (Motschulsky) from India, become frugivorous and feed on the leaves and fruits of banana trees (*Musa* spp., Musaceae).

4- *Mireditha* Reitter, 1912.

One species of this Asiatic genus has been found at high altitude in China on *Saussurea* (Asteraceae). The flowers of *Saussurea* are sweet-smelling and attract insects.

5- *Nodina* Motschulsky, 1858.

Asiatic genus, mostly Indo-Malayan. Relatively polyphagous. The following plants have been recorded: *Rosa, Prunus* (Rosaceae), *Vitis* (Vitaceae), *Melastoma* (Melastomataceae), *Mangifera* (Anacardiaceae), *Imperata, Oryza, Zea* (Poaceae), *Castanea* (Fagaceae), *Aralia* (Araliaceae), *Glochidion, Mallotus* (Euphorbiaceae), *Picea* (Pinaceae), *Cajanus, Crotalaria* (Fabaceae), *Garcinia*

(Clusiaceae), *Hydnocarpus* (Flacourtiaceae), *Nephelium* (Sapindaceae), *Rhodomyrtus* (Myrtaceae), *Cinnamomum* (Lauraceae), *Secamone* (Asclepiadaceae), *Citrus* (Rutaceae) and *Paulownia* (Scrophulariaceae). These selections indicate a rather broad polyphagy. The preferred foods are Rosaceae, Melastomataceae and Vitaceae which could mean that there is a tendency towards mostly edible, non-toxic species and avoidance of unpalateable plants. The selection of *Secamone* (Asclepiadaceae) is interesting because this could indicate a latent attraction.

6- *Epinodostoma* Bryant and Gressitt, 1957
Genus from Fiji Islands. One species, *E. alocasiae* Gressitt feeds on *Alocasia macrorrhiza* (Araceae).

7- *Pseudolpus* Jacoby, 1884.
One species has been captured on *Elettaria cardamomum* (Zingiberaceae) in New Guinea.

8- *Sedlacekia* Gressitt, 1969.
On *Pandanus* (Pandanaceae) in New Guinea.

9- *Colposcelis* Dejean, 1837 (= *Pagria* Lefèvre)
This genus from the tropical zone of the Old World extends northwards to Japan. It is essentially associated with the Fabaceae. Most of the species feed on *Phaseolus*, *Desmodium*, *Microlespedeza*, *Faba*, *Arachis*, *Erythrophleum*, *Glycine*, *Calopogonium*, *Pueraria*, *Vigna*, *Centrosema*, *Crotalaria*, *Dolichos*, *Pisum*, *Trifolium*, *Voandzeia*, *Cajanus*, etc. Several species, e.g. *C. aeneicollis* (Lefèvre), *C. flavopustulata* (Baly), *C. graphica* (Weise), and *C. kanarensis* (Jacoby), are pests of crops in Indonesia, Malaysia, Philippines and Bangladesh. Other species have been recorded from cotton (*Gossypium*, Malvaceae) and Fabaceae in East Africa (e.g. *C. gossypii* Bryant). Other species have also been collected from *Vitex* (Verbenaceae) and *Sporobolus* (Poaceae). Most of the captures in Zaire on Naucleaceae, Euphorbiaceae, Verbenaceae and Combretaceae are doubtful. On Fabaceae, the eggs are laid at the base of the plant on the ground. The larvae feed on the roots and the adults feed on the leaves after emergence.

10- *Metachroma* Chevrolat, 1837.
Neotropical genus, mostly polyphagous, species of which can become pests of crops, such as cotton, avocado, guava, rose apple, pineapple, banana, sapotillas and various other fruit trees. Various species have been recorded from *Gossypium* (Malvaceae), *Bucida* (Combretaceae), *Coffea* (Rubiaceae), *Eugenia*, *Eucalyptus*, *Psidium* (Myrtaceae), *Phaseolus*, *Inga*, *Pithecellobium*, *Crotalaria* (Fabaceae), *Citrus* (Rutaceae), *Prunus*, *Malus*, *Rosa*, *Pyrus* (Rosaceae), *Quercus* (Fagaceae), *Salix*, *Populus* (Salicaceae), *Oenothera* (Onagraceae), *Capsicum*

(Solanaceae), *Asclepias* (Asclepiadaceae), *Rauwolfia* (Apocynaceae), *Zinnia, Pluchea, Lactuca* (Asteraceae), *Hibiscus* (Malvaceae), *Saccharum* (Poaceae), *Persea* (Lauraceae), *Carya* (Juglandaceae), *Achras, Micropholis, Calocarpum, Pouteria* (Sapotaceae), *Scirpus, Cladium* (Cyperaceae), *Musa* (Musaceae), *Ananas* (Bromeliaceae), *Manihot* (Euphorbiaceae), *Rhizophora* (Rhizophoraceae), *Pinus* (Pinaceae) and *Taxodium* (Taxodiaceae). Curiously, the plants which are recorded the most often are *Quercus, Bucida, Gossypium* and various fruit trees. Several species feed on Coniferae, i.e. *M. orientale* (Jacoby), *M. interruptum* (Say) and *M. mexicanum* Jacoby. Records from Asclepiadaceae and Apocynaceae are interesting and deserve further study.

11- *Rhyparida* Baly, 1861.

Oriental genus of economic importance, feeding on various crops, including peanuts and cocoa. The host-plants of only 33 species are known from about 180, i.e. 18.3%. The species appear to be mostly polyphagous. Many species in Asia have adapted for feeding on cacao (*Theobroma cacao*, Sterculiaceae) from South America and coffee (*Coffea* sp., Rubiaceae) from Africa, which indicates great plasticity. Various species have been recorded from *Theobroma, Commersonia* (Sterculiaceae), *Ixora, Coffea* (Rubiaceae), *Artocarpus, Morus* (Moraceae), *Elettaria* (Zingiberaceae), *Arachis, Phaseolus* (Fabaceae), *Ceiba* (Bombacaceae), *Gossypium, Hibiscus* (Malvaceae), *Nicotiana* (Solanaceae), *Nephelium* (Sapindaceae), *Casuarina* (Casuarinaceae), *Pipturus* (Urticaceae), *Melastoma* (Melastomataceae), *Pandanus* (Pandanaceae), *Mangifera* (Anacardiaceae), *Colocasia* (Araceae), *Eucalyptus* (Myrtaceae), *Citrus* (Rutaceae), *Oryza, Themeda, Heteropogon, Imperata, Zea, Saccharum, Chrysopogon, Dichanthium* (Poaceae), *Musa* (Musaceae), *Elaeis, Cocos* (Arecaceae), *Couthovia* (Strychnaceae), *Agathis* (Araucariaceae), *Dillenia* (Dilleniaceae) and *Bruguiera, Rhizophora* (Rhizophoraceae). *Rhyparida morosa* Baly larvae bore into sugar cane roots in Australia.

12- *Micromolpus* Gressitt, 1969.

Polyphagous. Recorded from Sapindaceae (*Dodonaea*), Tiliaceae (*Triumfetta*) and Urticaceae (*Pipturus*) in New Guinea.

13- *Phainodina* Gressitt, 1969.

One species, *P. alpiniae* Gressitt has been recorded from *Alpinia* sp. (Zingiberaceae) in New Guinea.

14- *Deretrichia* Weise, 1913.

Two species have been recorded feeding on introduced cocoa (*Theobroma*, Sterculiaceae) in New Guinea.

15- *Vitibia* Fairmaire, 1882.

On *Commersonia, Melochia* (Sterculiaceae), *Excoecaria* (Euphorbiaceae),

Thespesia, *Gossypium* (Malvaceae) and *Agapetes* (Ericaceae). Endemic to Fiji. Apart from the affinities between Sterculiaceae and Malvaceae, there are no close relationships between the other food-plant families.

16- *Labasa* Bryant, 1925.
On *Pipturus* (Urticaceae), *Merremia* (Convolvulaceae) and *Trichospermum* (Tiliaceae). Endemic to Fiji Islands.

17- *Stygnobia* Weise, 1922.
Stygnobia cauta Weise has been recorded from *Hibiscus tiliaceus* (Malvaceae) in Samoa. The host-plants of *S. leveri* (Bryant) from Fiji are more numerous, i.e. *Alphitonia* (Rhamnaceae), *Glochidion* (Euphorbiaceae), *Couthovia* (Strychnaceae) and *Pipturus* (Urticaceae). Polyphagous.

18- *Rhyparidella* Gressitt, 1969.
New Guinea. One species, *R. casuarinae* Gressitt has been recorded from *Rosa* (Rosaceae), while another one, *R. cacaona* Gressitt, from New Britain, feeds on *Laportea* (Urticaceae), *Hibiscus esculentus* (Malvaceae) and *Theobroma cacao* (Sterculiaceae). Polyphagous.

19- *Eurydemus* Chapuis, 1874.
Eurydemus grandis (Baly) has been captured in Fiji on *Terminalia* (Combretaceae), *Phaleria* (Thymelaeaceae) and *Flacourtia* (Flacourtiaceae). *Eurydemus gussfeldi* (Karsch) from East Africa, has been recorded from *Gossypium* (Malvaceae) and *Nicotiana* (Solanaceae). Polyphagous.

20- *Zohranus* Aslam, 1968.
One species in Senegal on Poaceae.

21- *Afroeurydemus* Selman, 1965.
Central Africa. The host-plants recorded from Zaire are doubtful (Selman, 1972) except perhaps for *Sterculia* (Sterculiaceae). The genus has adapted to cacao (*Theobroma*, Sterculiaceae) an introduced plant, in the Côte d'Ivoire. Polyphagous.

22- *Menius* Chapuis, 1874.
Recorded from *Pennisetum* (Poaceae), *Gossypium* (Malvaceae), *Ipomoea* (Convolvulaceae), *Theobroma* (Sterculiaceae), *Dioscoreophyllum* (Menispermaceae), *Tephrosia* (Fabaceae), *Piliostigma* (Caesalpiniaceae), *Albizia* (Mimosaceae), *Coffea* (Rubiaceae) and even Arecaceae. Polyphagous.

23- *Microsyagrus* Pic, 1952.
African genus. Quoted from cotton (*Gossypium*, Malvaceae) and from various other plants such as Verbenaceae and Bignoniaceae which appear to be accidental captures and need to be reconfirmed. Also on Poaceae.

24- *Proliniscus* Selman, 1965.

African. Recorded from Poaceae and *Stereospermum* (Bignoniaceae) in Zaire.

25- *Syagrus* Chapuis, 1874.

Five species of *Syagrus* feed on cotton in South Africa and Madagascar but mostly *S. rugifrons* Baly. The eggs are deposited singly or in small clusters near the roots of cotton or other Malvaceae upon which the hatching larvae feed. Pupation occurs within the ground. Aestivation is also underground (Broodryk, 1965). Otherwise, the various species of *Syagrus* occur on *Leucaena, Acacia* (Mimosaceae), *Vigna, Glycine, Tephrosia* (Fabaceae), *Hibiscus, Gossypium, Urena, Sida, Theopena, Abutilon* (Malvaceae), *Heliotropium* (Boraginaceae), *Pennisetum, Oryza* (Poaceae) and *Elaeis* (Arecaceae). *Syagrus rosae* (Bryant) has also been captured on *Rosa, Canna, Gardenia* and *Gerbera* flowers, but probably without any clear trophic relationships. The majority of the species are associated with Malvaceae including cotton. Several species live as probable symphiles within the stipular thorns of East African *Acacia* together with *Crematogaster* ants (Selman, 1963).

26- *Sarum* Selman, 1965.

Nothing certain is known about this African genus. *Sarum baerti* Selman has been recorded from Apocynaceae, Fabaceae and Ixonanthaceae but these records are doubtful and need verification.

27- *Meniellus* Weise, 1903.

One species from Zaire (*M. kohlschutteri* Weise) has been recorded from *Combretum* (Combretaceae) and *Nauclea* (Naucleaceae) but these records also need verification.

28- *Massartia* Selman, 1965.

Appears to be associated with *Combretum* (Combretaceae) in Zaire (*M. minimus* (Burgeon). Another one, *M. nigrita* Selman is attracted to flowers of Rubiaceae (Selman, 1972).

29- *Paraivongius* Pic, 1936.

Many host-plants have been recorded in Africa but probably only Rubiaceae (*Coffea*), Tiliaceae (*Grewia*) and Sterculiaceae (*Theobroma*) are reliable. Polyphagous.

30- *Gaberella* Selman, 1965.

One species from Zaire recorded without any certainty from *Erythrophleum* (Caesalpiniaceae).

31- *Rhembastus* Harold, 1877.

Quoted on Mimosaceae (*Acacia*), Fabaceae (*Tephrosia*), Asteraceae

(*Helianthus*), Rubiaceae (*Coffea)*, Rosaceae (*Malus*) and Malvaceae (*Gossypium, Hibiscus*) in Africa. *Rhembastus pomorum* Bryant has adapted to introduced appletrees in the Transvaal.

32- *Mandolia* Selman, 1965.
On *Isoberlinia* (Caesalpiniaceae) in Zaire, but the record is doubtful and needs verification.

33- *Bedelia* Lefèvre, 1875.
On *Tamarix* (Tamaricaceae) and *Populus* and *Salix* (Salicaceae) in Central Asia.

34- *Chloropterus* Morawitz, 1861.
In dry areas of Europe, Asia and Africa. On *Tamarix* (Tamaricaceae) in North Africa and Central Asia and on *Limonium* (Plumbaginaceae) and *Salicornia* and *Atriplex* (Chenopodiaceae) in Asia.

35- *Phytorus* Jacoby, 1884.
Mostly polyphagous in the Far East. It has adapted to *Theobroma cacao* (Sterculiaceae) in Asia. Also on *Thea* (Theaceae), *Bouea, Mangifera* (Anacardiaceae), *Dahlia, Lactuca* (Asteraceae), *Brassica* (Brassicaceae), *Ricinus*, *Manihot*, *Mallotus* (Euphorbiaceae), *Rosa*, *Fragaria* (Rosaceae), *Ceiba*, *Ochroma*, *Eriodendron* (Bombacaceae), *Eugenia*, *Eucalyptus*, *Rhodomyrtus* (Myrtaceae), *Trema* (Ulmaceae) and *Hibiscus* (Malvaceae).

36- *Cleoporus* Lefèvre, 1884.
In Japan and China on Rosaceae (*Rosa, Potentilla, Fragaria*, *Malus, Prunus*, *Pyrus*, *Sanguisorba*), Hamamelidaceae (*Loropetalum*), Asteraceae (*Artemisia*), Rhamnaceae (*Zizyphus*), Salicaceae (*Salix)*, Caprifoliaceae (*Viburnum*), Fagaceae (*Quercus*, *Castanea*) and Myrtaceae (*Eucalyptus, Eugenia*). Elsewhere, *Cleoporus* have been recorded from Euphorbiaceae (*Glochidion*, *Phyllanthus*), Malvaceae (*Hibiscus*) and Sterculiaceae (*Theobroma*), etc. Polyphagous.

37- *Cleorina* Lefèvre, 1885.
On *Canna* (Cannaceae), *Clerodendrum* (Verbenaceae), *Mangifera* (Anacardiaceae), *Vernonia* (Asteraceae), *Eucalyptus* (Myrtaceae), *Lithocarpus* (Fagaceae), *Rubus* (Rosaceae), Zingiberaceae and Fabaceae, etc. Polyphagous. Asia and Australia.

38- *Stethotes* Baly, 1867.
On *Elettaria* (Zingiberaceae), *Pandanus* (Pandanaceae) and *Pipturus* (Urticaceae) in New Guinea.

39- *Typophorus* Chevrolat, 1837.

American genus, mostly polyphagous. Recorded from a very large number of plants including: *Convolvulus*, *Ipomoea* (Convolvulaceae), *Capsicum* (Solanaceae), *Vitis* (Vitaceae), *Vachellia* (Mimosaceae), *Glycine, Phaseolus* (Fabaceae), *Malvastrum, Gossypium* (Malvaceae), *Cordia* (Ehretiaceae), *Maranta* (Marantaceae), *Dioscorea* (Dioscoreaceae), *Persea* (Lauraceae) and *Parthenium* (Asteraceae), etc. The larvae, usually root feeders, feed on the rhizomes of various *Ipomoea*, and the adults on the leaves. The larvae have also been found on the roots of various Poaceae, *Zea*, *Saccharum* and *Panicum*. Other accidental food-plants for the adults which have been recorded in the literature include *Citrus* (Rutaceae), *Coffea, Gardenia* (Rubiaceae), *Foeniculum* (Apiaceae), *Heliconia* (Musaceae) and *Melicocca* (Sapindaceae).

40- *Paria* LeConte, 1858.

The genus *Paria,* which is related to *Typophorus*, is also polyphagous. The plants listed as food are the following: *Rosa*, *Potentilla, Rubus, Sorbus, Prunus, Fragaria, Malus, Pyrus,* (Rosaceae), *Iva, Solidago*, *Aster*, *Eupatorium* (Asteraceae), *Hypericum* (Hypericaceae), *Fraxinus* (Oleaceae), *Juglans, Carya* (Juglandaceae), *Solanum* (Solanaceae), *Vitis* (Vitaceae), *Comptonia* (Myricaceae), *Quercus* (Fagaceae), *Ipomoea* (Convolvulaceae), *Salix, Populus* (Salicaceae), *Platanus* (Platanaceae), *Cornus* (Cornaceae), *Pycnanthemum, Salvia* (Lamiaceae), *Gossypium* (Malvaceae), *Phaseolus, Trifolium* (Fabaceae), *Amaranthus* (Amaranthaceae), *Rumex* (Polygonaceae), *Taxodium* (Taxodiaceae), *Juniperus* (Pinaceae) and *Canna* (Cannaceae). *Paria* adults have also been captured on various Poaceae. The larvae have been collected on the roots of Rosaceae and Convolvulaceae. The eggs are deposited at the base of the plants. Adults adapted for feeding on various species of conifers (e.g. *Juniperus, Taxodium*) and those of other species, are pests of cotton and sweet potato.

41- *Cleoparida* Gressitt, 1967.

Two species of this genus have been recorded on *Freycinetia* (Pandanaceae) and on unidentified Urticaceae in the Solomon Islands. Probably polyphagous.

2.15.4. Tribe Colasposomini

42- *Odontionopa* Chevrolat, 1837.

One species has been found in South Africa eating *Prunus* leaves (Rosaceae). *Prunus* is introduced to South Africa. The natural host-plants remain unknown.

43- *Obelistes* Lefèvre, 1885.

Mentioned by Selman (1972) on Annonaceae (*Annona*) and *Erythrophleum* flowers (Caesalpiniaceae), in Zaire but these records are doubtful and need verification.

44- *Phascus* Lefèvre, 1884.

In Zaire, this genus feeds on *Eriosema* (Fabaceae), *Erythrophleum, Piliostigma* (Caesalpiniaceae) and on *Canthium* (Rubiaceae), Poaceae and Cyperaceae.

45- *Lefevrea* Jacoby, 1897.

African genus on *Piliostigma* (Caesalpiniaceae), *Cussonia* (Araliaceae) and *Crossopteryx* (Rubiaceae). Other data from Zaire are very doubtful (Selman, 1972).

46- *Colasposoma* Laporte, 1833.

Genus from the Old World which is essentially polyphagous. Recorded from *Tephrosia, Glycine* (Fabaceae), *Acacia* (Mimosaceae), *Coffea* (Rubiaceae), *Ipomoea, Convolvulus, Calystegia* (Convolvulaceae), *Citrus* (Rutaceae), *Eucalyptus* (Myrtaceae), *Prunus* (Rosaceae), *Lophira* (Ochnaceae), *Gossypium* (Malvaceae), *Zinnia* (Asteraceae), *Agave* (Agavaceae), *Morus* (Moraceae) and *Zizyphus* (Rhamnaceae). Also mentioned in South Africa, on *Cupressus* (Cupressaceae), *Solanum, Nicotiana* (Solanaceae) and *Commersonia* (Sterculiaceae). The presence of *Colasposoma* on *Citrus* in India seems rather exceptional, but another species has also been captured on the same plant in South Africa. The genus contains species which are potential pests of cotton, coffee, sweet potato and several other cultivated plants. In East Africa, *Colasposoma coffeae* Kolbe larvae have been found on *Coffea* roots (Le Pelley, 1959). The eggs are laid in the ground. The adults have also been captured on various Poaceae and *Pinus* (Pinaceae). An interesting phenomenon is the association of Convolvulaceae/Asclepiadaceae with at least one species, *Colasposoma dauricum* Mannerheim, which feeds on both *Ipomoea* and *Cynanchum*. However, the tendency of the genus is essentially towards the Convolvulaceae.

47- *Trichostola* Chapuis, 1874.

Appears polyphagous in the Mascarene Islands, but the observations must be reviewed and confirmed by laboratory rearing (Jolivet, 1979b). Several species have been captured on *Dodonaea* (Sapindaceae), *Rumex* (Polygonaceae) and *Elaeodendron* (Celastraceae).

48- *Melindea* Lefèvre, 1884.

Recorded from Africa on *Combretum* (Combretaceae), *Dombeya* (Sterculiaceae), *Sesbania* (Fabaceae) and also on Poaceae (*Pennisetum*) in Senegal.

49- *Euryope* Dalman, 1824.

Euryope is a genus which feeds exclusively on Asclepiadaceae. *Leptadenia* is a common host-plant in the Ethiopian tropical zone. The genus *Leptadenia* are creepers in the subdesertic areas (Sahel etc.) and four different species have been recorded as host-plants. *Euryope* also feeds on *Asclepias* in South Africa. References from Fabaceae are mistakes.

50- *Bechyneia* Jolivet, 1950.
This genus from South Africa has the same biology as *Euryope*. Probably synonymous with the previous one.

2.15.5. Tribe Eumolpini

51- *Nycterodina* Bechyne, 1951.
Bechyne and Bechyne (1966) recorded *N. bordoni* Bechyne "on flowers" in Venezuela, which is rather vague.

52- *Spintherophyta* Dejean, 1837.
American. Polyphagous. On *Viburnum* (Caprifoliaceae), *Boehmeria* (Urticaceae), *Condalia* (Rhamnaceae), *Lantana* (Verbenaceae), *Comptonia* (Myricaceae), *Rubus* (Rosaceae), *Corylus* (Corylaceae), *Quercus* (Fagaceae), *Azalea* (Ericaceae), *Baccharis* (Asteraceae), *Desmodium* (Fabaceae), *Prosopis* (Mimosaceae), *Salix* (Salicaceae), *Pinus* (Pinaceae) and *Juniperus* (Cupressaceae).

53- *Antitypona* Weise, 1921.
Neotropical. One species has been recorded from *Theobroma cacao* (Sterculiaceae) in Venezuela.

54- *Sphaeropsis* Lefèvre, 1876.
Bechyne and Bechyne (1966) recorded species "on flowers" in Venezuela but provided no other details.

55- *Nodonota* Lefèvre, 1885.
American. Mostly polyphagous. Recorded from 40 plant families and 92 genera, including the following: *Dahlia, Solidago, Artemisia, Vernonia, Chrysanthemum, Ambrosia, Helianthus, Hieracium, Cosmos, Baccharis, Verbesina, Cynara, Pluchea, Zinnia, Parthenium, Sclerocarpus* (Asteraceae), *Coffea* (Rubiaceae), *Solanum* (Solanaceae), *Alternanthera, Amaranthus, Celosia, Gomphrena* (Amaranthaceae), *Cucumis* (Cucurbitaceae), *Cydonia, Potentilla, Cerasus, Prunus, Pyrus, Malus, Rubus, Rosa, Fragaria* (Rosaceae), *Persea* (Lauraceae), *Salix, Populus* (Salicaceae), *Polygonum, Antigonon, Coccoloba, Colubrina* (Polygonaceae), *Vitis* (Vitaceae), *Trifolium, Glycine, Cajanus, Indigofera, Phaseolus, Vigna, Medicago, Vicia, Robinia* (Fabaceae), *Acacia, Pithecellobium, Vachellia, Prosopis* (Mimosaceae), *Alnus* (Betulaceae), *Theobroma* (Sterculiaceae), *Gossypium* (Malvaceae), *Croton, Ricinus* (Euphorbiaceae), *Ipomoea, Porana* (Convolvulaceae), *Annona* (Annonaceae),

Borago (Boraginaceae), *Brassica* (Brassicaceae), *Campsis* (Bignoniaceae), *Canna* (Cannaceae), *Chlorophora* (Moraceae), *Coriandrum, Daucus, Petroselinum* (Apiaceae), *Dianthus* (Caryophyllaceae), *Linum* (Linaceae), *Fuchsia* (Onagraceae), *Lagerstroemia* (Lythraceae), *Mangifera* (Anacardiaceae), *Duranta, Lantana, Verbena* (Verbenaceae), *Viola* (Violaceae), *Melicocca, Serjania* (Sapindaceae), *Oxalis* (Oxalidaceae), *Passiflora* (Passifloraceae), *Psidium* (Myrtaceae) and *Salvia* (Lamiaceae). Several species are pests of cotton, cocoa, grape and coffee while others feed on Poaceae (*Zea, Oryza, Triticum, Saccharum, Ammophila*) upon which they (mostly the larvae) produce severe damage. Several *Nodonota* feed on flowers and fruits of *Citrus* (Rutaceae).

56- *Isolepronota* Bechyne, 1949.
Neotropical. The trophic preferences are uncertain but species have been recorded (some doubtfully) from *Rosa*, *Eriobotrya* (Rosaceae) and *Eucalyptus* and *Psidium* (Myrtaceae) in Brazil.

57 - *Lepronota* Chapuis, 1874 .
Polyphagous. In Brazil, on *Lantana* (Verbenaceae), *Cydonia, Malus* (Rosaceae), *Solanum* (Solanaceae), *Erythrina* (Fabaceae) and *Psidium* (Myrtaceae).

58- *Agbalus* Chapuis, 1874.
Polyphagous. One species on soyabean (*Glycine max*, Fabaceae), various Melastomataceae, *Eucalyptus* (Myrtaceae) and rice (*Oryza sativa,* Poaceae) in Guyana.

59- *Pathius* Aslam, 1968.
African. On *Acacia* (Mimosaceae) and on Cyperaceae.

60- *Chrysolampra* Baly, 1859.
In China, only one species has been observed on its host-plant, namely *C. splendens* Baly, which feeds on *Prunus* (Rosaceae). Also recorded from *Cunninghamia* (Taxodiaceae).

61- *Chalcolema* Jacoby, 1890.
One Chinese species from *Cinnamomum* (Lauraceae).

62- *Taimbezinhia* Bechyne, 1954.
On leaves and fruits of *Theobroma cacao* (Sterculiaceae) in Brazil.

63- *Hylax* Lefèvre, 1884.
On *Cydonia* (Rosaceae) in Brazil.

64- *Iphimeis* Baly, 1864.
Neotropical. On *Nerium oleander* (Apocynaceae) introduced from the Medi-

terranean area. Also on *Citrus* (Rutaceae), *Solanum* (Solanaceae), *Acacia* (Mimosaceae), *Mucuna, Platycyamus, Glycine* (Fabaceae), *Coffea* (Rubiaceae), *Myrcia* (Myrtaceae), *Passiflora* (Passifloraceae), *Chorisia* (Bombacaceae), *Rosa* (Rosaceae), *Vitis* (Vitaceae) and *Cabralea* (Meliaceae).

65- *Costalimaita* Bechyne, 1954.
Polyphagous. From Venezuela and Brazil, on *Gossypium* (Malvaceae), *Psidium, Eugenia* and *Eucalyptus* (Myrtaceae), *Mangifera, Anacardium* (Anacardiaceae), *Musa* (Musaceae), *Phaseolus* (Fabaceae), *Persea* (Lauraceae) and *Vitis* (Vitaceae). Also on *Panicum* (Poaceae).

66- *Neoiphimeis* Bechyne, 1954.
On *Thea sinensis* (Theaceae) in Peru.

67- *Chalcophana* Chevrolat, 1843.
Neotropical. On Asteraceae: *Helianthus*, *Arctium*, etc.

68- *Otilea* Lefèvre, 1877.
On cocoa (*Theobroma*, Sterculiaceae) in Surinam.

69- *Corysthea* Baly, 1865.
Neotropical. On various plants, including rice, *Hibiscus* (Malvaceae), etc.

70- *Euphrytus* Jacoby, 1881.
On *Rosa* (Rosaceae) in Brazil and on *Poponax* and *Prosopis* (Mimosaceae) in Mexico.

71- *Eucolaspinus* Lea, 1916.
Genus from the Pacific. In New Zealand, one species has been recorded from Fabaceae and fruit trees (Rosaceae) by Tillyard (1926). Another species, *E. brunneus* (Fabricius) also from New Zealand, feeds on young buds of *Eucalyptus* (Myrtaceae) during summer and spring, although the insect actually appears to be polyphagous, since it also feeds on *Pinus* (Pinaceae) and on various fruit trees. Another species from New Zealand, *E. sculptus* (Brown), feeds on *Rubus australis* (Rosaceae) and *Aristotelia serrata* (Elaeocarpaceae)(Hudson, 1934). In Fiji, Bryant and Gressitt (1957) recorded two species from *Scaevola* (Goodeniaceae), *Tarenna* (Rubiaceae) and *Commersonia* (Sterculiaceae).

72- *Geloptera* Baly, 1861.
Australia. On *Citrus* (Rutaceae).

73- *Coytiera* Lefèvre, 1975.
Neotropical. On Convolvulaceae (*Ipomoea*) in Mexico.

74- *Prionodera* Chevrolat, 1837.
Neotropical. In Mexico, on *Baccharis* (Asteraceae).

75- *Promecosoma* Lefèvre, 1877.
Neotropical. Polyphagous. On *Eucalyptus* (Myrtaceae) introduced to Argentina. Also on *Verbesina, Eupatorium, Parthenium* (Asteraceae), *Vachellia, Prosopis* (Mimosaceae) in Mexico.

76- *Metazyonycha* Chevrolat, 1837.
On *Verbesina* (Asteraceae) in Mexico.

77- *Sternocolaspis* Bechyne, 1950.
Polyphagous in Brazil. On *Persea* (Lauraceae), *Gossypium* (Malvaceae), *Prunus, Malus, Pyrus* (Rosaceae), *Anacardium, Mangifera, Lithraea* (Anacardiaceae), *Ipomoea* (Convolvulaceae), *Theobroma* (Sterculiaceae), *Diospyros* (Ebenaceae), *Citrus* (Rutaceae), *Eucalyptus* (Myrtaceae), *Phaseolus* (Fabaceae), *Fuchsia* (Onagraceae), *Ricinus, Jatropha* (Euphorbiaceae), *Capsicum* (Solanaceae), *Vitis* (Vitaceae), etc. Also on Poaceae including sugarcane (*Saccharum*).

78- *Colaspis* Fabricius, 1801.
Enormous American genus of 250 species. Polyphagous. The adults feed on cotton, bananas and *Eucalyptus* and the larvae are root-feeders. On *Musa, Heliconia* (Musaceae), *Eucalyptus, Psidium, Myrcia* (Myrtaceae), *Machaerium, Phaseolus, Vigna, Pisum, Glycine, Arachis, Vicia, Lespedeza, Desmodium, Trifolium, Medicago, Meibomia* (Fabaceae), *Pithecellobium, Inga, Prosopis, Vachellia, Mimosa* (Mimosaceae), *Adipera, Cassia* (Caesalpiniaceae), *Solanum, Capsicum, Datura* (Solanaceae), *Gossypium, Hibiscus, Malachra* (Malvaceae), *Cissus, Vitis* (Vitaceae), *Brassica* (Brassicaceae), *Borreria, Hamelia* (Rubiaceae), *Helianthus, Ambrosia, Verbesina, Zinnia, Vernonia, Ximesia, Aster, Tagetes, Dahlia, Tithonia* (Asteraceae), *Linum* (Linaceae), *Papaver* (Papaveraceae), *Persea* (Lauraceae), *Piper* (Piperaceae), *Ipomoea* (Convolvulaceae), *Scaparia* (Scrophulariaceae), *Terminalia* (Combretaceae), *Theobroma* (Sterculiaceae), *Thymus, Coleus* (Lamiaceae), *Tropaeolum* (Tropaeolaceae), *Ficus* (Moraceae), *Litchi* (Sapindaceae), *Pyrus, Rosa* (Rosaceae), *Roystonea, Cocos* (Arecaceae), *Cattleya* (Orchidaceae), *Manihot, Croton* (Euphorbiaceae), *Achras* (Sapotaceae), *Lantana* (Verbenaceae), *Carya* (Juglandaceae), *Petiveria* (Phytolaccaceae), *Araujia* (Asclepiadaceae), *Anacardium* (Anacardiaceae), *Antigonon, Coccoloba, Polygonum* (Polygonaceae), *Rhododendron, Azalea, Vaccinium* (Ericaceae), *Oxalis* (Oxalidaceae), *Annona* (Annonaceae), *Mirabilis, Boerhavia* (Nyctaginaceae), *Begonia* (Begoniaceae), *Daucus* (Apiaceae), *Quercus* (Fagaceae), *Urera* (Urticaceae), *Heliotropium* (Boraginaceae), *Jussiaea, Fuchsia* (Onagraceae), *Beta* (Chenopodiaceae), *Camellia* (Theaceae), *Phlox* (Polemoniaceae), *Citrullus* (Cucurbitaceae), *Cyperaceae, Poaceae,* including

Paspalum, Panicum, Bambusa, Zea, Oryza and *Saccharum*. The genus is of great economic importance, because the larvae feed upon the roots of cultivated Poaceae while the adults feed on leaves, flowers and/or fruits of such plants as cotton, cocoa, alfalfa (lucerne), peanuts, potatoes, egg plants and bananas. Several species of the genus are also recorded from Coniferae, i.e. *Colaspis pini* Barber and others on *Pinus* (Pinaceae). Other species feed on *Kallistroemia* (Zygophyllaceae), *Citrus* (Rutaceae), *Ilex* (Aquifoliaceae), *Geranium* (Geraniaceae) and various Araceae. As is usual for the Eumolpinae, the larva lives within the ground (Blake, 1974) and during 9 months, feeds on grass roots (e.g. *Phleum, Paspalum)*, soyabean, banana rootlets and potato tubers. In the Northern hemisphere, the larva hibernates and the first pupae appear in June and the adult several weeks later. Many phyllophagous species also feed on fruits and cause scarring on bananas. Certain species do however, pollinate the flowers.

79- *Percolaspis* Bechyne, 1957.
On *Pennisetum, Panicum, Paspalum, Setaria* (Poaceae) and *Theobroma* (Sterculiaceae) in Brazil. The larvae feed on grass roots (Ferronatto, 1988).

80- *Freudeita* Bechyne, 1950.
Neotropical. On various shrubby *Solanum* (Solanaceae) in Brazil.

81- *Alethaxius* Lefèvre, 1885.
Neotropical. On *Citrus* (Rutaceae), *Annona* (Annonaceae), *Cydonia* (Rosaceae), *Thymus* (Lamiaceae) and various Poaceae *(Oryza, Panicum)*.

82- *Rhabdopterus* Lefèvre, 1885.
Neotropical. Polyphagous. On *Prosopis* (Mimosaceae), *Cajanus, Eysenhardtia* (Fabaceae), *Cephalanthus* (Naucleaceae), *Polygonum* (Polygonaceae), *Camellia* (Theaceae), *Rosa, Fragaria, Spiraea* (Rosaceae), *Salix, Populus* (Salicaceae), *Lonicera* (Caprifoliaceae), *Platanus* (Platanaceae), *Mangifera* (Anacardiaceae), *Persea* (Lauraceae), *Hydrangea* (Hydrangeaceae), *Musa* (Musaceae), *Pelargonium* (Geraniaceae), *Vitis* (Vitaceae), *Catasetum* (Orchidaceae), *Gossypium, Callirhoe* (Malvaceae), *Lantana,* (Verbenaceae), *Quercus* (Fagaceae), *Ulmus* (Ulmaceae), *Theobroma* (Sterculiaceae), *Commelina* (Commelinaceae), *Vaccinium, Rhododendron* (Ericaceae), *Ilex* (Aquifoliaceae), *Myrtus* (Myrtaceae), *Fuchsia* (Onagraceae), *Ixora* (Rubiaceae) and various Poaceae. The larvae feed on the roots of *Vaccinium* and other plants.

83- *Eumolpus* Weber, 1801.
Neotropical. Large species with bright colours of blue, green, red and purple, with polyphagous tendencies. On *Sida, Gossypium* (Malvaceae), *Solanum* (Solanaceae), *Beta* (Chenopodiaceae), *Leonurus* (Lamiaceae), Bignoniaceae, *Nerium* (Apocynaceae) and various Poaceae, including rice. Also on *Asclepias curassavica* (Asclepiadaceae)(Padre Moure, Vasconcellos and others, in litt.).

84- *Cleptor* Lefèvre, 1885.
Australia. One species, *C. inermis* Lefèvre, feeds as an adult on floral structures (i.e. flower parts and/or nectar) of *Pultenaea* (Fabaceae) and *Leptospermum* (Myrtaceae)(Hawkeswood & Jolivet, 1988).

85- *Edusella* Chapuis, 1874.
Australia. On *Acacia* in eastern Australia (Van den Berg, 1982), *Eucalyptus* (Myrtaceae) (Froggatt, 1907) and on *Prasophyllum* (Jones, 1972).

86- *Abirus* Chapuis, 1874.
Tropical Asia. On *Broussonetia, Ficus, Morus* (Moraceae) and *Ulmus* (Ulmaceae). In Japan, Indonesia and Vietnam, the genus has only been collected on Moraceae and Ulmaceae which indicates oligophagy.

87- *Abiromorphus* Pic, 1924.
China. On *Prunus* (Rosaceae), *Salix, Populus* (Salicaceae) and *Zizyphus* (Rhamnaceae).

88- *Tymnes* Chapuis, 1874.
American. On *Rubus* (Rosaceae), *Vitis* (Vitaceae) and *Ostrya* (Betulaceae) in the USA.

89- *Platycorynus* Chevrolat, 1837.
Temperate and tropical areas in the Old World. Like *Euryope*, it feeds on Asclepiadaceae and Apocynaceae (Jolivet, 1971-1982), but is rarely observed on other families. In temperate and tropical Asia and South Africa on *Trachelospermum* (Apocynaceae) and *Tylophora, Antherandra, Calotropis, Asclepias* (Asclepiadaceae). Other captures are purely accidental but in the Côte d'Ivoire, the senior author observed *Platycorynus azureus* (Sahlberg) on *Mallotus oppositifolius* (Euphorbiaceae).

90- *Chrysochares* Morawitz, 1861.
Palaearctic. Oligophagous. On *Cynanchum, Antitoxicum* (Asclepiadaceae) and *Trachomitum, Poacynum, Apocynum* (Apocynaceae). Also on *Calystegia* (Convolvulaceae).

91- *Chrysochus* Chevrolat, 1837.
Holarctic. On *Vincetoxicum, Cynanchum, Asclepias, Metaplexis, Antitoxicum* (Asclepiadaceae) and *Apocynum, Nerium, Trachomitum* (Apocynaceae). The larva lives 20 cm deep on the roots of the host-plants. The larva of *C. auratus* (Fabricius) feeds on the roots of *Asclepias* and *Apocynum cannabinum* (Peterson, 1960). Probably all genera of the section Corynodites (i.e. *Platycorynus, Chrysochares* and *Chrysochus*) live on Gentianales. Also on leaves of *Ipomoea batatas* (Convolvulaceae). Wyniger (1962) has also found the larvae on sweet

potato-tubers. Pupation occurs within the ground and the eggs are laid on the soil surface.

92- *Endocephalus* Chevrolat, 1837.
Neotropical. On *Vernonia* (Asteraceae) and *Cordia* (Ehretiaceae).

93- *Phanaeta* Lefèvre, 1878.
Neotropical. One species, *P. elongata* (Suffrian) on *Persea* (Lauraceae) in Cuba.

94- *Colaspoides* Laporte, 1833.
Cosmopolitan. Polyphagous. On *Litchi* (Sapindaceae), *Theobroma, Guazuma* (Sterculiaceae), *Rosa* (Rosaceae), *Prosopis* (Mimosaceae), *Bauhinia* (Caesalpiniaceae), *Liquidambar* (Altingiaceae), *Thea, Schima* (Theaceae), *Cornus* (Cornaceae), *Viburnum* (Caprifoliaceae), *Castanea* (Fagaceae), *Guarea* (Meliaceae), *Vaccinium* (Ericaceae), *Clerodendrum* (Verbenaceae), *Cinchona, Wedlandia, Hedyotis* (Rubiaceae), *Cannabis* (Cannabaceae), *Loropetalum* (Hamamelidaceae), *Symplocos* (Symplocaceae), *Anacardium, Mangifera* (Anacardiaceae), *Persea* (Lauraceae), *Phaleria* (Thymelaeaceae), *Syzygium, Eugenia, Psidium, Eucalyptus* (Myrtaceae), *Chrysobalanus* (Chrysobalanaceae), *Gossypium, Cienfuegosia* (Malvaceae) and *Terminalia* (Combretaceae). Many species are adapted for feeding on cotton in South America, on *Thea* in China and on cacao, *Eucalyptus* and various fruit trees in the tropics.

95- *Aulacia* Baly, 1867.
Indo-Malayan. One species on *Clidemia hirta* (Melastomataceae), a plant which is introduced to Malaysia.

96- *Heminodes* Jacoby, 1895.
Indian. On *Hibiscus* (Malvaceae) in Bangladesh.

2.15.6. Tribe Habrophorini

97- *Psathyrocerus* Blanchard, 1851.
Chile. On *Nothofagus pumilio* (Fagaceae).

2.15.7. Tribe Adoxini

98- *Scelodonta* Westwood, 1837.
Oligophagous. Old World Tropics. All species live on Vitaceae in Asia and Africa. Other records are doubtful. Recorded on *Leea* (Leeaceae), *Antidesma* (Stilaginaceae), *Vitis, Cissus, Ampelopsis, Cayratia* (Vitaceae), *Clerodendrum* (Verbenaceae), *Phyllanthus* (Euphorbiaceae), *Mangifera* (Anacardiaceae) and *Lophira* (Ochnaceae).

99- *Scelolanka* Medvedev, 1974.
On *Phyllanthus* (Euphorbiaceae) in Vietnam.

100- *Graphops* LeConte, 1884.
North American. Polyphagous. On *Gaura*, *Oenothera* (Onagraceae), *Hypericum* (Hypericaceae), *Cyrilla* (Cyrillaceae), *Fragaria* (Rosaceae), *Rudbeckia* (Asteraceae) and *Salsola* (Chenopodiaceae). Larvae feed on roots of *Oenothera* in Missouri (Riley & Enns, 1979).

101- *Aoria* Baly, 1863.
Tropical Asia. On *Cayratia, Tetrastigma, Vitis* (Vitaceae). Also captured on *Castanea* (Fagaceae), *Salix* (Salicaceae) and *Actinidia* (Actinidiaceae).

102- *Lefevrella* Jacobson, 1894.
Central Asia. On *Cerasus* (Rosaceae).

103- *Malegia* Lefèvre, 1883.
On *Pennisetum* (Poaceae) in Senegal, probably feeding on pollen.

104- *Microlypesthes* Pic, 1936.
China. On *Lespedeza* (Fabaceae).

105- *Pseudomalegia* Jacoby, 1897.
South Africa. On *Acacia* (Mimosaceae).

106- *Aulexis* Baly, 1863.
Far East. On *Lindera*, *Cinnamomum*, *Phoebe* (Lauraceae) and on Rubiaceae.

107- *Xanthonia* Baly, 1863.
America and Asia. On *Viburnum* (Caprifoliaceae) and *Picea* (Pinaceae) in China, on *Callicarpa* (Verbenaceae) in Vietnam, on *Fagus*, *Quercus* (Fagaceae) and *Ulmus* (Ulmaceae) in America and on *Morus* (Moraceae) in Japan.

108- *Lypesthes* Baly, 1863.
Asia. Polyphagous. On *Cinnamomum*, *Machilus* (Lauraceae), *Aesculus* (Hippocastanaceae), *Carpinus* (Carpinaceae), *Castanea*, *Quercus* (Fagaceae), *Corylopsis* (Hamamelidaceae), *Juglans, Pterocarya* (Juglandaceae), *Malus, Prunus, Pyrus* (Rosaceae), *Styrax* (Styracaceae), *Zelkova* (Ulmaceae) and *Cryptomeria* (Taxodiaceae).

109- *Fidia* Walsh, 1867.
America. On *Vitis, Parthenocissus, Ampelopsis* (Vitaceae). Accidental captures of the adults have been made on *Encelia* (Asteraceae) and *Cercis* (Caesalpiniaceae). The larvae have been found on *Vitis* roots in the USA.

110- *Brevicolaspis* Laporte, 1833.
On *Diplothemium, Cocos, Elaeis, Attalea* (Arecaceae) in Brazil. The adult gnaws the underside of the compound leaves of Arecaceae.

111- *Lio* Monros, 1949.
On *Myrtus luma* (Myrtaceae) in Chile.

112- *Hyperaxis* Gemmimger and Harold, 1874.
Asia. In the temperate zone, on *Quercus* (Fagaceae) and *Thea* (Theaceae) and in the tropics on *Callicarpa* (Verbenaceae) and *Melastoma* (Melastomataceae).

113- *Demotina* Baly, 1863.
In tropical Asia and the Pacific, on *Theobroma, Commersonia* (Sterculiaceae), *Morus* (Moraceae), *Melastoma, Clidemia* (Melastomataceae), *Psidium, Syzygium, Decaspermum* (Myrtaceae), *Macaranga, Glochidion* (Euphorbiaceae), *Maesa* (Myrsinaceae), *Quercus, Castanopsis* (Fagaceae), *Thea* (Theaceae), *Crotalaria* (Fabaceae), *Alphitonia* (Rhamnaceae), *Zingiberaceae, Rubus* (Rosaceae), *Grewia, Trichospermum* (Tiliaceae) and *Nephelium* (Sapindaceae). In the temperate zones, i.e. China and Japan, *Demotina* feed mostly on *Quercus, Castanopsis, Thea, Rubus* and *Morus*.

114- *Ischyromus* Jacobson, 1893.
Central Asia. On *Haloxylon* and *Salsola* (Chenopodiaceae).

115- *Andosia* Weise, 1896.
Central Asia. On *Artemisia* (Asteraceae) and *Salicornia* (Chenopodiaceae).

116- *Andosiomorpha* Lopatin, 1981.
Iran. On *Pistacia* (Pistaciaceae) and *Acer* (Aceraceae).

117- *Parnops* Jacobson, 1894.
Central Asia. On *Pyrus* (Rosaceae), *Salix, Populus* (Salicaceae).

118- *Eryxia* Baly, 1865.
African. Polyphagous. On *Hibiscus, Gossypium* (Malvaceae), *Canthium* (Rubiaceae), *Lagerstroemia* (Lythraceae), *Pennisetum* (Poaceae).

119- *Trichochrysea* Baly, 1861.
Asia. In the temperate and semi-tropical areas on *Cerasus, Prunus, Pyrus, Malus, Crataegus* (Rosaceae), *Castanea, Quercus* (Fagaceae), *Salix* (Salicaceae), *Albizia* (Mimosaceae), *Lespedeza* (Fabaceae), *Dahlbergia* (Gesneriaceae) and *Macaranga, Mallotus* (Euphorbiaceae). *Macaranga* is an ant-plant associated with *Crematogaster*.

120- *Nerissus* Chapuis, 1874.
Africa. On *Cissus* (Vitaceae) and *Theobroma* (Sterculiaceae). Other records from Moraceae, Annonaceae, Combretaceae and Verbenaceae are doubtful.

121- *Casmena* Chapuis, 1874.
Africa. On Arecaceae in Sierra Leone and Poaceae in Zaire.

122 - *Lymidus* Fairmaire, 1901.
Described from La Reunion. Recorded on *Theobroma cacao* (Sterculiaceae) in São Tomé (Blunk, 1954).

123- *Pseudocolaspis* Laporte, 1833.
Africa and Asia. Polyphagous. Often observed on flowers of Rubiaceae and Cistaceae.Recordedfrom*Pennisetum* and other Poaceae, *Combretum, Terminalia* (Combretaceae), *Heliotropium* (Boraginaceae), *Cissus* (Vitaceae), *Syzygium* (Myrtaceae) and *Allium* (Alliaceae). In India, one species has been captured on *Malus, Pyrus* (Rosaceae), *Dahlia* (Asteraceae) and *Castanea* (Fagaceae).

124 - *Macrocoma* Chapuis, 1 874 .
Old World. Polyphagous. On *Coffea* (Rubiaceae), *Gossypium* (Malvaceae), *Eryngium* (Apiaceae), *Adhatoda* (Acanthaceae), *Aesculus* (Hippocastanaceae), *Castanea* (Fagaceae), *Urelytrum, Sorghum* (Poaceae), *Vitex* (Verbenaceae), *Pulicaria* (Asteraceae), *Tamarix* (Tamaricaceae) and *Alhagi* (Fabaceae). Also on flowers of Poaceae and *Neoboutonia* (Euphorbiaceae).

125- *Bromius* Chevrolat, 1837.
Holarctic. Oligophagous on Vitaceae (*Vitis, Ampelopsis*) and Onagraceae (*Epilobium, Clarkia, Chamaenerion*).

126- *Anidania* Reitter, 1889.
Palaearctic. On Asteraceae in Central Asian steppes.

127- *Adoxinia* Reitter, 1889.
Central Asia. On *Haloxylon, Aellenia, Salsola* (Chenopodiaceae).

128- *Bromiodes* Jacoby, 1895.
India at high altitude. *Bromiodes squamosus* Bryant has been captured on *Pyrus* (Rosaceae) in Simla (Bryant, 1923).

129- *Acrothinium* Marshall, 1865.
Asia. On *Ampelopsis, Vitis, Parthenocissus* (Vitaceae), *Mentha* (Lamiaceae), *Commelina* (Commelinaceae), *Corylopsis* (Hamamelidaceae) and *Salix* (Salicaceae).

130- *Myochrous* Erichson, 1847.
America. Polyphagous. The larvae of *Myochrous denticollis* (Say) feed on the roots of corn (*Zea*, Poaceae). On *Xanthosoma* (Araceae), *Gossypium*, *Malva*, *Hibiscus* (Malvaceae), *Eucalyptus* (Myrtaceae), *Nicotiana, Capsicum, Solanum* (Solanaceae), *Oryza*, *Saccharum*, *Panicum*, *Paspalum*, *Zea*, *Sorghum, Alopecurus*, *Phleum* (Poaceae), *Musa* (Musaceae), *Baccharis*, *Pluchea*, *Xanthium, Ambrosia*, *Helenium*, *Silphium* (Asteraceae), *Glycine*, *Lespedeza*, *Melilotus* (Fabaceae), *Taxodium* (Taxodiaceae), *Vitis* (Vitaceae), Cucurbitaceae, *Salix* (Salicaceae), *Alnus* (Betulaceae) and *Pyrus* (Rosaceae).

131- *Glyptoscelis* Chevrolat, 1837.
America. Polyphagous. On *Cajanus, Crotalaria, Glycyrrhiza, Arachis, Medicago* (Fabaceae), *Prosopis* (Mimosaceae), *Citrus* (Rutaceae), *Cyperus* (Cyperaceae), *Mangifera* (Anacardiaceae), *Oryza, Saccharum, Zea* (Poaceae), *Gossypium*, *Sphaeralcea* (Malvaceae), *Salix* (Salicaceae), *Vitis* (Vitaceae), *Pyrus*, *Malus*, *Prunus* (Rosaceae), *Libocedrus, Juniperus* (Cupressaceae), *Pinus, Picea, Tsuga* (Pinaceae), *Sequoia* (Taxodiaceae), *Artemisia* (Asteraceae), *Carya* (Juglandaceae), *Cordia* (Ehretiaceae) and *Psidium* (Myrtaceae). The genus often feeds on Coniferae in North America.

132- *Pachnephoptrus* Reitter, 1892.
Azerbaidjan at the Iranian border. On *Astralagus* (Fabaceae).

133- *Pachnephorus* Chevrolat, 1837.
Cosmopolitan. Polyphagous. On *Saccharum, Zea*, *Triticum*, *Sorghum* (Poaceae), *Inula*, *Guizotia*, *Siegesbeckia*, *Cirsium*, *Carduus*, *Helianthus*, *Artemisia* (Asteraceae), *Sesamum* (Pedaliaceae), *Hibiscus, Gossypium* (Malvaceae), *Cocos* (Arecaceae), *Foeniculum* (Apiaceae), *Limonium* (Plumbaginaceae), *Arachis* (Fabaceae), *Polygonum* (Polygonaceae), *Ipomoea* (Convolvulaceae) and *Tamarix* (Tamaricaceae).

134- *Chalcosicya* Blake, 1930.
America. Polyphagous. On *Casuarina* (Casuarinaceae), *Coccoloba* (Polygonaceae), *Colubrina* (Rhamnaceae), *Conocarpus* (Combretaceae), *Croton* (Euphorbiaceae), *Dioscorea* (Dioscoreaceae), *Gossypium* (Malvaceae), *Ipomoea* (Convolvulaceae) and *Phaseolus* (Fabaceae).

135- *Colaspina* Weise, 1893.
Europe. Polyphagous. On *Artemisia*, *Helichrysum* (Asteraceae), *Helianthemum* (Cistaceae), *Scrophularia* (Scrophulariaceae), *Quercus* (Fagaceae), *Genista* (Fabaceae) and *Bromus* (Poaceae).

136- *Colaspidea* Laporte, 1833.
Holarctic. Polyphagous. On *Erica* (Ericaceae), *Cistus* (Cistaceae), *Ros-*

marinus (Lamiaceae), *Ulex* (Fabaceae), *Quercus* (Fagaceae), *Pelargonium* (Geraniaceae), *Artemisia* (Asteraceae), *Malus* (Rosaceae) and *Vitis* (Vitaceae).

137- *Apolepis* Baly, 1863.
Indo-Malayan. On *Mangifera* (Anacardiaceae), *Theobroma* (Sterculiaceae), *Thea*, *Camellia* (Theaceae) and *Nephelium* (Sapindaceae). In the case of the fruit trees, the adults feed on leaves and fruits.

138- *Aphilenia* Weise, 1889.
Central Asian steppes. On *Calligonium* (Polygonaceae) and *Haloxylon* (Chenopodiaceae).

139- *Atomyria* Jacobson, 1894.
Central Asia. On *Populus* (Salicaceae).

140- *Parascela* Baly, 1878.
China. On *Prunus* (Rosaceae) and *Populus* (Salicaceae).

Concluding comments on Eumolpinae:
140 genera of Eumolpinae from a total of 437 described (i.e. 30.6%) have their host-plants more or less determined. If we consider the number of species for which the food selection is known and the total number of described eumolpine species (which is approximately 3,500) we can say that less than 10% of the species have a biology which is more or less known. Among the genera recorded here, 54.6% are polyphagous, 26.9% are oligophagous and 18.4% have a food selection not clearly defined. However, most of the genera and their species have very large polyphagous tendencies; the record is with the genus *Nodonota* with more than 92 host-plant genera identified.

The plant families selected are distributed as follows: Dicotyledonae (93), Monocotyledonae (14) and Gymnospermae (Coniferae)(4). There are thus 111 plant families presently known to be selected by the Eumolpinae.

The reason why we do not know better the biology of these insects, is that there are little or no coherent observations from large areas of the world such as Madagascar, Indonesia, Malaysia, Australia and continental Africa (except for the southern regions). In addition, practically nothing has been recorded from the Solomon Islands, New Caledonia, Vanuatu (New Hebrides), and many other areas such as India where the published data are often erroneous. It is without doubt that the Holarctic zone is the best known, along with China, Vietnam, Thailand and certain areas of tropical America, although there are still immense gaps in knowledge about the beetle fauna of these regions. Without all these necessary data yet to be obtained, it is impossible to determine the selection patterns/trends within the subfamily.

The larvae of Eumolpinae have a morphological and biological unity in that they are blind, endogenous[1] and are root-feeders but their food selections can be only explained when sufficient data are available. Only one thing is absolutely certain, i.e. the tendency of certain genera (9 or 6.6%) towards Asclepiadaceae and Apocynaceae (Gentianales). However, there is certainly still much more to be observed and recorded. Are we seeing here a primitive element from which the polyphagy of particular genera has been derived, or was this polyphagy primitive? The most common eumolpine selections are the Vitaceae, Convolvulaceae, Euphorbiaceae, Malvaceae, Rosaceae, Sterculiaceae and Rubiaceae. The primitive genera have special selections, e.g. *Hornibius* and *Psathyrocerus* on *Nothofagus* (Fagaceae) and *Stenomela* on *Gunnera* (Gunneraceae). *Psathyrocerus* and *Hemidacne* are very primitive and are both devoid of the medio-cubital patch (this patch is only slightly visible in *Hemidacne*).

There is probably a close relationship between Gentianales and Convolvulaceae since several eumolpines, e.g. *Colasposoma dauricum* Mannerheim, display a restricted selection for both plant groups. However, there do not appear to be any close relationships between Gentianales and Geraniales (Euphorbiaceae) but such dual selections exist in the genus *Platycorynus*. Several scientists in Brazil have partly confirmed our theory about the Apocynaceae and Asclepiadaceae concerning certain South American Eumolpinae, e.g. the case of *Eumolpus* feeding on *Asclepias*.

The selection of *Eucalyptus*, a tree which has been introduced to tropical America, can be explained since several species of Eumolpinae commonly feed there on native *Psidium* and *Eugenia* (Myrtaceae). Similar comments can be made about the selection of Sterculiaceae, Malvaceae, Tiliaceae and Bombacaceae (Malvales) because there is a strong link between them, but the fact that some eumolpines have a broad polyphagy, should also be taken into consideration.

The host-plants recorded from Zaire in Africa must be treated with the utmost caution because very often they are not very reliable. All eumolpine larvae have been observed feeding on roots within the soil. In the case of *Syagrus*, which are asumed to be myrmecophilous in the stipular thorns of *Acacia* in east Africa, nobody has observed the larvae which are probably reared within the ant-nests. Eumolpine larvae are extremely difficult to rear in the soil and do not complete their life-cycle on naked roots.

The aposematic bright colours of many Eumolpinae adults are often associated with the toxicity of their host-plants. Genera which are very well protected by their toxins, such as *Chrysochares, Chrysochus, Platycorynus, Eumolpus, Iphimeis, Acrothinium, Trichochrysea* and *Colasposoma*, possess the brightest colours, i.e. green, purple, red, blue. Others such as *Euryope* are brown with black spots and are clearly visible on the green of the leaves. The black colour of certain *Abirus* species can be protective as for *Timarcha*. Several other genera are green, greenish, blackish, brownish, or covered with strong, whitish

[1] Exceptionally, some primitive genera (*Hornibius*) have borer larvae in buds.

hairs. Some species do not appear to be protected and perhaps they are able to escape from predators by being small and inconspicuous. The small green or blue *Basilepta* on *Rubus* in Asia, do not have any serious predators but they drop and feign death at the slightest approach.

Eumolpine blood is toxic like the blood of Chrysomelinae and Alticinae, but reflex bleeding is rare or absent. They fly clumsily and rarely, except in the tropics, where they use their flight to escape from predators. Many species, even entire genera, are brachypterous and apterous.

Most of the known Eumolpinae are polyphagous and this is the reason why they have adapted so readily to introduced crops such as cocoa, tea, coffee, grapes and bananas in other continents. Several *Colaspis* leaf-feeders rapidly became fruit (banana) feeders in America, attacking the fruit epidermis.

Sinigrin or mustard oil appears to be repulsive to the Eumolpinae so that attacks upon cabbages and related plants are very exceptional. Few species have been recorded on Apiaceae and Cucurbitaceae.

Many Eumolpinae, either by choice or by secondary adaptation, become pests of fruit trees in temperate areas (e.g. pear, apple, peach, quince, plum, cherry) or in the tropics (e.g. mango, litchi, custard apple, guava, avocado, banana, *Sapotilla, Eugenia*). The Eumolpinae also attack strawberry, pineapple and *Rubus*. At least 28 genera have been recorded from cotton (*Gossypium*)(20.7%), 12 genera from coffee (*Coffea*)(8.8%), 21 genera on cocoa (*Theobroma*)(15.5%) and 7 genera on tea (*Thea*)(5.2%). Twelve genera (8.8%) contain numerous species adapted to Coniferae. Eighteen genera have chosen *Eucalyptus* and other Myrtaceae. Many species feed on alfalfa, peanuts, *Acacia* and various Fabaceae, citrus trees, potato, egg-plant, numerous wild or cultivated grasses (Poaceae)(e.g. rice, corn, maize, sugarcane, sorghum) and sedges (Cyperaceae). The damage caused by the adults on Poaceae (i.e. leaves, ears) is generally weak, but the larvae feed on roots and can be serious pests. Many species feed on grape vines and certain genera are specialized on Vitaceae and related plants.

Finally, the Eumolpinae is one of the largest subfamilies of Chrysomelidae, with a total of 437 genera and 3,500 species, but is also one of the poorest known biologically. Evolutionary trends in selection will only be elucidated once the food-plants of most of the genera have been determined.

References listed in the bibliography pertaining to Eumolpinae are as follows:- Alibert (1951), Araujo & Silva (1968), Balsbaugh (1982), Bechyne & Bechyne (1966), Blake (1950, 1974), Blunk (1954), Bosq (1942), Boving & Craighead (1931), Broodryk (1965), Bryant (1923), Bryant & Gressitt (1957), Chapin (1979), Costa-Lima (1936, 1955), Ferronatto (1988), Froggatt (1907), Gardner (1934), Gressitt (1963, 1966-1969), Hawkeswood & Jolivet (1988), Hocking (1970), Hudson (1934), Jerez & Ibarra-Vidal (1992), Jolivet (1971c, 1979b, 1982, 1987b), Jones (1972), Kalshoven *et al.* (1981), Kelly (1915), Kurcheva (1958, 1967, 1975), Le Pelley (1959), Lopatin (1984), Medvedev & Dap (1983), Monros (1952b), Ostmark (1975), Paterson (1931-1932, 1936), Peterson (1960), Riley & Enns (1979), Salt (1928), Selman (1962, 1963, 1972), Staines (1991), Stehr (1991), Tillyard (1926), Wyniger (1962).

SECTION IV: TRICHOSTOMA

2.16. *Chrysomelinae*

The Chrysomelinae are generally very selective in their host-plant choices and very often these choices are at the genus level, but also at the species level, e.g. *Timarcha, Chrysolina, Calligrapha* (Jolivet & Petitpierre, 1973, 1976a,b, 1981). There are no adequate classifications of this rather heterogeneous subfamily and the old classification of Weise (1916) is far from perfect. The genus *Timarcha* is a tribe in itself and this has been mentioned many times by Sharp and Muir (1912), Paterson (1931), Chen (1934), Powell (1941) and Jolivet (1957). Chen (1934) proposed separation of the Timarchini from the Entomoscelini but this solution is only approximate. The classification of Daccordi in Seeno and Wilcox (1982) is rather satisfactory despite several taxonomic points which remain unclarified. Daccordi highlights the uniqueness of the Timarchini (ring-like tegmen) compared to the rest of the Chrysomelini, which are divided into very differing subtribes. This classification satisfies most of the specialists working on these groups, and has merit in separating *Xenomela, Oreomela, Cyrtonastes* and others from *Timarcha,* to which they have only a superficial resemblance through general convergence. No other genus except for *Timarcha* have been found to possess a ring-like tegmen among the Chrysomelinae, a character which, for some specialists, qualifies *Timarcha* for the status of subfamily.

The Doryphorina, a Neotropical subtribe, have a biological feature in common amongst its genera, i.e. the total absence of gregarines inside the gut. The various species have very different trophic preferences even within a single genus, e.g. *Leptinotarsa.* The most common tendency is towards Solanaceae, but Asteraceae, Apocynaceae, Zygophyllaceae and Asclepiadaceae are frequent selections. The Solanaceae are the most diversified in the Neotropical world and their center of origin is probably the American continent and this is the reason why the "solanophagous" species are the most common there.

The host-plants of *Barymelina* from Madagascar are not known.

2.16.1. Tribe Timarchini

1- *Timarcha* Latreille, 1829.

There are some differences between *Timarcha* sensu stricto and the smaller *Timarchostoma* Motschulsky, 1860. These subgenera prefer the Rubiaceae (*Galium*) in northern Europe and feed in southern Europe and North Africa upon a certain number of other Rubiaceae, and according to the species, on Plantaginaceae, Brassicaceae, Scrophulariaceae, Asteraceae and Dipsacaceae (Tiberghien, 1972; Jolivet & Petitpierre, 1973). The botanical relationships are evident between the Plantaginaceae, Rubiaceae and the Dipsacaceae, less evident with the Scrophulariaceae and not at all with the Asteraceae and

Brassicaceae. The basic tendency is clearly towards Rubiaceae/Plantaginaceae. One of us (PJ) has tested specimens of *Timarcha tenebricosa* Fabricius on various tropical Rubiaceae such as *Mitracarpus* and *Borreria*. Both plants belong to a different tribe from that of *Galium* and *Rubia* and are rejected by this insect.

2- *Metallotimarcha* Motschulsky, 1860.
On *Galium* (Rubiaceae) in northern Europe.

3- *Americanotimarcha* Jolivet, 1948.
On *Rubus* and *Fragaria* (Rosaceae) on the west coast of the USA.

2.16.2 Tribe Chrysomelini

4- *Cosmogramma* Erichson, 1847.
On *Triumfetta bartramia* (Tiliaceae) in Brazil (PJ, pers. obs.).

5- *Zygogramma* Chevrolat, 1837.
On various Asteraceae in the USA and Mexico (Moldenke, 1971). In Brazil and Argentina, on various Malvaceae, i.e. *Zygogramma novemstriata* Stål, *Z. satrapa* Stål, *Z. malvae* Stål, Z. *sextaeniata* Stål, and *Z. virgata* Stål. *Z.* (*Tritaenia*) *cognata* Stål has been captured on *Solanum tuberosum* (Solanaceae) in Venezuela. *Zygogramma* is an artificial and composite genus, poorly interpreted. The species feeding on Asteraceae are very selective in their host-plant choice.

6- *Calligrapha* Chevrolat, 1837.
Mostly in North America on Asteraceae, Convolvulaceae, Salicaceae, Tiliaceae, Corylaceae, Ulmaceae, Betulaceae, Carpinaceae, Rosaceae, Cornaceae, Malvaceae and Sterculiaceae. Several of these families are related (Malvales) but the others show much less affinity to these. The species of *Calligrapha* are very specific. Most of them fly well and feed on trees and shrubs, whereas *Chrysolina* remain on low herbaceous plants.

7- *Deuterocampta* Chevrolat, 1837.
Recorded from *Anacardium* (Anacardiaceae) in Brazil and on *Heliotropium* (Boraginaceae) in Argentina (Bosq, 1942) which may be accidental.

8- *Leucocera* Chevrolat, 1837.
On Malpighiaceae (*Malpighia*) and Euphorbiaceae (*Phyllanthus*) in the West Indies. The two families are closely related (Geraniales).

9- *Strichosa* Chevrolat, 1843.
Very few data available on this Chilean genus. On *Berberis buxifolia* (Berberidaceae).

10- *Elytrosphaera* Chevrolat, 1843.

One species has been observed by one of us (PJ) in the Atlantic primary forest on *Adenostemma brasilianum* (Asteraceae). This appears to be the normal food-plant of the subgenus or perhaps of the whole genus, together with other Asteraceae at high altitudes (Jolivet, 1987a). *Elytrosphaera* formerly frequenting the clearings of the primary forests in middle (700 m) and high altitudes have become endangered species. Several other species, formerly classified within the genus *Grammodesma*, feed on the same plants (Asteraceae).

11- *Leptinotarsa* Stål, 1859.

This genus includes 41 species in the USA and Central and South America. It has been recently studied by Hsiao (1986, 1988) and Jacques (1988). The staff of Wageningen University have published many papers on the biology of *L. decemlineata* (Say) which has invaded Europe to the Ural Mountains and probably beyond. The dispersion of this species is very rapid because it flies readily and displays migratory flights. The specificity of the various species for Solanaceae, Asteraceae and Zygophyllaceae, is very restricted and very often only certain species of *Solanum* are selected. The adaptation of *L. decemlineata* from *Solanum rostratum* to *S. tuberosum* has been very rapid and remarkable because on certain other species of *Solanum* and on certain other Solanaceous genera, the beetle is unable to complete all of its life-cycle (Hsiao, 1981; Jacques, 1988). Some Panamanian species on *Solanum* shrubs with xerophytic leaves.

12- *Eugonycha* Chevrolat, 1843.

One species, *Eugonycha melanostoma* Stål on arborescent *Solanum* (Solanaceae) at Serra do Japi, Brazil (PJ).

13- *Proseicela* Chevrolat, 1837.

Incorrectly recorded on *Schizolobium paralybum* (Caesalpiniaceae) in Ecuador. On *Solanum* shrubs in Brazil (Medeiros, 1991). Viviparous.

14- *Labidomera* Chevrolat, 1837.

On *Asclepias* and *Cynanchum* (Asclepiadaceae) in the USA and on *Witheringia* (Solanaceae) in Costa Rica.

15- *Platyphora* Gistel, 1857.

On arborescent species of *Solanum* in tropical America. Most of the species are viviparous. At least one species shows cycloalexy among its larvae. One species, *P. langsdorfi* Germar has been recorded on *Ipomoea* (Convolvulaceae) by Costa-Lima (1936). Not all *Platyphora* feed on Solanaceae; however, the ones that do, are very specific in their plant choice and generally accept only one, two or three *Solanum* species. Some species on Asteraceae.

16- *Doryphora* Illiger, 1807.
On *Prestonia isthmica* (Apocynaceae) in Costa Rica (Eberhard, 1981).

17- *Ambrostoma* Motschulsky, 1860.
On *Ulmus* (Ulmaceae) in Mongolia and China.

18- *Crosita* Motschulsky, 1860.
Crosita from Central Asia feed on *Artemisia* (Asteraceae) while the subgenus *Bittotaenia* appears to feed on Lamiaceae. *Crosita afghanica* Lopatin has been recorded from *Nepeta* (Lamiaceae) in Central Asia.

19- *Gnathomela* Jacobson, 1895.
On Apiaceae in Central Asia (Lopatin, 1984).

20- *Chrysolina* Motschulsky, 1860.
The enormous genus *Chrysolina* has been divided by Bechyne, Daccordi and others into 56 or 57 subgenera. The genus is absent from New Guinea, Australia and tropical America; otherwise it is practically cosmopolitan. The host-plant selections arranged subgenus by subgenus or species by species, have been reviewed in detail by Jolivet and Petitpierre (1976a) and Jolivet *et al.* (1986). The oligophagous species are very strict in their food choices and most of the groupings appear natural. *Chrysolina* species either fly poorly or not at all, although there are some exceptions, e.g. *C. americana* (L.) and *C. aurichalcea* (Mannerheim), where flight still occurs among certain populations of those species. The wings of *Chrysolina* are generally reduced or absent. The wing muscles are also reduced and most of the species feed on low herbs where they rely on their cryptic or nocturnal habits for protection. The following botanical families arranged in decreasing order of importance, are chosen among *Chrysolina:* Lamiaceae, Asteraceae, Hypericaceae, Apiaceae, Scrophulariaceae, Plantaginaceae, Ranunculaceae and Brassicaceae. If the Central Asian species of *Crosita* and the alpine *Oreina* are considered subgenera of *Chrysolina*, which is rather logical, then the Asteraceae become the dominant host-plant family, with the Apiaceae also important. There are still many gaps to be filled because only 15 subgenera are well-known trophically, i.e. 44.7% of the described species. Several rare species from Mongolia, Canada and Alaska have been recorded from Salicaceae, Chenopodiaceae, *Betula* (Betulaceae), etc., but these records are very doubtful. One must differentiate between a resting place and actual feeding. Gaps in knowledge mostly concern the diets of the montane species from tropical Asia, Sri Lanka, Indonesia and South Africa. There are no *Chrysolina* known from Madagascar.

21- *Oreina* Chevrolat, 1837.
Alpine and montane genus with several incursions into lowlands. On many Asteraceae and Apiaceae. It is not known for certain whether the species feed-

ing on Asteraceae also feed on Apiaceae and vice versa. Laboratory tests will be necessary to clarify this matter. Various biotypes even exist among the species feeding on Asteraceae, such as in *O. cacaliae* (Schrank), which can be divided into *Senecio*-feeding groups and *Petasites*-feeding ones (Kreslavsky *et al.*, 1976). Most of the subgenera appear to be somewhat natural if the host-plant choices are examined (Jolivet *et al.*, 1986).

22- *Agrosteomela* Gistel, 1857.
Takizawa (1977) described the larva of this insect and discovered its host-plant, *Eupatorium* (Asteraceae).

23- *Promechus* Boisduval, 1835.
These large and splendid beetles are often common with their larvae along New Guinean rivers on arborescent Araliaceae, i.e. *Schefflera* and *Boerlagiodendron*. They have also been found on Pittosporaceae and Cunoniaceae by Gressitt and Hart (1974). These two families are related but are not closely related to Araliaceae.

24- *Ceralces* Gerstaecker, 1855.
Africa. On Lamiaceae and *Clerodendrum* (Verbenaceae). There are various reports from *Manihot* (Euphorbiaceae), cocoa (*Theobroma*, Sterculiaceae) and *Coffea* (Rubiaceae) but these are all accidental.

25- *Sphaeratrix* Gistel, 1848.
Reported to feed on *Maytenus* (Celastraceae) in East Africa but probably doubtful.

26- *Colaphellus* Weise, 1916.
On almost all Brassicaceae.

27- *Colaspidema* Laporte, 1833.
The subgenus *Colaspidema* sensu stricto appears to feed on Fabaceae and the subgenus *Colaphomega* prefers Scrophulariaceae and Plantaginaceae, closely related families.

28- *Gastrophysa* Chevrolat, 1837.
All the species feed on Polygonaceae (*Rumex, Polygonum*, *Rheum*, etc.), but they appear very host-specific at the species level.

29- *Phaedon* Latreille, 1829.
The genus *Phaedon* has been divided into 8 subgenera, but their food choice differs depending on the species. There are also essential differences between the Holarctic and tropical species. In the temperate zone, the following food-plants have been recorded: Ranunculaceae (*Ranunculus*), Brassicaceae

(*Cochlearia, Stellaria, Brassica, Capsella, Rorippa, Nasturtium, Raphanus, Lepidium, Barbarea, Sisymbrium, Sinapis, Armoracia, Strigosella, Cardamine*), Scrophulariaceae (*Veronica)*, Lamiaceae (*Galeopsis, Mentha*), Callitrichaceae (*Callitriche*). Several tropical *Phaedon* feed on *Cordia* (Ehretiaceae), *Baccharis, Senecio, Bidens, Flaveria, Tessaria, Tagetes* (Asteraceae), *Gossypium* (Malvaceae), *Prosopis* (Mimosaceae) and *Lantana* (Verbenaceae). The subgenus *Paraphaedon* feeds in Europe on various Apiaceae (*Anthriscus, Ferula, Apium, Daucus, Heracleum, Selinopsis, Smyrnium*). The subgenus *Sternoplatys* feeds on *Malachium* and *Stellaria* (Caryophyllaceae) and on *Dasyphora* and *Potentilla* (Rosaceae). Several *Phaedon* feed on *Rubus* (Rosaceae) in South-East Asia.

30- *Sclerophaedon* Weise, 1882.
Sclerophaedon feeds on *Stellaria* (Caryophyllaceae) and *Caltha* (Ranunculaceae) in Europe.

31- *Cyrtonus* Latreille, 1829.
West Mediterranean genus. Nocturnal. Feeds on small, low growing Asteraceae (i.e. *Lappa, Hyoseris, Artemisia, Hieracium*) and rests under stones during the day.

32- *Hydrothassa* Thomson, 1866.
All species feed on Ranunculaceae (*Ranunculus, Caltha*) in Europe and America.

33- *Prasocuris* Latreille, 1802.
Holarctic. Feeds on plants from various families, i.e. *Cicuta, Helosciadium, Sium, Phellandrium, Oenanthe, Callitriche, Angelica, Levisticum* (Apiaceae), *Caltha* (Ranunculaceae), *Anagallis* (Primulaceae), *Veronica* (Scrophulariaceae), *Myosotis* (Boraginaceae) and *Callitriche* (Callitrichaceae). *Prasocuris* are rather host-specific at the species level.

34- *Plagiodera* Chevrolat, 1837.
Almost cosmopolitan. Normally occurs on *Salix* and *Populus* (Salicaceae) in the temperate Northern zone, but also in Chile (*P. erythroptera* Blanchard) (Zapata, 1970; Bosq, 1942) and in Brazil. One South American species, *P. viridipennis* Stål has been recorded from *Salix humboldtiana* and another one on *Tibouchina*, but the latter record is doubtful. A South African species, *P. cuprea* Baly feeds on *Gymnosporia* (Celastraceae).

There are several reports of the genus from the Philippines and India on *Flacourtia* and *Xylasma* (Flacourtiaceae).

35- *Chrysomela* Linnaeus, 1758.
The genus *Chrysomela* is divided into several subgenera. Food-plants are *Alnus, Betula* (Betulaceae), *Populus, Salix* (Salicaceae), *Carpinus* (Carpinaceae), Salicales and Fagales.

36- *Linaeidea* Motschulsky, 1860.
On Betulaceae (*Alnus, Betula*) and Carpinaceae (*Carpinus*).

37- *Gastrolina* Baly, 1859.
Asia. Feeds on *Juglans* and *Pterocarya* (Juglandaceae) and also on *Alnus* (Betulaceae).

38- *Gastroliniodes* Chûjô and Kimoto, 1960.
Japan. On *Carpinus* (Carpinaceae) and *Corylopsis* (Hamamelidaceae).

39- *Agasta* Hope, 1909.
Asia. On *Casearia* (Flacourtiaceae).

40- *Phaedonia* Weise, 1898.
Phaedonia feeds on *Sesbania* and *Indigofera* (Fabaceae) in Africa and on *Desmodium*, *Pueraria*, *Glycine* and *Alysicarpus* (Fabaceae) in Asia. It has adapted easily to soya bean *(Glycine max)* in Indonesia.

41- *Mesoplatys* Baly, 1875.
Also on arborescent Fabaceae in Africa (i.e. *Indigofera*, *Aeschynomene*, *Sesbania* and *Erythrina*) and *Lagerstroemia* (Lythraceae).

42- *Ageniosa* Weise, 1908.
Wrongly recorded on *Rosa* (Rosaceae) in South Africa. Larva and adult feed on *Plectranthus* spp. (Lamiaceae) (Grobbelaar & Jolivet, 1993, pers. obs.).

43- *Cyclomela* Baly, 1856.
Biology unknown, but the beetle is found among shrubs of the primary rain forest in Queensland (Hawkeswood, 1987).

44- *Starycea* Selman, 1977.
One species, *Starycea acaciae* (Lea), has been recorded from *Acacia* (Mimosaceae) in Australia.

45- *Dicranosterna* Motschulsky, 1860.
On *Acacia* (Mimosaceae) in Australia.

46- *Trochalodes* Weise, 1901.
On *Acacia* (Mimosaceae) and *Eucalyptus* (Myrtaceae) in Australia.

47- *Lioplacis* Chevrolat, 1843.
On *Baccharis* (Asteraceae) in Brazil.

48- *Paropsis* Olivier, 1807.

Most of the species of *Paropsis* in Australia and Papua New Guinea feed on young leaves and stems of various *Eucalyptus* species (Myrtaceae). Adults and larvae strip the leaves of *Eucalyptus* forests in Australia from the end of spring until the beginning of winter (Carne, 1966) and the trees suffer and sometimes die after the attack. It appears that there is some specificity in nature but many species will accept almost all *Eucalyptus* species in the laboratory. In Papua New Guinea, one of us (PJ) observed some *Paropsis* beetles feeding on very young leaves of *Eucalyptus*; this preference for tender leaves is shared with *Chrysomela* which prefer young shoots of *Populus* (Salicaceae) in Europe. A few *Paropsis* accept or prefer *Acacia* (Mimosaceae). Apart from *Eucalyptus* and *Acacia*, several *Paropsis* feed on *Leptospermum, Melaleuca* and *Tristania* among the Myrtaceae. *Paropsis charybdis* Stål has been accidentally introduced into New Zealand where it attacks *Eucalyptus*.

49- *Procrisina* Aslam, 1968.

On *Leptospermum* (Myrtaceae) in Australia.

50- *Paropsisterna* Motschulsky, 1860.

On *Eucalyptus* in Australia.

51- *Chrysophtharta* Weise, 1901.

On various Myrtaceae and *Acacia* (Mimosaceae) in Australia.

52- *Paropsides* Motschulsky, 1860.

On Rosaceae (*Pyrus, Crataegus)* in the Far East and India and on Myrtaceae (*Eugenia*) in New Guinea *(Aparopsis).*

53- *Trachymela* Weise, 1908.

On Myrtaceae and *Acacia* (Mimosaceae) in Australia. One species, *T. tincticollis* Blackburn has been introduced into South Africa, where it attacks *Eucalyptus.*

54- *Pyrgoides* Aslam, 1968.

On *Acacia* (Mimosaceae) and *Eucalyptus* (Myrtaceae) in Australia.

Concluding remarks on Paropsini:

Several Paropsini feed on *Eucalyptus* but also on *Melaleuca* and *Leptospermum* (Myrtaceae), while others exclusively feed on the section obliquae of the genus *Eucalyptus*. On *Eucalyptus* (subgenus *Symphyomyrtus)* in Tasmania occurs *Chrysophtharta agricola* (Chapuis), *C. variicollis* (Chapuis) and *Paropsis charybdis* Stål and on *Acacia dealbata, Pyrgoides orphana* (Erichson). The Paropsini live on Myrtaceae and Mimosaceae in the Australo-Papuan zone and on Rosaceae in the Far East. The selections are somewhat homogeneous for the group as a whole.

55- *Gonioctena* Chevrolat, 1837.

Takizawa (1976) has recorded the host-plants of several subgenera. The general tendency, despite many gaps, appears to be towards the Fabaceae, Salicaceae, Ulmaceae, Betulaceae, Carpinaceae, Fagaceae and Rosaceae. *Gonioctena* sensu stricto has been found on *Alnus* (Betulaceae), *Salix*, *Populus* (Salicaceae), *Carpinus* (Carpinaceae), *Prunus, Sorbus* (Rosaceae), *Fagus* (Fagaceae), *Corylus* (Corylaceae) and *Sambucus* (Sambucaceae). Subgenus *Spartophila* has been found on at least 12 genera of Fabaceae including *Genista* and *Sarothamnus*, subgenus *Spartoxena* on *Genista, Ulex* (Fabaceae) and *Salix* (Salicaceae), subgenus *Brachyphytodecta* on various Fabaceae, subgenus *Sinomela* on *Celtis* (Ulmaceae) and (?) Convolvulaceae, subgenus *Spartomena* on Salicaceae, Fabaceae and Rosaceae, subgenus *Goniomena* on Salicaceae and Rosaceae and subgenus *Machomena* on *Genista* (Fabaceae). *Gonioctena* often exhibits variation and the production of coloured mutations is well-known (Zulueta, 1925, 1929). The differentiation of subgenera is acceptable, but does not always correspond to the food selections.

56- *Cercyonops* Jacobson, 1900.

On Fabaceae in Central Asia (Altai).

57- *Centroscelis* Chevrolat, 1837.

Also on Fabaceae in Africa.

58- *Phratora* Chevrolat, 1837.

Phratora feeds on *Salix*, *Populus* (Salicaceae) and *Betula* (Betulaceae) in the case of *P. hudsonia* Brown and others from the Arctic. Several *Salix* feeders, such as *P. polaris* Schneider, also feed on *Polygonum* (Polygonaceae) in the laboratory (Morris, 1970).

59- *Stethomela* Baly, 1856.

On *Acacia* (Mimosaceae) and *Melastoma* (Melastomataceae) in Queensland, Australia.

60- *Calomela* Hope, 1840.

Australia. Mostly on Mimosaceae *(Acacia)*. Also recorded from Proteaceae and Pittosporaceae but these records are doubtful as food plants.

61- *Macelola* Selman, 1975.

Australia. This beetle has been recorded from dry sclerophyll forest but the host-plant remains unknown.

62- *Novacastria* Selman, 1983.

On *Nothofagus* (Fagaceae) in tropical and subtropical rainforests of Australia. The potential hosts have been tested (De Little, 1985). The insect accepts the European *Castanea* but rejects *Fagus sylvatica* and *Quercus robur*; the complete life-cycle is only successful on *Nothofagus moorei* and *N. cunninghamii.*

63- *Augomela* Baly, 1856.
On *Acacia* (Mimosaceae) and *Maytenus* (Celastraceae) in Queensland, Australia (Hawkeswood, 1988).

64- *Platymela* Baly, 1856.
On Sapindaceae (*Cupaniopsis*) in Queensland (Hawkeswood, 1988).

65- *Lycaria* Stål, 1857.
On Lauraceae in Vietnam.

66- *Oidosoma* Quedenfeldt, 1891.
On *Clerodendrum* (Verbenaceae) and (?) Pedaliaceae and (?) Sesamiaceae in Africa. The last two families are doubtfully recorded as hosts.

67- *Chalcolampra* Blanchard, 1853 (= *Phola* Weise, 1890).
Unidentified blue-flowering plant in Queensland (Hawkeswood, 1988). Also on *Parahebe* (Scrophulariaceae) and *Prostanthera* (Lamiaceae) (Reid, 1993).

68- *Phyllocharis* Dalman, 1824.
On *Clerodendrum* (Verbenaceae) in South-East Asia, Australia and New Guinea.

69- *Kuramela* Gressitt, 1964.
On *Citrus* (Rutaceae) in New Guinea but may be an accidental record.

70- *Johannica* Blackburn, 1887.
On *Pandorea* (Bignoniaceae) in Queensland (Reid, 1991) and may feed in the wild on *Cissus* (Vitaceae).

71- *Acacicola* Lea, 1902.
On *Acacia* (Mimosaceae) in New South Wales, Australia.

72- *Allocharis* Sharp, 1882.
Larvae and adults of *A. marginata* Sharp feed on *Veronica salicifolia* (Scrophulariaceae) along the rivers in New Zealand.

73- *Oreomela* Jacobson, 1895.
Asia. Found by Lopatin (1984) on *Nepeta* (Lamiaceae).

74- *Iscadida* Chevrolat, 1843.
Little is known of the host-plants of the timarchoid genera of the *Iscadida*

group in South Africa. They rest on leaves during the day in semi-arid areas and feed during the night. They feed on *Rhoicissus* (Vitaceae) a bush and a poor climber. The adult females drop their eggs to the ground and the very active larvae are probably root feeders (Grobbelaar & Jolivet, unpub. pers. obs.).

75- *Potaninia* Weise, 1889.

On Urticaceae (*Boehmeria*, *Maoutia*). All other references to host-plants are incorrect. South Asia and Far East.

76- *Entomoscelis* Chevrolat, 1837.

Mostly on Brassicaceae (*Sisymbrium, Erysimum, Brassica, Thlaspi, Cakile, Crambe, Diplotaxis, Sinapis, Descurainia, Capsella, Lepidium, Eruca, Cochlearia* and *Raphanus*), but also recorded from Fumariaceae (*Fumaria)*, Ranunculaceae (*Adonis*) and Asteraceae (*Cirsium*, *Carduus* and *Petasites).*

77- *Microtheca* Stål, 1860.

On Brassicaceae (*Brassica, Sinapis, Nasturtium, Lepidium, Barbarea, Raphanus*) in South America. Introduced to the USA. The larvae feed near roots and pupate in the ground inside a cocoon (Jolivet, 1950). The adults feed on the pollen of many flowers, including Rosaceae.

78- *Lamprolina* Baly, 1855.

According to Froggatt (1907) and Hawkeswood (1986, 1987) adults feed on foliage of *Bursaria* and *Pittosporum* (Pittosporaceae) in Queensland, Australia. The orange larvae have black spines and live on the young summit leaves. They consume the mesophyll of the adaxial surface of the leaves.

2.16.3. Tribe ?

79- *Cadiz* Andrews and Gilbert, 1992.

North America. The taxononic position of this newly created genus has not been resolved but it has affinities with the Chrysomelinae (Andrews & Gilbert, 1992). Adults and larvae feed on *Coldenia plicata* (Boraginaceae).

Concluding remarks on Chrysomelinae:

If, according to Monros, the specialization of Chrysomelidae sens. lat. upon flowering plants dates from the mid-Jurassic to Cretaceous, then there are presently few groups, such as the Aulacoscelinae and archaic Sagrinae (*Carpophagus)* which are known to be feeders upon Gymnospermae. The food selections of all other Chrysomelidae not associated with Angiospermae appear to have resulted from secondary adaptations.

All Chrysomelinae feed on Dicotyledonae and none have been recorded from Monocotyledonae. Many of these relationships show a high degree of co-evolution. It appears unlikely that the Chrysomelinae have been derived from the Neotropical Aulacoscelinae; there is certainly a similarity in anatomy, i.e. wings, genitalia, but the host-plants and the general habitus are quite different.

There are some specializations exhibited by certain genera which are sometimes associated with apterism, towards the habit type of the host-plant. Several genera, e.g. *Chrysolina*, *Timarcha* and *Elytrosphaera*, frequent low-growing herbs, while others inhabit shrubs, e.g. *Chrysomela, Promechus, Platyphora* and *Mesoplatys,* or even trees, e.g. *Paropsis* and related genera. Several genera such as *Calligrapha* exhibit all these different adaptations according to the species.

Of 174 described genera of Chrysomelinae, the host-plants of at least 76 genera (c. 43%) are known in detail or only superficially. Overall, the host-plant data are insufficient and no host-plants have been recorded for many cryptic species which live under stones during the day, often in the mountains of Central Asia, or in open country in South Africa. As per usual, nothing is known of the biology of the Chrysomelinae of Madagascar.

Host-plant specificity is widespread among the Chrysomelinae both at the generic and specific levels. This appears to be the general rule not only for the well-known genera such as *Timarcha*, *Chrysolina*, *Gonioctena*, *Phratora*, *Leptinotarsa*, *Plagiodera* and *Paropsis*, but also for other less-known genera.

No genus appears to be really polyphagous. At least 40 plant families are selected by the Chrysomelinae. These are Salicaceae, Juglandaceae, Ranunculaceae, Ulmaceae, Corylaceae, Carpinaceae, Polygonaceae, Brassicaceae, Fumariaceae, Pittosporaceae, Hamamelidaceae, Rosaceae, Fabaceae, Mimosaceae, Euphorbiaceae, Callitrichaceae, Celastraceae, Tiliaceae, Malvaceae, Sterculiaceae, Hypericaceae, Myrtaceae, Melastomataceae, Araliaceae, Apiaceae, Cornaceae, Primulaceae, Asclepiadaceae, Convolvulaceae, Solanaceae, Scrophulariaceae, Bignoniaceae, Ehretiaceae, Vitaceae, Boraginaceae, Verbenaceae, Lamiaceae, Plantaginaceae, Rubiaceae, Dipsacaceae and Asteraceae. Other recorded families are doubtful hosts in the light of present knowledge. About 200 plant genera have been found among these plant families, without counting the genera tested in the laboratory. The plants most often selected are Salicaceae, Betulaceae, Polygonaceae, Brassicaceae, Fabaceae, Malvaceae, Myrtaceae, Apiaceae, Solanaceae, Plantaginaceae, Rubiaceae and Asteraceae.

In the case of the Paropsini, and several others, the selection appears to be more ecological than botanical/chemical, i.e. it is linked with the most common and readily available host-plants in the area (*Acacia* and *Eucalyptus*). This is not an isolated phenomenon.

The karyotypes of certain genera such as *Chrysolina* are sometimes in accordance with the plant choice, i.e. with the natural taxonomy. With other genera such as *Timarcha*, there are no clear-cut relationships between host-plants and chromosomes. The basic number, as defined by the remarkable researches of Petitpierre (1976 and seq.) appears to be 2n=20 (*Timarcha*) and the most common is 2n=24 (*Chrysolina, Timarcha*). However, it is still too early to reliably trace any parallelisms between chromosomes, morphology and food preferences for the subfamily as a whole.

The evolution of feeding preferences can sometimes be deduced from the data already known for a genus or subgenus. It is hoped that future researches will be rapid in order to fill the gaps in knowledge concerning the biology and host-plants of the other 100-odd genera before they disappear from the surface of the planet. Many of these genera will certainly not survive the effects of insecticides, fertilizers, herbicides and the destruction of the world's forests. The forest reserves appear to be getting comparatively smaller and smaller, so that the beetles, like other insects (even the smallest ones), face more and more difficulty in surviving. This is not to die forever.

References listed in the bibliography pertaining to Chrysomelinae are as follows:- Andrews & Gilbert (1992), Bosq (1942), Bourdouxhe & Jolivet (1981), Breden & Wade (1985, 1988), Brown (1945), Carne (1966), Chen (1934, 1964), Costa-Lima (1936), Costa *et al.* (1983), De Little (1985), Dimock & Tingey (1986), Eberhard (1981), Froggatt (1907), Gressitt & Hart (1974), Gressitt & Kimoto (1961-1963), Gritsenko *et al.* (1986), Hare & Kennedy (1986), Harrison (1987), Hawkeswood (1983, 1986, 1987, 1988), Hsiao (1974, 1981, 1986, 1988), Hsiao & Fraenkel (1968), Hsiao & Hsiao (1983), Jacques (1985, 1988), Jayanth & Nagarkatti (1987), Jolivet (1950-1991c), Jolivet & Petitpierre (1973, 1976a, b, 1981), Jolivet *et al.* (1986, 1988), Jolivet & Van Parys (1977), Kalshoven (1951, 1981), Kasuya (1985), Kimoto (1964-1966), Kovalev & Medvedev (1983), Kovtun (1966), Kreslavsky *et al.* (1976), Kulik & Timoshin (1986), Larsson & Ohmart (1988), Lopatin (1984), Maulik (1926, 1945, 1950), McFadyen (1987), Medeiros (1991), Medvedev & Dap (1982, 1983), Medvedev & Roginskaya (1988), Moldenke (1971), Morris (1970), Paterson (1930, 1931), Petitpierre (1976), Powell (1941), Reid (1991, 1993), Sharp & Muir (1912), Takizawa (1970-1978b), Tieberghien (1971, 1972), Vasconcellos-Neto & Jolivet (1988, 1989), Vinogradova (1988), Vlasova (1978), Wade & Breden (1986), Weise (1916), Xue (1986), Zaitsev (1983, 1986), Zulueta (1925,1929).

2.17. Galerucinae

Almost all Galerucinae feed on Dicotyledonae but many adults also consume the pollen of Poaceae flowers. Pollen of Cucurbitaceae are also eaten by many species. *Coelomera*, on *Cecropia* (Cecropiaceae), not only feeds on pollen (eventually), but also on the müllerian bodies, normally considered to be a mutualistic adaptation between *Cecropia* and *Azteca* ants. *Coelomera* also suck the nectar from the female flowers, feed on the pearl-bodies of the leaves and consume the internal stem tissue (only at the larval stage). A few species feed on Monocotyledonae.

There are actually more than 5800 described species of Galerucinae, and there are probably many more, especially in the forest canopy niche. Their larval biology is very diverse. Some larvae are subterraneous and feed on roots like Eumolpinae (e.g. *Phyllobrotica*, *Platyxantha*, *Luperus*, *Aulacophora*, *Craniotectus*, *Diabrotica* and *Cerotoma*). These larvae are normally polyphagous and some can cause severe damage to crops. Other larvae, such as those of *Exosoma*, feed inside the bulbs of Liliaceae and Amaryllidaceae. Some larvae of *Monoxia* are leaf miners. Larvae of *Galeruca*, *Agelastica*, *Arima*, *Galerucella*, *Monocesta* and *Lochmaea*, are free-living on plants like the adults.

Pupation occurs within the soil inside a cavity (*Aulacophora*) or inside bark crevices at the base of the trees or inside the buds or directly on the leaves. Eggs are laid on the leaves or inside the soil crevices and are often attached by a sticky secretion or faecal mass.

The genus *Coelomera* in Brazil sometimes displays a peculiar behaviour. Not all *Coelomera* species have the same biology, but at least three or four species, including *C. ruficornis* Baly, lay their eggs into *Cecropia* stems at an internode. The female pierces the prostoma the previous day. The eggs are very well protected here in the same niche where the eggs of the *Azteca* ants, e.g. *A. alfari* Emery, are laid (Andrade, 1981). This is an example of parasitism of a mutualistic relationship between a tree and an ant (Jolivet, 1987d-1988c). The instinct involved in the piercing of the stem must be the product of a long evolution. However, this case of maternal care is probably not unique among the Chrysomelidae and must exist elsewhere, including amongst the *Cecropia* biocoenosis itself (Jolivet, 1988c).

Galerucine larvae are often gregarious on leaves, e.g. *Galerucella* and *Coelomera*. *Cerotoma* larvae are known to destroy the nitrogen-fixing nodules on the roots of legumes. Many chrysomelid subterranean larvae feed on *Rhizobium* nodules of a large number of legumes (Siddappaji & Gowda, 1980). Generally, *Diabrotica* and *Cerotoma* larvae are attracted to Cucurbitaceae roots by feeding stimulants, mostly the cucurbitacins (Nishida *et al.*, 1986).

Hibernation may occur at the egg, larval or adult stage according to the climate or the species. Reflex bleeding is rare but sometimes occurs in the adult or larval stages. The three larval stages of *Diabrotica* exhibit reflex bleeding when disturbed. This phenomenon is limited to the intersegmental membrane separating the last two abdominal segments. *Diabrotica* adults do not display reflex bleeding and their blood coagulates slowly in contrast to that of the larvae. Among *Coelomera*, adults display a femoro-tibial haemorrhage and the larvae only exhibit what appears to be enteric or buccal discharge. Several larvae, e.g. those of *Agelastica*, possess repugnatory glands. Defence of Galerucinae is undertaken by reflex immobilization, reflex bleeding, buccal or enteric discharge, fluid toxicity, aposematic colours and/or larval cycloalexy. Galerucines do not jump and their hind femorae are generally normal in size (i.e. not expanded). They have few predators, but are often heavily infected with parasitoids.

Diabrotica is an enormous genus of which the various species feed on 280 plant species distributed among 29 families; however, the genus appears to have a basic and primitive feeding relationship with the Cucurbitaceae, since this is now well-developed for the Old World *Aulacophora*. From the Cucurbitaceae diet, the genus became polyphagous, attacking monoculture crops. In America, it is common to find local species on Cucurbitaceae flowers, like *D. speciosa*

Germar from Brazil. A good summary of the biology and host-plants of *Diabrotica* is provided by Krysan and Miller (1986).

Diabrotica larvae often have a different diet from those of their adults. Like Eumolpinae, the larvae often feed on grass roots. The adults also have the tendency to feed on flowers and pollen. In the USA, *Diabrotica* mostly feed on corn (*Zea*, Poaceae), Cucurbitaceae, Fabaceae and sweet potato (*Ipomoea*, Convolvulaceae).

136 genera of Galerucinae are listed here from over 489 described (i.e. 27.8%). 83 genera are mono- or oligophagous (61.4%) and 52 are primarily or secondarily polyphagous (38.5%). Fundamentally, the Galerucinae are feeders upon Cucurbitaceae, Fabaceae or Verbenaceae and have later adapted to other families. The Poaceae provide pollen for adult nutrition and roots for the larvae but these plants do not appear to be directly attacked by the adults. More than 100 plant families have been recorded as potential food for the Galerucinae. Among the 17 monocotyledonous families selected, the Amaryllidaceae, the Zingiberaceae and the Liliaceae (sens. lat.) are common hosts, but often these are only eaten by the larvae.

We follow here the classification of Seeno and Wilcox (1982) but there are many analogies between galerucine and alticine larvae and transitional genera exist. Galerucine larvae are often elongated and parallel (orthosomatic), but sometimes are slightly bent (cyphosomatic). *Coelomera* larvae possess a very large anal shield, a kind of horny plaque, which is often ciliated and which serves for protection. The shield is much smaller among *Luperus* and is present among Neotropical species. The reduced 10th anal segment which is situated under the 9th, is a pygopod and is used for hooking onto the substrate and for moving forward. Ocellae are either absent (subterraneous larvae) or present, one on each side of the head. The antennae are mono- or biarticulate and not triarticulate as in Chrysomelinae. The larvae possessing defensive glands such as those of *Agelastica,* display tubercles on abdominal segments 1-8 above each stigmata. Galerucine larvae generally measure from 4 to 18 mm long, like those of Alticinae.

Laboissiere (1934), after Mayet (1907), described the biology of the larvae of *Exosoma lusitanica* (L.) which live inside *Narcissus* bulbs (Amaryllidaceae). The larvae, sometimes present in large numbers, live inside the same bulb inside a decomposed magma. The larva uses its excreta for cementing the pupal case. *Exosoma* larvae measure from 15 to 18 mm long and are blind. When hatching, the adult diet completely changes from that of the larvae and becomes relatively polyphagous on hosts such as *Urospermum* (Asteraceae), *Vitis* (Vitaceae), *Ocotea* (Lauraceae), *Crotalaria, Medicago* (Fabaceae), *Rhododendron* (Ericaceae), *Linum* (Linaceae), *Gossypium, Hibiscus* (Malvaceae), *Salix* (Salicaceae) and *Alnus* (Betulaceae). Larvae have also been found inside the stems of *Asparagus*, *Allium*, *Muscari* and *Hemerocallis* (Liliaceae sens. lat.).

The Galerucinae are generally winged and fly well and because of this, they feed on bushes and huge trees, even in the forest canopy. Several species are brachypterous, such as *Ruwenzoria viridis* Laboissiere, which feeds on tree heaths in the Ruwenzori in Africa at 3000 meters altitude. Certain species are micropterous or apterous, such as some *Galeruca* species, and feed on low herbs. Apterism is relatively uncommon among galerucines. Sometimes the elytra regress with the wings so that these species (e.g. *Arima*) resemble small *Meloe* beetles (Meloidae).

Certain galerucine genera lay their eggs in clusters, while others such as *Galeruca*, glue about 70 eggs into a kind of ootheca which is orange in colour. The larva of *Lochmaea crataegi* (Forster) develops inside the fruit pulp of *Crataegus* (Rosaceae). Several *Longitarsus*, e.g. *L. nigripennis* (Motschulsky) (Alticinae) also develop inside berries (i.e. *Piper*, Piperaceae).

2.17.1. Tribe Oidini

1- *Oides* Weber, 1801.

Old World. Polyphagous. On Poaceae, Fabaceae, Vitaceae, Rubiaceae, Cucurbitaceae, Schisandraceae, Euphorbiaceae, Menispermaceae, Sterculiaceae, Malvaceae, Rosaceae and Rutaceae. Actually, the greatest number of captures have been made from Vitaceae (i.e. *Vitis*, *Cayratia*, *Tetrastigma* and *Cissus*) in the Far East, which are undoubtedly the normal host-plants for the genus. The attraction towards the pollen of the Poaceae does not represent any special relationship. In India, Pillai and Rajamma (1987) have observed *Oides affinis* Jacoby feeding on leaves and pseudostems of *Amorphophallus campanulatus* (Araceae).

2.17.2. Tribe Galerucini

2- *Monocesta* Clark, 1865.

America. Oligophagous. Few host-plants are known for the Neotropical species. On *Ulmus* (Ulmaceae) and *Corylus* (Corylaceae). Probably also on *Crataegus* (Rosaceae) in the USA. *Monocesta depressa* Clark feeds in Panama on *Cecropia*. So do *Syphaxia* in Peru (Jolivet, in. litt.).

3- *Coelomera* Chevrolat, 1837.

Neotropical. Whether they are myrmecophilous or not, the 35 described species of *Coelomera* all feed on *Cecropia* (Cecropiaceae). Several species, e.g. *Coelomera helenae* Jolivet are rarer and are often monophagous. Pupation occurs within the soil and the larvae are all gregarious.

4- *Dircema* Clark, 1865.

On *Cecropia* also in Amazonia.

5- *Derospidea* Blake, 1931.

America. On *Citrus* and *Zanthoxylum* (Rutaceae) in the USA.

6- *Trirhabda* LeConte, 1865.
On Asteraceae (i.e. *Solidago, Ambrosia, Cirsium, Artemisia, Baccharis, Bahia, Encelia, Brickellia, Franseria, Aplopappus, Chrysothamnus* and *Gutierrezia*) in North America. Several other reports need to be confirmed.

7- *Miraces* Jacoby, 1888.
On *Colubrina* (Rhamnaceae) in Puerto Rico.

8- *Hemiphracta* Weise, 1902.
On *Ficus* (Moraceae) in Africa.

9- *Mombasa* Fairmaire, 1884
On *Coffea* (Rubiaceae) in East Africa.

10- *Clitena* Baly, 1864.
Appears to feed by ecological selection on several forest trees in East Africa, i.e. Aceraceae, Salicaceae and Betulaceae.

11- *Periclitena* Weise, 1902.
On *Cordia* (Ehretiaceae) in India.

12- *Clitenella* Laboissiere, 1927.
On *Celtis* (Ulmaceae) in Taiwan and China. Also recorded from *Rubus* (Rosaceae) but this may be accidental.

13- *Sastra* Baly, 1865.
One species on *Trema orientalis* (Ulmaceae) in New Guinea.

14- *Sastroides* Jacoby, 1884.
On *Phyllanthus* (Euphorbiaceae) in Asia.

15- *Polysastra* Shute, 1983.
Adapted to crops in New Guinea such as *Theobroma* (Sterculiaceae), *Coffea* (Rubiaceae), *Curcuma* and *Elettaria* (Zingiberaceae) which may indicate polyphagy. The native host-plants are unknown.

16- *Poneridia* Weise, 1908
On *Ficus* (Moraceae) in New Guinea and Australia. The reference of *Hevea* (Euphorbiaceae) is doubtful.

17- *Malacotheria* Fairmaire, 1881.
Fiji. On *Pipturus* (Urticaceae)

18- *Diorhabda* Weise, 1883.
Genus associated with *Tamarix* (Tamaricaceae) in Europe, Asia and North Africa (to Sahel). Common in Central Asia and Mongolia.

19- *Galerupipla* Maulik, 1936.
Asia. On various families, i.e. Asteraceae, Rosaceae and Onagraceae. Also on *Trapa natans* (Trapaceae).

20- *Atysa* Baly, 1864.
On Lauraceae in China and elsewhere. Other reports from Convolvulaceae and Fagaceae are doubtful.

21- *Chujoa* Gressitt and Kimoto, 1963.
On *Fagus* (Fagaceae) in Japan.

22- *Galerucella* Crotch, 1873.
About 40 species in the Old World. The species are mono- or polyphagous, but the genus shows rather broad selections, i.e. Trapaceae, Rosaceae, Rubiaceae, Ehretiaceae, Myrtaceae, Malvaceae, Pontederiaceae, Ericaceae, Betulaceae, Corylaceae, Asteraceae, Cabombaceae, Lythraceae, Onagraceae, Lamiaceae, Salicaceae, Polygonaceae, Meliosmaceae, Hydrocharitaceae, Primulaceae, Ulmaceae, Hippocastanaceae, Nymphaeaceae, Alismataceae, Grossulariaceae, Scrophulariaceae, Cornaceae, Vitaceae, Moraceae, Caprifoliaceae and Rhamnaceae. Also on floating ferns, Salviniaceae. *Galerucella nymphaeae* L. is oligophagous and feeds on *Nuphar lutea*, *Nymphaea alba* and *N. tetragona* in Fennoscandinavia, whereas *G. sagittariae* (Gyllenhal) is polyphagous on various Rosaceae (i.e. *Comarum*, *Rubus*, *Fragaria* and *Alchemilla*) and also on *Rumex* (Polygonaceae). *G. sagittariae* rejects the Nymphaeaceae (Hippa & Koponen, 1986).

23- *Hoplostines* Blackburn, 1890.
On *Urtica* and *Dendrocnide* (Urticaceae) in Queensland, Australia.

24- *Tricholochmaea* Laboissiere, 1932.
Holarctic. On Betulaceae, Rosaceae, Ericaceae, Salicaceae and Grossulariaceae. The species are somewhat specific in their food choice.

25- *Pyrrhalta* Joannis, 1866.
Palaearctic. The species of *Pyrrhalta* are specific in their plant choice like the previous genus. On Ulmaceae, Sambucaceae, Caprifoliaceae, Lythraceae, Salicaceae, Aceraceae, Corylaceae, Betulaceae, Rosaceae and Hamamelidaceae. Certain species feed on *Viburnum* (Caprifoliaceae) while others prefer *Ulmus* (Ulmaceae).

26- *Xanthogaleruca* Laboissiere, 1934.
Holarctic. On *Ulmus* (Ulmaceae).

27- *Lochmaea* Weise, 1883.
Holarctic. On trees belonging to Betulaceae, Salicaceae, Rosaceae and Fagaceae. Also on Ericaceae and Cucurbitaceae. Mikheev and Kreslavsky (1980, 1981, 1984, 1986) and Kreslavsky *et al.* (1981) have studied sympatric races of *Lochmaea capreae* L. in Russia; one race feeds on birch (*Betula*), while the other one feeds on willow (*Salix).* All differences in the food preferences of adults of either race were shown to be associated with a gene responsible for physiological peculiarities of the larvae. Willow-race monophagy may develop from birch-race polyphagy due to a single mutation resulting in the inability to develop on birch. Other species exhibit monophagy or oligophagy.

28- *Megaleruca* Laboissiere, 1922.
Africa. Mostly on Piperaceae and Ulmaceae. Also recorded on Pinaceae in South Africa.

29- *Schematiza* Chevrolat, 1837.
Neotropical. 37 species have been described. Two are known to be specific on *Cordia* (Ehretiaceae). *Schematiza cordiae* Barber has been introduced to Mauritius and elsewhere to control a pest species of *Cordia.*

30- *Yingaresca* Bechyne, 1956.
America. On *Cordia* (Ehretiaceae) or *Lantana* (Verbenaceae), two related families.

31- *Metrogaleruca* Bechyne and Bechyne, 1969.
One species on *Cordia* (Ehretiaceae) in tropical America.

32- *Erynephala* Blake, 1936.
On *Chenopodiaceae* in the USA.

33- *Monoxia* LeConte, 1865.
North America. On Asteraceae, Chenopodiaceae, Fabaceae and Amaranthaceae. Other families recorded are improbable. The larvae are leaf miners.

34- *Ophraella* Wilcox, 1965.
North America. On various Asteraceae (*Solidago*, *Iva, Eupatorium*, *Ambrosia*, *Chrysopsis*, *Aster*, *Parthenium, Helianthus, Ratibida, Coreopsis* and *Xanthium*). Also one species on Lythraceae.

35- *Neolochmaea* Laboissiere, 1939.
Recorded on *Rosa* (Rosaceae), *Nicotiana* (Solanaceae) and *Lantana* (Verbenaceae) in Brazil. Very improbable or the genus may be composite. In

Amazonia, *Neolochmaea tropica* Jacoby appears to only feed on *Borreria verticillata* (Rubiaceae) (Jolivet, 1988b).

36- *Galeruca* Muller, 1764.
Mostly Palaearctic. Polyphagous with some preferences according to the species. Divided into well-defined subgenera. Several species in the USA, native or introduced. Feeds on low herbs, because they are flightless. On Brassicaceae, Lamiaceae, Asteraceae, Alliaceae, Plantaginaceae, Rubiaceae, Dipsacaceae, Polemoniaceae, Chenopodiaceae, Ranunculaceae, Polygonaceae, Fabaceae, Convolvulaceae, Caryophyllaceae, Apiaceae, Rosaceae and Solanaceae. *Galeruca laticollis* Sahlberg and *G. monticola* Kiesenwetter mostly feed on Ranunculaceae. Most of the species prefer Asteraceae, Dipsacaceae and Brassicaceae.

37- *Theone* Gistel, 1857.
On *Artemisia* (Asteraceae) in Asia.

38- *Pallasiola* Jacobson, 1925.
Northern Asia. On *Artemisia* (Asteraceae).

39- *Arima* Chapuis, 1875.
Western Europe. Polyphagous on Rubiaceae, Asteraceae, Lamiaceae and Dipsacaceae. Apterous.

40- *Apterogaleruca* Chûjô, 1962.
Larva and adult on *Gonostegia* (Urticaceae) in Taiwan.

41- *Nyctiphantus* Semenov, 1902.
On *Calligonum* (Polygonaceae) in Asia.

42- *Geinella* Strand, 1935.
On *Rhodiola* (Crassulaceae) and *Polygonum* (Polygonaceae) in China.

43- *Geinula* Ogloblin, 1936.
On *Salvia* (Lamiaceae) and *Artemisia* (Asteraceae) in Sichuan, China.

44- *Apophylla* Thomson, 1858.
Old World tropics. On Poaceae, Ehretiaceae, Lamiaceae, Convolvulaceae and Acanthaceae (Tubiflorae). The captures from Poaceae in Africa are related to pollen feeding.

2.17.3. Tribe Metacyclini

45- *Sesselia* Laboissiere, 1931.
In Africa on Poaceae and Cyperaceae and on other flowers, where beetles feed on petals and pollen.

46- *Escourtiana* Jacoby, 1900.
Two species in Africa on Poaceae (pollen).

47- *Pseudoides* Jacoby, 1892.
Two species in Vietnam and in Thailand on *Cassia* (Caesalpiniaceae).

48- *Chapuisia* Duvivier, 1885.
On *Prunus* (Rosaceae) in East Africa.

49- *Exora* Chevrolat, 1837.
On *Theobroma* (Sterculiaceae) and *Coffea* (Rubiaceae) in America.

50-*Malacorhinus* Jacoby, 1887.
On *Acacia* (Mimosaceae) in the USA.

51- *Chthoneis* Baly, 1864.
One Cuban species on *Cordia* (Ehretiaceae).

2.17.4. Tribe Sermylini

52- *Dercetina* Gressitt and Kimoto, 1963.
One species in Assam on *Mitrogyna* (Rubiaceae) and another one on *Sagittaria* (Alismataceae). Also several observations have been made on *Solanum* (Solanaceae) and *Musa* (Musaceae). This may indicate polyphagy but the first observation appears to be the only reliable one.

53- *Arthrotus* Motschulsky, 1857.
On *Trema* (Ulmaceae), *Litsea* (Lauraceae), *Pterocarya* (Juglandaceae), *Pyracantha* (Rosaceae), *Alnus* (Betulaceae), *Ormosia* (Fabaceae) and *Liquidambar* (Altingiaceae). Polyphagous and adapted to forest trees. In southern China on *Metasequoia* (Taxodiaceae), *Salix* (Salicaceae) and possibly *Brassica* (Brassicaceae) and *Dioscorea* (Dioscoreaceae)(?). Several of these captures must only be accidental.

54- *Bonesia* Baly, 1865.
On *Theobroma* (Sterculiaceae) an introduced tree in Africa.

55- *Sermylassa* Reitter, 1912.
On *Galium* spp. (Rubiaceae). Monophagous in Europe.

56- *Aplosonyx* Chevrolat, 1837.
East Asia. On Araceae, Poaceae and Vitaceae. The real host-plant is certainly *Cissus* (Vitaceae).

57- *Sphenoraia* Clark, 1865.
On Polygonaceae, Convolvulaceae, Rubiaceae and Pinaceae in Asia. Probably *Oldenlandia* (Rubiaceae) is the normal host-plant.

58- *Agelasa* Motschulsky, 1860.
East Asia. On *Actinidia* (Actinidiaceae). Recorded also from Salicaceae and Hydrangeaceae.

59- *Hamushia* Chûjô, 1956.
On *Stellaria* (Caryophyllaceae) in Japan.

60- *Gallerucida* Motschulsky, 1860.
Asiatic genus recorded from Rosaceae, Polygonaceae, Vitaceae, Saxifragaceae, Fabaceae, Onagraceae and Araceae. The species appear to be monophagous, although the genus as a whole displays diversification in its diet.

61- *Meristoides* Laboissiere, 1929.
On *Alocasia* (Araceae) in South-East Asia.

62- *Meristata* Strand, 1935.
Asia. On fruit trees such as *Pyrus* and *Malus* (Rosaceae) and on *Carduus* (Asteraceae) in India.

63- *Agelastica* Chevrolat, 1837.
Palaearctic. Oligophagous. On Betulaceae *(Alnus)*, Corylaceae (*Corylus*), Salicaceae (*Salix, Populus*), Rosaceae (*Cerasus, Prunus, Pyrus, Malus, Rosa*), Carpinaceae (*Carpinus*). Essentially on Fagales and Rosales. The choice of *Salix* is very exceptional.

64- *Morphosphaera* Baly, 1861.
On *Ficus* (Moraceae) in Asia. A marked exception is *M.ginkgoae* Gressitt and Kimoto, which has been recorded from China (Huoeh) on *Ginkgo biloba* (Ginkgoaceae), a tree which is still growing wild (as a living fossil) in isolated areas of China (Chekiang).

65- *Miltina* Chapuis, 1875.
Vietnam. On *Morinda* (Rubiaceae).

2.17.5. Tribe Luperini

66- *Aulacophora* Chevrolat, 1837.

Old World. Essentially on Cucurbitaceae (leaves, flowers, pollen): *Cucumis, Cucurbita, Citrullus, Luffa, Lagenaria, Momordica, Trichosanthes, Benincasa* and *Melothrix*. Several purported cases of allotrophy on low herbs need to be verified. In India, one species has been recorded on *Calotropis* (Asclepiadaceae), a highly toxic plant, but this is unlikely. Adults of *Aulacophora* are found on flowers of various plants and the larvae feed on roots like *Diabrotica.*

67- *Lamprocopa* Hincks, 1949.

Africa. Despite unconfirmed reports on Poaceae (pollen), Fabaceae and Solanaceae, *Lamprocopa* mostly feed on Cucurbitaceae.

68- *Leptaulaca* Weise, 1902.

On various plants, i.e. Poaceae, Rubiaceae, Dioscoreaceae, Convolvulaceae, Oxalidaceae and Sterculiaceae, including cocoa (*Theobroma*) and coffee (*Coffea*). Normal host is most probably Cucurbitaceae.

69- *Pseudocophora* Jacoby, 1884.

On Cucurbitaceae (*Thladiantha, Gynostemma*) in Vietnam.

70- *Agetocera* Hope, 1840.

On Cucurbitaceae (*Gynostemma*) in Vietnam. The observation of one species (*A. discedens* Weise) on *Rubus* appears to be accidental (Takizawa, 1978a).

71- *Paragetocera* Laboissiere, 1929.

On Cucurbitaceae in Taiwan.

72- *Diacantha* Chevrolat, 1837.

In Africa on Cucurbitaceae (*Cucurbita, Cucumis)*. Also recorded on various other plant families, i.e. Onagraceae, Solanaceae, Sterculiaceae, Poaceae, Convolvulaceae, Rubiaceae, Fabaceae, Malvaceae and Lamiaceae. Most of these probably represent spatial findings without actual feeding being observed.

73- *Paridea* Baly, 1886.

On Cucurbitaceae (*Gynostemma, Trichosanthes*) in Asia.

74- *Asbecesta* Harold, 1877.

Africa. On Cucurbitaceae (*Cucumis*, *Cucurbita* and *Lagenaria*). Like most Luperini, the larvae feed on roots.

75- *Prosmidia* Weise, 1901.

Polyphagous on Fabaceae, Rubiaceae and Boraginaceae in Africa.

76- *Laetana* Baly, 1864.
On cotton in West Africa (Malvaceae). The natural host-plant is unknown.

77- *Diabrotica* Chevrolat, 1837.
Like the Aulacophorina in the Old World, which have an attraction towards the Cucurbitaceae, the Diabroticina in the New World display the same tendency. However, *Diabrotica* have adapted their feeding upon more than 500 plant species, including many crops. In Latin America, *Diabrotica* frequently occur on pumpkin flowers, e.g. *D. speciosa* (Germar). Like the Eumolpinae, the larvae of *Diabrotica* feed on the roots of Cucurbitaceae and Poaceae and are broadly polyphagous. *Diabrotica* adults have been found on a multitude of plants, i.e. Cyperaceae, Poaceae, Musaceae, Alliaceae, Commelinaceae, Liliaceae, Agavaceae, Iridaceae, Amaryllidaceae, Typhaceae, Cucurbitaceae, Begoniaceae, Urticaceae, Rubiaceae, Ehretiaceae, Asteraceae, Convolvulaceae, Sapindaceae, Brassicaceae, Fabaceae, Verbenaceae, Malvaceae, Phytolaccaceae, Rosaceae, Solanaceae, Anacardiaceae, Sterculiaceae, Orchidaceae, Pedaliaceae, Chenopodiaceae, Rutaceae, Moraceae, Amaranthaceae, Caricaceae, Vitaceae, Polygonaceae, Annonaceae, Euphorbiaceae, Balsaminaceae, Lauraceae, Mimosaceae, Passifloraceae, Myrtaceae, Apiaceae, Boraginaceae, Scrophulariaceae, Combretaceae, Lamiaceae and Hippocastanaceae. *Diabrotica* are mostly famous as pests of beans, peanuts, corn, cabbage, soya bean, potatoes, tobacco, tomatoes, sugarcane, cereals, coffee, gourds, fruit trees, pumpkins and other crops. Their polyphagy results from secondary adaptations. They are recorded from 11 monocot families and 45 dicot families, surpassing the record for the genus *Nodonota* (Eumolpinae). At Easter Island in the Pacific, *Diabrotica viridula* (Fabricius), which is probably introduced, feeds on local corn (Olalquiaga, 1980).

78- *Cochabamba* Bechyne, 1955.
On *Citrus* (Rutaceae) in Argentina which represents a secondary adaptation to an introduced tree. Polyphagous.

79- *Paranapiacaba* Bechyne, 1958.
Polyphagous in Brazil and Argentina. On Cucurbitaceae, Verbenaceae, Rutaceae, Fabaceae, Solanaceae, Caryophyllaceae and Poaceae.

80- *Amphelasma* Barber, 1947.
On *Prosopis* (Mimosaceae) in Mexico.

81- *Acalymma* Barber, 1947.
New World. On Cucurbitaceae, Malvaceae, Poaceae, Apiaceae, Passifloraceae, Combretaceae, Brassicaceae, Asteraceae, Solanaceae, Fabaceae and Phytolaccaceae. Larvae feed on Cucurbitaceae. Secondarily polyphagous. *Citrullus, Echinocystis, Cucurbita, Cucumis* and *Sechium* are the genera of Cucurbitaceae consumed.

82- *Paratriarius* Schaeffer, 1906.
America. On Cucurbitaceae (*Cucumis*), Poaceae, Solanaceae, Commelinaceae and Asteraceae.

83- *Isotes* Weise, 1922.
Neotropical. On Cucurbitaceae (*Cucurbita*) and Fabaceae.

84- *Gynandrobrotica* Bechyne, 1955.
Neotropical. On Cucurbitaceae (*Citrullus*) and Fabaceae. One species, *G. fenestrata* (Baly), feeds on *Cecropia santanderensis* (Cecropiaceae) in Venezuela.

85- *Neobrotica* Jacoby, 1887.
America. Polyphagous. On Fabaceae, Musaceae, Annonaceae, Solanaceae, Arecaceae, Cannaceae, Bignoniaceae, Convolvulaceae, Sapindaceae, Poaceae, Sterculiaceae, Malvaceae, Asteraceae, Bromeliaceae and Verbenaceae.

86- *Cerotoma* Chevrolat, 1837.
America. On Liliaceae, Poaceae, Begoniaceae, Fabaceae, Solanaceae, Cucurbitaceae, Rosaceae, Convolvulaceae, Rubiaceae, Malvaceae, Brassicaceae, Asteraceae, Chenopodiaceae, Urticaceae, Polemoniaceae, Celastraceae and Rutaceae. Polyphagous like *Diabrotica*.

87- *Phyllechthris* Dejean, 1837.
On Fabaceae with one observation on Pinaceae in the USA.

88- *Trichobrotica* Bechyne, 1956.
Neotropical. On Fabaceae and Asteraceae.

89- *Luperosoma* Jacoby, 1891.
America. On Fabaceae and Asteraceae.

90- *Ectmesopus* Blake, 1940.
On Fabaceae in Puerto Rico.

91- *Medythia* Jacoby, 1887.
On Fabaceae and Poaceae in Asia.

92- *Madurasia* Jacoby, 1896.
On Fabaceae in Asia.

93- *Xingeina* Chen, 1987.
On *Potentilla* (Rosaceae) in China (Yunnan and Sichuan) up to 4400 m altitude.

94- *Shaira* Maulik, 1936.
On *Potentilla* (Rosaceae) in Yunnan and Sichuan (China) up to 4200 m altitude.

95- *Marseulia* Joannis, 1866.
Asia Minor. On Poaceae, Fabaceae, Chenopodiaceae, Linaceae, Brassicaceae, Liliaceae, Apiaceae, Asteraceae, Arecaceae etc.

96- *Microlepta* Jacoby, 1886.
South-East Asia. On Sterculiaceae, Fabaceae, Hypericaceae.

97- *Prasyptera* Baly, 1878.
On Cucurbitaceae (*Cucurbita*). Has adapted to *Theobroma* (Sterculiaceae) in New Guinea.

98- *Eumelepta* Jacoby, 1892.
On Fabaceae in Vietnam.

99- *Craniotectus* Laboissiere, 1932.
Malaysia. On Fabaceae, Malvaceae and Rosaceae. Larvae endogenous on roots.

100- *Synetocephalus* Fall, 1910.
On Rosaceae in the USA and also recorded from *Aesculus* (Hippocastanaceae).

101- *Pseudoluperus* Beller and Hatch, 1932.
On Fabaceae and Apocynaceae in the USA.

102- *Triarius* Jacoby, 1887.
On Fabaceae and Solanaceae in the USA.

103- *Scelolyperus* Crotch, 1874.
North America, Asia. Recorded from various plants, i.e. Anacardiaceae, Staphyleaceae, Rhamnaceae, Juglandaceae and Berberidaceae.

104- *Clerotilia* Jacoby, 1885.
On Rhamnaceae in Japan.

105- *Phyllobrotica* Chevrolat, 1837.
Holarctic. On Lamiaceae, Primulaceae and perhaps Araceae.

106- *Euliroetis* Ogloblin, 1936.
On Fabaceae in Japan.

107- *Xenarthracella* Laboissiere, 1940.
On *Clerodendrum* (Verbenaceae) in the Ruwenzori of Africa at 2000 m altitude.

108- *Hoplasoma* Jacoby, 1884.
On *Clerodendrum*, *Premma* and *Vitex* (Verbenaceae) in Asia. Other records from Asteraceae, Fagaceae and Ulmaceae are doubtful.

109- *Mimastra* Baly, 1865.
South-East Asia. On Moraceae, Ulmaceae, Rosaceae, Euphorbiaceae, Fabaceae, Tiliaceae, Apocynaceae, Chenopodiaceae, Sterculiaceae and Rubiaceae. Feeds on tree leaves. Also recorded from a fern, *Adiantum* (Adiantaceae).

110- *Haplosomoides* Duvivier, 1890.
On *Clerodendrum* (Verbenaceae) in Asia.

111- *Hallirhotius* Jacoby, 1888.
Africa. On Fabaceae. Adults also feed on the pollen of Poaceae and Cyperaceae.

112- *Exosoma* Jacoby, 1903.
On Lauraceae, Fabaceae, Ericaceae, Linaceae, Malvaceae, Liliaceae, Amaryllidaceae, Alliaceae, Salicaceae, Betulaceae, Asteraceae, Vitaceae, Asclepiadaceae, Apiaceae, Chenopodiaceae, Cucurbitaceae and Poaceae. Adults are polyphagous and often found on Asteraceae flowers. The larvae live inside the bulbs of Amaryllidaceae, Liliaceae, etc. Pupation also occurs within the bulbs.

113- *Liroetis* Weise, 1889.
On Salicaceae and Fagaceae in the Far East.

114- *Taomueria* Chen and Jiang, 1986.
On *Rosa* (Rosaceae) in China.

115- *Siemssenius* Weise, 1922.
On Caprifoliaceae and Altingiaceae (*Liquidambar*) in Asia.

116- *Cneorane* Baly, 1865.
On Fabaceae (e.g. *Lespedeza, Desmodium, Canavalia* and *Centrosema*). There is a doubtful observation on Taxodiaceae in Asia.

117- *Cassena* Weise, 1892.
Asia, Australia, New Guinea. On Fabaceae (*Phaseolus, Glycine, Mucuna, Erythrina*) and various Malvaceae, Rosaceae, Sterculiaceae, etc.

118- *Niasia* Jacoby, 1889.
On *Derris* (Fabaceae) in Asia.

119- *Cneoranidea* Chen, 1942.
On *Corylus* (Corylaceae) in China.

120- *Ootheca* Dejean, 1837.
On Fabaceae, Pedaliaceae, Malvaceae, Cucurbitaceae, Solanaceae, Rubiaceae and Lauraceae. Basically feeds on cultivated legumes in Africa.

121- *Euluperus* Weise, 1886.
On Rosaceae in Europe. Other data on Salicaceae, Apiaceae and Poaceae are doubtful.

122- *Calomicrus* Stephens, 1831.
Old World. On Fabaceae and Cistaceae and on flowers of Ranunculaceae and Betulaceae.

123- *Eusattodera* Schaeffer, 1906.
On Rosaceae in the USA.

124- *Luperodes* Motschulsky, 1858.
Almost cosmopolitan. On Fabaceae, Verbenaceae, Malvaceae, Lythraceae, Onagraceae, Poaceae, Hippocastanaceae, Rosaceae, Asteraceae, Lamiaceae, Scrophulariaceae, Myrtaceae, Ranunculaceae, Brassicaceae, Betulaceae, Salicaceae, Polygonaceae, Balsaminaceae, Chenopodiaceae, Caryophyllaceae, Apiaceae, Hydrophyllaceae, Solanaceae, Plantaginaceae, Cucurbitaceae, Liliaceae and Juglandaceae. Several species, e.g. *L. quaternus* (Fairmaire) appear to occur only on Fabaceae in Africa. The Fabaceae dominate but some species are not selective and become broadly polyphagous (secondary polyphagy).

125- *Metrioidea* Fairmaire, 1881.
Almost cosmopolitan; not found in Europe. Adults on Poaceae, Malvaceae, Cactaceae, Ranunculaceae and Rosaceae. Also on flowers of Asteraceae.

126- *Monolepta* Chevrolat, 1837.
This enormous, cosmopolitan genus contains at least 600 species. A very broad range of host-plants are selected: Gnetales, Poaceae, Araceae, Dioscoreaceae, Arecaceae, Bromeliaceae, Verbenaceae, Rutaceae, Fabaceae, Malvaceae, Hypericaceae, Betulaceae, Apocynaceae, Proteaceae, Vitaceae, Anacardiaceae, Lauraceae, Euphorbiaceae, Cucurbitaceae, Bombacaceae, Anacardiaceae, Rosaceae, Barringtoniaceae, Rubiaceae, Asteraceae, Chenopodiaceae, Brassicaceae, Polygonaceae and Juglandaceae. Several lar-

vae, e.g. those of *M. apicicornis* Thomson, mine the roots of oil palm, *Elaeis guineensis* (Arecaceae) at the Côte d'Ivoire. The corresponding adults live on the palm leaves. Many *Monolepta* adults feed on Poaceae pollen. Some species, e.g. *M. erythrocephala* L., are monophagous on *Polygonum*, but cases of allotrophy on *Fragaria* are well-known. Other species have been recorded from Cyperaceae, Coniferae, Amaranthaceae, Hydrangeaceae, Salicaceae, Caprifoliaceae, Sterculiaceae, Rhamnaceae, Polemoniaceae, Sonneratiaceae, Styracaceae, Ulmaceae, Moraceae and Altingiaceae.

127- *Lilophaea* Bechyne, 1958.
On Rosaceae in Venezuela.

128- *Atrachya* Dejean, 1837.
Polyphagous on many wild and cultivated plants, i.e. Salicaceae, Rosaceae, Moraceae, Fabaceae, Poaceae, Cucurbitaceae, Solanaceae and Asteraceae. Also appears to feed on *Metasequoia* (Taxodiaceae) in China.

129- *Candezea* Chapuis, 1879.
On Moraceae, Fabaceae (including *Glycine max*) in Queensland. Also on cotton (Malvaceae) in Mozambique.

130- *Paleosepharia* Laboissiere, 1936.
On Taxodiaceae, Altingiaceae, Cornaceae and Fagaceae in Asia.

131- *Macrima* Baly, 1878.
On Betulaceae (*Alnus*) in Asia and on *Crotalaria* (Fabaceae) in New Guinea.

132- *Bonesioides* Laboissiere, 1925.
On Fabaceae and Malvaceae in Africa.

133- *Strobiderus* Jacoby, 1884.
On Fabaceae and Convolvulaceae in Africa and Asia.

134- *Cerophysa* Chevrolat, 1837.
On Fabaceae, Anacardiaceae and Ulmaceae in Asia and the Pacific.

135- *Hoplosaenidea* Laboissiere, 1933.
On Ulmaceae and Rutaceae in Asia.

136- *Luperus* Muller, 1764.
Luperus larvae are gregarious and subterranean and feed on the rootlets of herbaceous plants (Fabaceae and Poaceae). The adults are polyphagous and feed on Poaceae, Liliaceae (flowers), Euphorbiaceae, Onagraceae, Salicaceae, Rosaceae, Fabaceae, Flacourtiaceae, Betulaceae, Apiaceae, Fagaceae,

Carpinaceae, Ulmaceae, Corylaceae, Asteraceae, Brassicaceae, Tamaricaceae, Ericaceae, Polygonaceae, Grossulariaceae, Pistaciaceae and Rhamnaceae. Several species such as *Luperus circumfusus* Marsham feed on Fabaceae (*Sarothamnus, Ulex, Genista, Spartium, Calycotome* etc.). Some others like *Luperus pinicola* Duftschmid feed on *Pinus, Larix* and *Picea* (Pinaceae), which is probably a recent adaptation. Other species are fruit tree pests (Rosaceae).

137- *Theopea* Baly, 1864.
In Malaysia on introduced cacao (*Theobroma*, Sterculiaceae). The native host-plants remain unknown.

138- *Ruwenzoria* Laboissiere, 1919.
On *Philippia* (Ericaceae) in the Ruwenzori, Africa, at 3000 m altitude. (Jolivet, 1966).

139- *Platyxantha* Baly, 1864.
On Moraceae and Sterculiaceae in Asia. Larvae feed on roots.

140- *Fleutiauxia* Laboissiere, 1933.
On Moraceae, Salicaceae, Juglandaceae, Rosaceae, Betulaceae and Fagaceae in the Far East.

141- *Diaphaenidea* Laboissiere, 1933.
On *Derris* (Fabaceae), an insecticidal plant, in Asia.

142- *Palpoxema* Baly, 1861.
On Meliaceae and Sterculiaceae (introduced cacao) in Malaysia.

143- *Megalognatha* Baly, 1878.
Africa. Polyphagous. On Rosaceae, Fabaceae, Lauraceae, Anacardiaceae, Poaceae, Myrtaceae, Rubiaceae, Malvaceae and Asteraceae. On *Eriobotrya* (Rosaceae) in South Africa. Grobbelaar (1993) has confirmed the great polyphagy of the genus. Probably the larvae are subterraneous root feeders.

Concluding remarks on Galerucinae:-
As noted previously, the Galerucinae exhibit well-defined trophic preferences on Cucurbitaceae, Fabaceae and perhaps Verbenaceae, Rosaceae, Ericaceae and certain Fagales. From these specific selections, a secondary polyphagy has sometimes developed, namely among *Diabrotica* and related genera which have adapted rapidly for feeding on tropical crops. Like Eumolpinae, the genera with subterraneous larvae, are generally polyphagous in the adult stage. Only *Aulacophora* has remained oligophagous on Cucurbitaceae. The cucurbitacins (tetracyclic tripertenes), are a class of bitter toxic chemicals almost always present among the Cucurbitaceae. It is evident they are very attractive to many Luperini.

The Galerucinae are found on low herbs (apterous and brachypterous species) as well as on shrubs and large trees. 118 plant families have been recorded as host-plants. These are distributed as follows:- Pteridophyta (ferns)-2, Gymnospermae (mostly Coniferae)-4, Monocotyledonae- 17, Dicotyledonae-95. It is evident the host-plant listings are far being complete since so many genera have not even been studied. Only one species has been captured on Cactaceae. Most of the *Galeruca* which feed on Asteraceae, Brassicaceae and Dipsacaceae are truly polyphagous, but some species are oligophagous. Many genera, purported to be polyphagous, are actually composed of oligo- and monophagous species. Some are really polyphagous. Those species feeding on flowers are able to affect pollination because they fly generally well and can transport pollen from flower to flower.

Although much more research is needed, it is likely that the present sampling is sufficient to determine the selection trends amongst the genera of the subfamily and that these will remain unchanged when more data is published on their host-plants.

References pertaining to Galerucinae listed in the bibliography are as follows:- Andrade (1981), Grobbelaar (1993), Hawkeswood (1987), Hippa & Koponen (1986), Jolivet (1966, 1987c,d, 1988a,b,c), Kalshoven *et al.* (1981), Kreslavsky *et al.* (1981), Krysan & Miller (1986), Laboissiere (1934), Maes & Staines (1991), Martorell (1976), Mayet (1907), Medvedev & Dap (1982), Messina (1982), Mikheev & Kreslavsky (1980, 1981, 1984, 1986), Nishida *et al.* (1986), Olalquiaga (1980), Palmer & Goeden (1991), Pillai & Rajamma (1987), Schaefer (1987), Seeno & Wilcox (1982), Shute (1983), Siddappaji & Gowda (1980), Staines (1986), Takizawa (1972, 1978a), Wilcox (1979).

2.18. Alticinae

Together with the Chrysomelinae and the Galerucinae, the Alticinae are part of Section 4 of the Chrysomelidae, known as the Trichostoma. However, they are more closely related to the Galerucinae than to the Chrysomelinae. Generally, species of Alticinae are not generalists and if their selections sometimes appear representative of polyphagy, it is only because of the loose, accidental feeding relationships with alternative plants as a result of climatic/environmental changes such as drought etc. This phenomenon was discovered by Peyerhimhoff (1915) in Algeria and was later confirmed by Furth (1983) in Israel, by Jolivet (1985) in the Cape Verde Islands, by Petitpierre (1985) in the Balearic Islands and by Doguet (1984) in Algeria. Petitpierre is probably correct in regarding these secondary plants as a refuge. Such refuges are sometimes necessary because of the temporary disappearances of the real and normal host-plant(s) as a result of adverse climatic changes such as exceptional drought. In Europe, *Quercus, Populus* and *Pistacia*, and in the tropics, fruit trees such as guavas, *Persea* and *Carica* can assist adults to survive these natural hazards. In general, the larvae do not feed on these secondary (alternative) host-plants. Succulence of several cultivated crops can also attract adult Alticinae. Generally, Alticinae are oligophagous and some species from particular genera feed exclusively on plants of closely related families, their selections being chemical, a phenomenon which is being studied more and more by specialists. Unfortunately, several genera remain rather artificial and this possibly explains the apparent incoherence of certain selections recorded for them.

Most of the Alticinae are terrestrial but several genera contain subaquatic species such as *Pseudolampsis* in Florida and perhaps *Agasicles* in South America.

Like the Eumolpinae and the Galerucinae, Alticinae can become serious agricultural pests. Recent papers by Furth (1976-1985) have summarized the host-plants of the Mediterranean genera. Serious outbreaks of pests are known in the Holarctic and in the tropics among the Alticinae, but mostly with cultivated crops.

The larvae of Alticinae are miners inside leaves (e.g. *Sphaeroderma, Dibolia, Febra, Mantura, Mniophila, Argopistes*, several *Phyllotreta*, including *P. nemorum* L. and others and *Argopus)*, fruits (*Longitarsus nigripennis* Motschulsky), stems, petioles, flowers and buds (e.g. *Chaetocnema, Psylliodes hyoscyami* (L.), *P. chrysocephala* L., *P. napi* (Fabricius), *Clitea*) or are subterranean and feed on roots (i.e. *Phyllotreta, Longitarsus, Podagrica, Nisotra, Aphthona*, several *Chaetocnema, Xenidea, Epitrix, Asorestia, Crepidodera, Batophila, Clavicornaltica, Psylliodes attenuata* (Koch), *P. affinis* (Paykull), *Erystus* and several *Altica*). Other larvae are free-living (e.g. *Podontia, Blepharida* and *Altica)* on the leaves of shrubs or small herbs. Some root feeders such as *Podagrica* and *Nisotra* only chew the superficial layers of the roots. *Phyllotreta vittula* Redt., most *Aphthona* and *Epitrix* and some *Longitarsus* are also superficial root feeders. Free larvae very often remain under the leaves away from light. Generally, these larvae are well-known scientifically and have been studied and described by various authors, but much work still remains to be undertaken.

As noted above, the root-feeders do not all feed the same way. For instance, the larva of *Longitarsus echii* (Koch) chew into the central area of the roots, while those of *L. aeneus* Kutsch. mostly feed on the cortex and on the rootlets (Wapshere, 1982). Most alticine larvae are subterranean and feed on or inside roots but the movement from leaves to stems and from stems to roots is often frequent (Smith, 1985). The leaf-feeding larvae, either free or mining, are somewhat cyphosomatic in shape, while root-feeding and subterranean larvae are always orthosomatic. The larvae of several Asian alticines, which live as adults and larvae in the ground (i.e. *Clavicornaltica)*, are still undescribed. It is very probable that they do not differ from the general alticine scheme. Other endogenous alticines occur in Africa and tropical America. Endogenous larvae are uniformly white except for the head, the prothorax and the 9th abdominal segment. Miners or external feeding larvae are more pigmented, mostly on the head. Like the Galerucinae (Jolivet, 1987c) the antennae of Alticinae are mono-

or bisegmented and ocelli are often absent. Larval characteristics are similar to those of the Galerucinae. Among the miners, their legs are very small and well separated. The abdomen is composed of 9 to 10 segments. Often the 10th segment is disk-shaped and is used as a pygopod. The 9th tergum can be apparent, rounded in shape or as a prominent, hardened plate bearing setae. The free and cyphosomatic larvae of *Blepharida* feed on *Rhus* leaves and the 10th segment bears a pygopod. The length of alticine larvae varies from 4 to 14 mm according to the species. Good descriptions of some larvae have been provided by Maulik (1926). Larvae of *Podontia*, a genus closely related to *Blepharida*, are free-living and are covered with a black shiny film composed of their own excreta. They feed on the margins of the leaves and can attain a large size. Pupation occurs in the ground within an earthen cell (Maulik, 1926; Kalshoven, 1951). The adult jumps or drops to the ground in response to danger.

Larvae of *Blepharida*, studied by Frost (1973) and Furth (1982) exhibit similar habits to those of *Podontia* and also pupate in the ground within a cell. The eggs are deposited in pads of 10 by the female and are agglutinated with fecal matter. The free larva, measuring up to 14 mm long, is yellow or green in colour with small stripes and has a moist cuticle similar to that of a slug and often has fecal filaments attached to the anus. These larvae feed on the margins of leaves, while most other alticine larvae either chew holes into the laminae of thin leaves or produce transparent windows through the thicker ones. Most alticine adults feed in a similar manner. Although most of the Alticinae are oligophagous, certain genera have become secondarily polyphagous like *Systena* in America.

It is very difficult to distinguish the larvae of Alticinae from those of the Galerucinae (Steinhausen, 1982). Many alticine larvae lack ocelli, while several others, i.e. *Dibolia*, *Mantura* and *Podagrica*, possess them.

According to Scherer (1982), most of the alticine species and genera occur in the Neotropical region (220 genera), while 65 live in Africa, 30 in the Palaearctic region, 45 in the Nearctic region and 100 in the Oriental region. The Australian fauna is mostly endemic with several immigrants from the north. The classification of the genera presented here follows that of Seeno and Wilcox (1982). Since the subfamily Alticinae contains 514 genera and subgenera with more than 7,000 species, it is impossible to list all of the host-plants for each species and the host-plants of many genera are still unknown. Roughly, the trophic selections of about 150 genera have been recorded (i.e. c. 29%). This sampling is sufficient to roughly deduce the selective tendencies of the most important genera.

Many recent papers have listed host-plants, but unfortunately this has not occurred in the past. As per usual, nothing is known about the food-plants of Alticinae from Madagascar and little is known from Australia. It is a great pity that such rich territories possessing luxuriant and original floras, have been so little explored in the domain of food-plant relationships. Certain aspects of the Alticinae host-plant relationships are discussed in Jolivet (1986b), but more recent papers by Matsuda, Nielsen, Begossi and Benson (in Jolivet *et al.*, 1988) have reviewed these in detail. Chemical aspects are not treated here in the present work. However, it should be noted that certain Alticinae are parasites of myrmecophilous plants and are relatively immune against attacks by the inhabitant ants. Others feed on epiphytic plants or frequent the forest canopy.

Host-plant data are very irregularly distributed for the various regions of the world. Even in the Neotropical and African regions the gaps in knowledge are considerable. Most of the Alticinae which live in the canopy of tropical forests, predominantly in South America, are still undescribed and unknown biologically. It is evident that killing and collecting them by the use of pyrethroids will not determine their insect/plant associations. We also fear that by the time technology allows us to reach the top of the forests, most of the rich fauna of Curculionidae and Chrysomelidae (especially Alticinae and Galerucinae) will have already disappeared together with the trees upon which they lived, because the forests are rapidly disappearing from Amazonia, West Africa and Asia through clearing for avaricious gain by minorities.

Alticinae are protected by their toxicity, reflex bleeding (often associated with aposematic colouration), escape mechanisms (jumping), flight and reflex immobilization. In some large species such as *Podontia*, the larva rolls itself into a ball like a wood-louse while the adult drops to the ground and does not even attempt to fly or jump.

Concerning their flight, it is evident that most of the Alticinae are able to fly high and well and can easily colonize the forest canopy as do the Galerucinae. However, many genera or species are brachypterous, micropterous or even apterous (Jolivet, 1959; Furth, 1979-1980; Shute, 1980) and so cannot fly. Such variations can occur even within a particular species and are also common among the Chrysomelinae. Sometimes those individuals which are flightless, have been considered by Peyerhimhoff as comprising a biological race, which feeds on a related but different host-plant to that of the flying race (Jolivet, 1959). They could perhaps be considered as emerging species, but all the data on them needs to be checked and their ecology and morphology restudied.

In *Longitarsus*, wing polymorphism and seasonal variation has been related in the past to differing host-plants. Sometimes different forms cohabit or sometimes they are separated in time and space. The humeral calus of the elytra is supposed to regress among apterous forms (Peyerhimhoff, 1911).

Apterous genera include *Trachytetra, Batophila, Orestia, Minotula* (insular), *Minota* (alpine), *Podagrica* (several species only), *Apteropeda, Mniophila, Ivalia, Kamala* and *Clavicornaltica* (endogenous and micropterous). The last genus also has reduced eyes, but strangely enough, possesses strong hind femorae, so that it is able to jump. For certain species, this probably means a short, epigeous life (Scherer, 1979). In any case, among the five Sri Lankan species of *Clavicornaltica* , only the males are winged, so this means a probable aerial dispersion during the mating period by this sex only. *Clavicornaltica* could have evolved by adaptation to an endogenous mode of life. Their closest relatives are probably *Kamala* or *Ivalia* which are both apterous but have normal eyes.

As mentioned previously, many plants act as secondary refuges for Alticinae in times of drought or during absence of the normal host-plant. This is the case of *Quercus* in temperate climates and with *Carica* and *Psidium* in the tropics. Many references to captures of Alticinae mention such plants which are evidently unrelated to the normal host-plant(s) or represent accidental captures at a resting stage or are false reports. We have been hesitant about systematically rejecting these findings but have retained them after the summary of the normal food-plants for each genus where applicable. Due to the unclear classification of the Alticinae, the genera have not been arranged into tribes (see also Seeno and Wilcox, 1982). Recent transfers between Alticinae and Galerucinae have not been included (e.g. *Microdonacia*, *Orthaltica*, etc., see Reid, 1992).

1- *Crimissa* Stål, 1858.

Neotropical. The species appear restricted to Anacardiaceae (*Anacardium*, *Mangifera*) and Bignoniaceae in Brazil. These two families are not closely related.[1]

2- *Podontia* Dalman, 1824.

Indo-Australian, but reaching Taiwan. It contains large-bodied species restricted to Anacardiaceae (*Spondias*, *Rhus*, *Mangifera*, *Toxicodendrum*), Burseraceae (*Canarium*) and Elaeocarpaceae (*Elaeocarpus*). These families are closely related. *Podontia* has also been recorded from Caesalpiniaceae, Moraceae (*Ficus*) and Theaceae (*Thea*) which are probably accidental records. *Podontia* larvae are covered with a black, shiny layer of their own feces. They feed on foliage and grow to a length of 14 mm before pupating in an earthen cell within the soil. The adults are able to leap and feign death when disturbed by dropping to the ground. The female lives for 3 months and produces about 500, very large (1.6 mm diameter) eggs readily found on the underside of the leaves in loose clusters covered with a dark substance (Kalshoven, 1951).

3 - *Ophrida* Chapuis, 1875.

Indo-Australian. On Burseraceae (*Boswellia*, *Garuga* and *Canarium*). Probably also on Apocynaceae.

4- *Blepharida* Chevrolat, 1837.

The sumac flea beetles feed on *Rhus* (Anacardiaceae) and have almost a cosmopolitan distribution. The free larva is slightly cyphosomatic. *Blepharida* have been recorded from *Rhus* in Africa, on *Rhus* and *Schinus* (Anacardiaceae) in Brazil and Guyana, on *Allophyllus* (Sapindaceae) in Ethiopia, on *Matayba* and *Allophyllus* (Sapindaceae) in the West Indies, on *Bursera* (Burseraceae) in Mexico, on *Schinus, Rhus* and *Cotinus* (Anacardiaceae) in North America and

[1] A genus close to *Crimissa*: *Procalus* Clark, 1865, has been recently studied by Viviane Jerez (1992) in Chile. The adults feed also on Anacardiaceae: *Lithrea caustica, Schinus montanus, Schinus polygamus* and *Schinus latifolius*. (See also Addenda, p. 213).

Rhus in Israel etc. Thus the genus is closely associated with Sapindaceae and Geraniales and has been restudied recently by Furth (1982-1992). According to Furth, the Neotropical species listed under *Blepharida* probably belong to *Notozona*.

5- *Diamphidia* Gerstaecker, 1855.

These African species are very poisonous. The pupae and larvae are used as arrow poison in the Kalahari of Africa (Jolivet, 1967). Many larvae of *Diamphidia* are preyed upon by *Lebistina* larvae (Coleoptera: Lebiidae). This is a real case of peckhamian or aggressive mimicry. All the species of *Diamphidia* feed on *Commiphora* (Burseraceae). The larvae feed freely on the leaves of the host-plant. Eggs are laid in batches of about 15 on the twigs and attached with excreta. Pupation occurs within a cocoon in the soil.

6- *Polyclada* Chevrolat, 1893 (= *Cladocera* Hope, 1840).

Similar to *Diamphidia*. They are also used as an arrow poison in the Kalahari. They feed on *Sclerocarya caffra* (Anacardiaceae) and are also preyed upon by *Lebistina* larvae. Adults may also feed on *Pseudospondias* (Anacardiaceae). Other references to *Dalbergia* (Fabaceae) in Djebel Marra (Sudan) and *Clerodendrum* (Verbenaceae) as host-plants are very doubtful. One of us (PJ) observed one species during 1962 on a small bush (Anacardiaceae) at 2000 m altitude in the Djebel Marra of Africa. This mountain reaches 3088 m altitude.

7- *Luperomorpha* Weise, 1887.

Indo-Australian. Several species on *Clerodendrum* (Verbenaceae) which is the natural host-genus. Very common in Vietnam on several *Clerodendrum* species, including *C. fragrans*. Also reported (as secondary host-plants or as accidental captures) on *Mangifera* (Anacardiaceae), *Ipomoea* (Convolvulaceae), *Solanum, Petunia, Capsicum, Lycopersicon* (Solanaceae), *Pisum, Glycine, Phaseolus, Robinia, Wisteria, Astralagus, Vigna, Lespedeza, Milletia, Cajanus* (Fabaceae), *Morus, Broussonetia* (Moraceae), *Chrysanthemum, Hemistepta* (Asteraceae), *Raphanus* (Brassicaceae), *Citrus, Zanthoxylum* (Rutaceae), *Ranunculus* (Ranunculaceae), *Scabiosa* (Dipsacaceae), *Castanea* (Fagaceae), *Buddleja* (Buddlejaceae), *Beta* (Chenopodiaceae), *Dioscorea* (Dioscoreaceae), *Iris* (Iridaceae) and *Cryptomeria* (Taxodiaceae).

8- *Luperaltica* Crotch, 1873.

North America. Feeds on *Ambrosia, Solidago, Gutierrezia, Eupatorium* (Asteraceae). *Lespedeza* (Fabaceae) is probably a secondary (or accidental) host-plant.

9- *Licyllus* Jacoby, 1885.

All the species of this genus have been reported on *Solanum* leaves (Solanaceae) in New Guinea, New Caledonia and Australia.

10- *Hespera* Weise, 1889.

Probably polyphagous. Certain Asian and African species can be monophagous. On *Rhus* (Anacardiaceae) in Africa. On *Berberis* (Berberidaceae), *Chenopodium* (Chenopodiaceae), *Artemisia* (Asteraceae), *Hippophae* (Elaeagnaceae), *Nepeta* and *Elssholzia* (Lamiaceae), *Polygonum* (Polygonaceae), *Hibiscus* (Malvaceae) and *Zea* (Poaceae) in Asia.

11- *Yunohespera* Chen *et al.*, 1984.

On *Rhododendron* (Ericaceae) in the Yunnan Mountains, China at 2920 m altitude (Chen *et al.*, 1984).

12- *Phyllotreta* Chevrolat, 1837.

Cosmopolitan. Most of the species (175) on Brassicaceae, Resedaceae, Cleomaceae, Limnanthaceae, Capparaceae and Tropaeolaceae, at least chemically related. Most of the larvae are root- or subterranean stem-feeders. However, certain species are miners inside stems and leaves in the larval stage. The following host-plants have been reported:- *Capparis, Cadaba, Crateva, Maerua* (Capparaceae), *Gaylusea, Reseda, Ochradenus, Randonia* (Resedaceae), *Tropaeolum* (Tropaeolaceae), *Gynandropsis, Polanisia, Cleome* (Cleomaceae), *Erucastrum, Rapistrum, Dentaria, Camelina, Arabis, Isatis, Bunias, Erysimum, Radicula, Thlaspi, Cochlearia, Hirschfeldia, Alyssum, Alliaria, Raphanus, Brassica, Sisymbrium, Nasturtium, Diplotaxis, Cardamine, Rorippa, Lepidium, Coronopus, Armoracia, Sinapis, Barbarea, Berteroa, Biscutella, Descurainia, Crambe, Cardaria, Cakile, Neslia, Schouwia, Eruca, Cheiranthus, Ochthodium, Farsetia, Moricandia, Psychine, Erucaria, Enarthrocarpus, Matthiola, Iberis, Reboudia, Lonchophora, Anchonium, Fibigia, Peltaria, Capsella, Hesperis, Dontostemon, Sophia, Chorispora, Draba, Senebiera, Malcolmia* and *Zilla* (Brassicaceae). All of these plant families contain sinigrin and this is the reason why *Phyllotreta* feed on these plants. The relationships between the Limnanthaceae and the Tropaeolaceae on one hand and the Rhoedales on the other is more chemical than botanical and is probably due to the thioglucosides producing isothiocyanates. The host-plants of 122 species of *Phyllotreta* are known from over 175 described species (i.e. 69.7%) which is rather remarkable and provides an excellent sample. The larvae normally feed on rootlets and germinating seeds. Furth (1979) has shown that *Phyllotreta* in Israel are divided into three groups according to their host-plants. This specificity is due to slight differences between the chemicals and inhibitory substances. Many secondary or accidental plants have been recorded for *Phyllotreta*. As mentioned previously they are either accidental or represent temporary refuges. Many of these plants have no close taxonomic connection with the Rhoeadales. They are *Quercus* (Fagaceae), *Carpinus* (Carpinaceae), *Ulmus* (Ulmaceae), *Populus* (Salicaceae), *Humulus, Cannabis* (Cannabaceae), *Gossypium, Hibiscus* (Malvaceae), *Solanum, Lycopersicon* (Solanaceae), *Nitraria* (Zygophyllaceae), *Phaseolus, Pisum, Glycine, Trifolium, Medicago, Vicia, Dolichos, Faba, Desmodium* (Fabaceae),

Acacia (Mimosaceae), *Zizyphus* (Rhamnaceae), *Tilia* (Tiliaceae), *Morus* (Moraceae), *Stellaria* (Caryophyllaceae), *Limonium* (Plumbaginaceae), *Ambrosia, Lactuca, Erigeron* (Asteraceae), *Carica* (Caricaceae), *Ricinus, Phyllanthus* (Euphorbiaceae), *Fraxinus* (Oleaceae), *Prunus, Pyrus, Fragaria, Malus* (Rosaceae), *Asclepias*, *Leptadenia* (Asclepiadaceae), *Allium* (Alliaceae) and *Juglans* (Juglandaceae). However, several species of *Phyllotreta* are monophagous, i.e. feeding on one species of plant only, e.g. *P. armoraciae* (Koch) which feeds in nature only on *Armoracia* (Brassicaceae). Among the leaf-miners, certain species are restricted to petioles or the central vein, while others mine the abaxial parenchyma. The transition from leaf-miner to stem-miner has been observed with *P. zimmermanni* (Crotch) (Smith, 1985) and from stem-miner to root-feeder with *P. armoraciae.* The majority of *Phyllotreta* larvae feed on stems or on roots, but there are more species on roots than on stems. Almost all *Phyllotreta* feed on Brassicaceae and related families, but very few are serious pests and they can be easily controlled by insecticides because they do not offer any resistance unlike caterpillars and other insects. Apterism is exceptional but occurs among several species feeding on *Iberis* (Brassicaceae) in Algeria. However, it is certain that several species normally live on Chenopodiaceae (*Atriplex, Suaeda, Salsola, Spinacia, Chenopodium, Beta)*, while others feed on *Plantago* (Plantaginaceae). Other species, such as *P. vittula* Redt. appear to be polyphagous but develop normally and mature their ovaries only on Poaceae and partially on Brassicaceae (Kostromitin, 1973-1982). The following Poaceae have been recorded for *P. vittula:- Triticum, Secale, Zea, Panicum, Dactylis, Hordeum, Avena, Setaria, Agropyrum, Bromopsis, Poa, Festuca, Koeleria, Sesleria, Bromus, Lolium, Brachypodium* and others. It is evident that *P. vittula* has also been collected on *Trifolium, Vicia* (Fabaceae), *Tropaeolum* (Tropaeolaceae), *Beta, Chenopodium, Atriplex* (Chenopodiaceae), *Sonchus, Scorzonera, Taraxacum* (Asteraceae) and *Cyperus* (Cyperaceae). Several Brassicaceae have also been recorded by Heikertinger (1925) as regular host-plants and it appears most likely that it is from Brassicaceae that *P. vittula* has adapted to Poaceae and others.

13- *Trachytetra* Sharp, 1886.

New Zealand. Appears to be polyphagous, feeding on several Sympetalae, *Coprosma* (Rubiaceae), *Leucopogon* and *Cyathodes* (Epacridaceae). However it has also been found on *Melicytus* (Violaceae) and *Pittosporum* (Pittosporaceae).

14- *Pleuraltica* Sharp, 1886.

New Zealand. Two species on *Coprosma* (Rubiaceae).

15- *Aphthona* Chevrolat, 1837.

Cosmopolitan. Very important genus with more than 200 species. Most of the host-plants are well-known. Feeds mostly on Euphorbiaceae, Geraniaceae, Cistaceae, Rosaceae, Linaceae, Iridaceae and Lythraceae. The Poaceae, Rutaceae and Lamiaceae can be considered secondary hosts. *Aphthona* are well-known on secondary and accidental hosts. Most of them are without any true trophic relationships, but sometimes actual feeding occurs on them. These are *Cyperus*, *Carex* (Cyperaceae), *Pennisetum, Sorghum, Zea, Poa, Oryza, Cynodon* (Poaceae)(mostly on flowers feeding on grass pollen), *Citrus* (Rutaceae), *Clerodendrum* (Verbenaceae), *Stigmaphyllon, Heteropteris* (Malpighiaceae), *Paederia, Coffea* (Rubiaceae), *Mangifera* flowers (Anacardiaceae), *Cucumis* (Cucurbitaceae), *Colubrina,* (Rhamnaceae), *Myrcia, Psidium* (Myrtaceae), *Sisymbrium, Brassica, Raphanus, Farsetia* (Brassicaceae), *Peganum* (Zygophyllaceae), *Cirsium, Helichrysum, Taraxacum, Bellis, Artemisia, Carduus, Eclipta, Grindelia* (Asteraceae), *Beta, Chenopodium* (Chenopodiaceae), *Lychnis* (Caryophyllaceae), *Ulmus* (Ulmaceae), *Morus* (Moraceae), *Viburnum* (Caprifoliaceae), *Vicia, Melilotus, Vigna, Pisum, Medicago, Cajanus, Trifolium, Phaseolus, Dolichos, Arachis, Crotalaria* (Fabaceae), *Acacia, Inga, Pithecellobium, Prosopis, Albizia* (Mimosaceae), *Plantago* (Plantaginaceae), *Convolvulus, Ipomoea* (Convolvulaceae), *Solanum, Nicotiana* (Solanaceae), *Evonymus* (Celastraceae), *Pistacia* (Pistaciaceae), *Gossypium, Abutilon, Hibiscus* (Malvaceae), *Paliurus* (Rhamnaceae), *Fagus, Quercus, Castanea* (Fagaceae), *Heliotropium* (Boraginaceae), *Salvia* (Lamiaceae), *Urtica* (Urticaceae), *Erica* (Ericaceae), *Populus* (Salicaceae), *Calotropis* (Asclepiadaceae), *Aristolochia* (Aristolochiaceae), *Edgeworthia* (Thymelaeaceae), *Mentha* (Lamiaceae), *Lythrum* (Lythraceae), *Bacopa* (Scrophulariaceae), *Pinus* (Pinaceae), *Elaeagnus* (Elaeagnaceae) and *Amygdalis, Prunus, Crataegus, Spiraea, Rosa, Filipendula, Rubus, Sanguisorba, Malus, Pourthiaea* (Rosaceae). In reality, most of the species, as mentioned above, feed on *Euphorbia, Phyllanthus, Croton, Mallotus, Macaranga* (an ant-tree), *Ricinus* (Euphorbiaceae), *Sesamum* (Pedaliaceae), *Helianthemum* (Cistaceae), *Linum* (Linaceae), *Geranium, Erodium* (Geraniaceae), *Iris* (Iridaceae) and *Sempervivum, Sedum* (Crassulaceae). Almost everywhere, even in Fiji, the chief host-plants are the Euphorbiaceae (mainly the genus *Euphorbia*). However, some species like *A. euphorbiae* (Schrank), which feeds mostly on *Euphorbia* and *Linum*, have also been recorded from Chenopodiaceae, Convolvulaceae, Solanaceae and Asteraceae. The Linaceae, Geraniaceae, Malpighiaceae, Zygophyllaceae, Rutaceae and Euphorbiaceae are closely related families (Geraniales). Many *Aphthona* species feed on pollen from several grass species but are not attracted to the plants themselves. Feeding on tender and edible leaves of families like Poaceae is only accidental.

The larva of *Aphthona nigrilabris* Duvivier mines the *Euphorbia* leaves in India (Zaka-ur-Rab, 1991).

16- *Tribolia* Chen, 1933.

Food selection of this genus is similar to that of the previous one. Mostly found on Euphorbiaceae in Vietnam.

17- *Trachyaphthona* Heikertinger, 1924.

Asia and the Pacific. On *Alpinia* (Zingiberaceae), *Costus* (Costaceae), *Calamus* (Arecaceae), *Viburnum*, *Weigela*, *Lonicera* (Caprifoliaceae), *Paederia* (Rubiaceae), *Edgeworthia* (Thymelaeaceae), *Psidium* (Myrtaceae) and several Poaceae. It appears that Caprifoliaceae and Rubiaceae are the normal host-plants in Japan.

18- *Zipangia* Heikertinger, 1924.

On *Viburnum, Lonicera, Weigelia* (Caprifoliaceae) and *Paederia* (Rubiaceae) in Asia. Also on *Calopogonium, Glycine* (Fabaceae) and *Elaeagnus* (Elaeagnaceae).

19- *Decaria* Weise, 1895.

On *Heliotropium* (Boraginaceae) in Kenya, but also recorded from flowers of *Cola* (Sterculiaceae) and *Ocimum* (Lamiaceae).

20- *Gabonia* Jacoby, 1893.

According to Seeno and Wilcox (1982), this genus is a synonym of *Jamesonia* Jacoby, 1895. It appears to be a forest dwelling genus adapted to various plants but the main host preferences have not yet been determined. On *Piper* (Piperaceae), *Theobroma, Cola* (Sterculiaceae), *Senecio, Helichrysum* (Asteraceae)(in mountainous areas to 3000 m), *Hibiscus* (Malvaceae), *Rosa* (Rosaceae), *Coffea* (Rubiaceae), *Persea* (Lauraceae) and *Jatropha* (Euphorbiaceae). It can also feed on introduced cocoa (*Theobroma*, Sterculiaceae) in Africa. The observation, provided by Boppré and Scherer (1981), that the males are attracted during the night to wilted leaves of *Heliotropium* (Boraginaceae), which contain pyralizidine alkaloids, does not indicate the real host-plant, but can be related to sexual behaviour. The montane *Senecio* species (Asteraceae), on which *Gabonia keniae* Bryant feeds in East Africa, contain similar chemicals.

21- *Lupraea* Jacoby, 1885.

Recorded on *Quercus* (Fagaceae) in the USA.

22- *Homoschema* Blake, 1950.

In the West Indies on Malpighiaceae (*Stigmatophyllum, Bunchosia, Malpighia, Banisteria, Heteropterys, Byrsonima*). Also on *Coffea* (Rubiaceae), *Clusia* (Clusiaceae), *Colubrina* (Rhamnaceae) and *Crotalaria* (Fabaceae).

23- *Heyrovskaya* Madar and Madar, 1968.

Canary Islands. On *Euphorbia* (Euphorbiaceae).

24- *Apraea* Baly, 1877.

The main host-plant of this West Indian genus has not been determined with any certainty. On *Quercus* (Fagaceae), *Dioscorea* (Dioscoreaceae), *Eugenia* (Myrtaceae), *Prosopis* (Mimosaceae), *Monarda* (Lamiaceae), *Serjania* (Sapindaceae) and several Asteraceae.

25- *Acanthonycha* Jacoby, 1891.
On *Passiflora* (Passifloraceae) in Argentina and on *Carpotroche* (Flacourtiaceae) in Brazil. These families are closely related and belong to the Parietales.

26- *Heikertingerella* Csiki, 1940.
On *Theobroma* (Sterculiaceae) in Brazil and *Tecoma* (Bignoniaceae) in Jamaica. These two families are closely related.

27- *Longitarsus* Berthold, 1827.
One of the most important alticine genera. There are 330 described and perhaps many more remain to be discovered. The trophic selections are well-known for most of the species and some are actually polyphagous. On Boraginaceae, Lamiaceae, Linaceae, Asteraceae, Plantaginaceae, Convolvulaceae, Thymelaeaceae, Scrophulariaceae, Solanaceae, Dipsacaceae and Ranunculaceae. Preferences for Boraginaceae, Lamiaceae and Asteraceae are common. It is a cosmopolitan genus, and like many genera of Alticinae, secondary host-plants are frequent. The larvae are root-feeders, but are sometimes reported to be leaf-miners. Several larvae mine the stems under the epidermal layers. The Verbenaceae are also a common choice in certain areas.

Apterous and micropterous forms are relatively common, often within a given species (Jolivet, 1959; Furth, 1979; Shute, 1980). The following host-plants have been reported for the various species of *Longitarsus* throughout the world:- *Ammobium, Santolina, Leucanthemum, Asteriscus, Anthemis, Chrysanthemum, Helichrysum, Cousinia, Othonnopsis, Hertia, Euryops, Matricaria, Bellis, Bidens, Arctium, Cichorium, Silybum, Centaurea, Taraxacum, Pollenis, Tussilago, Erigeron, Blumea, Petasites, Pyrethrum, Marribium, Artemisia, Achillea, Eupatorium, Kleinia, Senecio, Cirsium, Carduus, Leontodon, Inula, Xanthium, Aster, Cinneraria, Sonchus, Mycelis, Crepis* (Asteraceae), *Plantago* (Plantaginaceae), *Asperugo, Pulmonaria, Alkanna, Lycopsis, Elizaldia, Onosma, Rindera, Trachystemon, Macrotamia, Borago, Heliophytum, Tournefortia, Trichodesma, Heliotropium, Echium, Cynoglossum, Anchusa, Myosotis, Symphytum, Brunnera, Solenanthus, Onosmodium, Pulmonaria, Nonnea, Cerinthe, Lithospermum, Mattia, Messerschmidtia* (Boraginaceae), *Phlomis, Molucella, Lamiastrum, Galeobdolon, Stachys, Mentha, Lycopus, Pogostemon, Clinopodium, Buglossoides, Nepeta, Marjorana, Ballota, Salvia, Prasium, Marrubium, Satureia, Prunella, Thymus, Glechoma, Eremostachys, Lavandula, Teucrium, Sideritis, Pevovskia, Isodon, Leonurus, Calamintha, Coridothymus, Rosmarinus, Origanum, Melissa, Betonica, Dracocephalum* (Lamiaceae), *Scabiosa, Cephalaria, Dipsacus, Succisa* (Dipsacaceae), *Lantana, Callicarpa,*

Phyla (Verbenaceae), *Calystegia, Ipomoea, Convolvulus* (Convolvulaceae), *Scrophularia, Rhinanthus, Isoplexis, Verbascum, Bacopa, Euphrasia, Veronica, Gratiola, Odontites, Orthanta, Pedicularis, Alectorolophus* (Scrophulariaceae), *Piper* (Piperaceae), *Daphne, Passerina, Thymelaea* (Thymelaeaceae), *Ranunculus, Thalictrum, Clematis, Pulsatilla* (Ranunculaceae), *Linum* (Linaceae), *Solanum, Atropa, Lycopersicon* (Solanaceae), *Cajanus, Vigna, Genista, Melilotus, Lathyrus, Phaseolus, Pueraria, Crotalaria, Lotus, Medicago, Onobrychis, Trifolium* (Fabaceae), *Prosopis* (Mimosaceae), *Lobelia* (Campanulaceae), *Cannabis* (Cannabaceae); perhaps also on *Psidium* (Myrtaceae), *Citrus* (Rutaceae), *Campanula* (Campanulaceae), *Ficus, Morus* (Moraceae), *Urtica* (Urticaceae), *Gossypium* (Malvaceae), *Epilobium* (Onagraceae), *Nasturtium* (Brassicaceae), *Rumex, Polygonum* (Polygonaceae), *Silene* (Caryophyllaceae), *Juncus* (Juncaceae), *Equisetum* (Equisetaceae) and *Eryngium, Sium* (Apiaceae). Probably many of the last-mentioned captures have been of insects feeding on pollen only. The following secondary (and accidental) plants have been recorded in the literature: *Bridelia* (Euphorbiaceae), *Camellia* (Theaceae), *Quercus* (Fagaceae), *Potentilla, Amygdalis, Pyrus, Malus, Fragaria* (Rosaceae), *Tabebuia* (Bignoniaceae), *Utricularia* (Lentibulariaceae), *Asystasia* (Acanthaceae), *Sueda, Atriplex, Beta* (Chenopodiaceae), *Opuntia* (Cactaceae), *Pennisetum, Oryza, Lolium, Poa, Avena* (Poaceae), *Lysimachia* (Primulaceae), *Galium, Asperula* (Rubiaceae), *Crateva* (Capparaceae), *Erodium* (Geraniaceae) and *Balanites* (Balanitaceae). The *Longitarsus* from the Island of St. Helena are endemic and specific to certain plants, i.e. *Longitarsus melissi* Wollaston feeds on *Sium kelenianum* (Apiaceae), *L. janulus* Wollaston on the cabbage tree, *Senecio prenanthiflora* (Asteraceae) and *L. helenae* Wollaston on *Lobelia scaevifolia* (Campanulaceae). The latter report is the only one available at the present time of a *Longitarsus* species feeding on *Lobelia* but this host-relationship must be rather common in the East African mountains, even though observations have yet to be made. Several *Longitarsus* are actually monophagous on *Piper, Plantago*, etc. Overall, the trophic selections of *Longitarsus* somewhat parallel the relationships within those plant families, e.g. between Plantaginaceae, Rubiaceae, Dipsacaceae (Plantaginales and Rubiales) and Lamiaceae, Verbenaceae, Boraginaceae, Scrophulariaceae, Lentibulariaceae, Acanthaceae, Gesneriaceae, Solanaceae, Convolvulaceae (Tubiflorae). The case with Asteraceae is somewhat different. It should be noted that the above-mentioned families differ considerably in their chemical composition but they must have a common attractant or at least a lack of repulsive substances. Relationships between Plantaginales (and certain Rubiales) and Tubiflorae are confirmed by several Chrysomelidae, such as *Timarcha* and *Chrysolina*.

28- *Metroserrapha* Bechyne, 1958.

This genus occurs only in the islands of Mauritius, Reunion and Rodriguez. It is apparently polyphagous since it has been collected from *Philippia montana* (Ericaceae), *Senecio* (Asteraceae), *Rumex* (Polygonaceae), *Menecylon* (Menecylaceae), *Ardisia* (Myrsinaceae) and *Ligustrum* (Oleaceae). As some of these host-plants, including *Rumex*, are introduced plants to these islands, it is difficult to determine the primary and secondary hosts.

29- *Glyptina* LeConte, 1859.
American. Recorded from *Cirsium*, *Solidago* (Asteraceae), *Euphorbia*, *Piscaria* (Euphorbiaceae), *Rubus*, *Potentilla* (Rosaceae), *Mentha* (Lamiaceae), *Thermopsis* (Fabaceae), *Solanum* (Solanaceae), *Gossypium* (Malvaceae), *Salsola* (Chenopodiaceae) and *Carya* (Juglandaceae). *Euphorbia* appears to be the normal host-plant (Riley & Enns, 1979, 1982).

30- *Platiprosopis* Chevrolat, 1834.
American. In Brazil on *Theobroma* (Sterculiaceae) and in Brazil, Honduras and Salvador on *Phaseolus*, *Canavalia* and *Mucuna* (Fabaceae).

31- *Leptophysa* Baly, 1877.
Neotropical. On flowers of *Cleome* (Cleomaceae) and on *Brassica* (Brassicaceae) in Venezuela and Brazil. It appears restricted to the Rhoeadales but has also been recorded on *Solanum* (Solanaceae), *Beta* (Chenopodiaceae), *Lamiaceae* and *Cordia* (Ehretiaceae). Also on *Adiantum* (Adiantaceae) and several other tree ferns in the West Indies.

32- *Systena* Chevrolat, 1837.
New World. It is the most polyphagous, if not the only true one, among the Alticinae. Probably, this polyphagy derives from the Asteraceae which appears to be selected very often. About 100 species comprise the genus. They can be very serious pests of crops. On *Trifolium*, *Medicago*, *Pisum*, *Cajanus*, *Canavalia*, *Cicer*, *Arachis*, *Glycine*, *Crotalaria*, *Vigna*, *Phaseolus*, *Vicia*, *Faba* (Fabaceae), *Prosopis* (Mimosaceae), *Solanum*, *Lycopersicon*, *Nicotiana*, *Capsicum* (Solanaceae), *Beta, Chenopodium, Kochia, Salsola, Spinacia* (Chenopodiaceae), *Lactuca, Pluchea, Synedrella, Calliopsis, Helichrysum, Baccharis, Verbesina, Bidens, Dahlia, Tagetes, Xanthium, Coreopsis, Galinsoga, Carduus, Ambrosia, Helenium, Cynara, Aster, Chrysanthemum, Eupatorium, Solidago, Sclerocarpus* (Asteraceae), *Amaranthus* (Amaranthaceae), *Buddleja* (Buddlejaceae), *Gossypium, Hibiscus, Abelmoschus* (Malvaceae), *Lantana, Valerianoides*, *Phyla, Stachytarpheta, Lippia, Verbena, Tectona* (Verbenaceae), *Arracacia*, *Daucus*, *Cuminum*, *Eryngium*, *Foeniculum* (Apiaceae), *Borreria*, *Diodia*, *Galium*, *Gardenia* (Rubiaceae), *Cucurbita*, *Cucumis* (Cucurbitaceae), *Saccharum, Zea, Sorghum, Lolium*, *Oryza*, *Panicum*, *Hordeum*, *Festuca*, *Euchlaena*, *Dactylis,* (Poaceae), *Portulaca* (Portulacaceae), *Melochia* (Sterculiaceae), *Ipomoea* (Convolvulaceae), *Allium* (Alliaceae), *Cornus* (Cornaceae), *Apocynum, Allamanda* (Apocynaceae), *Vitis* (Vitaceae), *Brassica, Raphanus, Nasturtium* (Brassicaceae), *Plantago* (Plantaginaceae), *Rumex, Rheum, Polygonum, Fagopyrum* (Polygonaceae), *Salvia, Origanum*, *Thymus* (Lamiaceae), *Pelargo-*

nium (Geraniaceae), *Prunus, Fragaria, Crataegus* (Rosaceae), *Sambucus* (Caprifoliaceae), *Oenothera* (Onagraceae), *Alnus, Betula* (Betulaceae), *Borago, Heliotropium* (Boraginaceae), *Liquidambar* (Altingiaceae), *Carya* (Juglandaceae), *Quercus* (Fagaceae), *Ulmus* (Ulmaceae), *Scaparia, Antirrhinum* (Scrophulariaceae), *Sesamum* (Pedaliaceae), *Taxodium* (Taxodiaceae), *Viola* (Violaceae), *Citrus* (Rutaceae), *Carica* (Caricaceae), *Morus* (Moraceae), *Ligustrum* (Oleaceae), *Linum* (Linaceae), *Manihot* (Euphorbiaceae), *Phlox* (Polemoniaceae), *Polyscias* (Araliaceae), *Spondias* (Anacardiaceae), *Tetragonia* (Tetragoniaceae), *Asparagus, Tulipa* (Liliaceae) and *Dianthus* (Caryophyllaceae). Since *Systena* are polyphagous, the above list of host-plants could be easily extended but fundamental alticine selections are all represented within this list. It appears that polyphagy is mostly associated with succulent crops.

33- *Prasona* Baly, 1861.
Neotropical. On *Helianthus* and *Centaurea* (Asteraceae).

34- *Cyrsylus* Jacoby, 1891.
On *Clerodendrum* (Verbenaceae) in Puerto Rico.

35- *Agasicles* Jacoby, 1904.
Neotropical. The genus has been used in the biological control of *Alternanthera* (Amaranthaceae) or alligator weed in the USA and in certain Asian countries. It appears host-specific on this plant in Argentina and Brazil. *Agasicles* has also been recorded on Polygonaceae. These two plant families are closely related.

36- *Cacoscelis* Chevrolat, 1837.
The genus appears to mainly feed on *Passiflora* (Passifloraceae), but has also been reported on *Cecropia* (Cecropiaceae) and on introduced *Eucalyptus* (Myrtaceae) in Brazil.

37- *Phrynocephala* Baly, 1861.
Neotropical. On *Prosopis* (Mimosaceae).

38- *Syphrea* Baly, 1876.
On Euphorbiaceae (*Ricinus, Croton* and *Crotonopsis*) in tropical America.

39- *Strabala* Chevrolat, 1837.
America. Mostly on *Passiflora* (Passifloraceae), but also on *Pisum, Phaseolus, Medicago* (Fabaceae), *Diodia* (Rubiaceae), *Capsicum, Lycopersicon, Solanum* (Solanaceae), *Musa* (Musaceae), Orchidaceae, *Zea* (Poaceae), *Ipomoea* (Convolvulaceae) and *Spinacia* (Chenopodiaceae). Passifloraceae is probably the preferred host family.

40- *Hermaeophaga* Foudras, 1859.

Europe, Asia Minor, India, Africa according to the definition of the genus. It appears to be associated with Euphorbiaceae (*Mercurialis, Acalypha, Croton, Ricinus, Crozophora, Andrachne*). Also reported (often accidental) on *Dioscorea* (Dioscoreaceae), *Capparis* (Capparaceae), *Corchorus* (Tiliaceae), *Gossypium* (Malvaceae), *Heliotropium* (Boraginaceae), *Celosia* (Amaranthaceae), *Sorghum* (Poaceae), *Dolichos, Cajanus* (Fabaceae), *Tecoma* (Bignoniaceae) and *Citrus* (Rutaceae).

41- *Orthocrepis* Weise, 1888.

In Tanzania, on *Tragia* (Euphorbiaceae) and in Vietnam on *Ficus* (Moraceae).

42- *Exoceras* Jacoby, 1891.

Found in the West Indies on the tree-fern *Adiantum* (Adiantaceae) but this is not necessarily the real host-plant.

43- *Lactina* Harold, 1875.

On Solanaceae (*Dunalia, Cestrum*) in the Neotropics.

44- *Docema* Waterhouse, 1877.

On *Heliotropium* (Boraginaceae) in the Galapagos islands.

45- *Altica* Geoffroy, 1762.

Cosmopolitan. About 236 species have been described. The normal (preferred) food-plants are the following:- *Eucharidium, Boisduvalia, Isnardia, Clarkia, Chamaenerion, Hartmannia, Epilobium, Oenothera, Jussiaea, Circaea, Godetia, Ludwigia, Fuchsia, Gaura* (Onagraceae), *Cuphea, Lythrum, Lagerstroemia, Ammannia* (Lythraceae), *Myriophyllum* (Haloragaceae), *Trapa* (Trapaceae), *Vitis, Ampelopsis, Parthenocissus* (Vitaceae). These families are closely related both chemically and taxonomically. There are also a large number of other plants (either chemically related or not closely related at all) which have been recorded as hosts for *Altica.* Most of these appear not to represent real host-plants but probably mostly accidental captures of resting beetles on non-hosts. These are as follows:- *Agrimonia, Malus, Rosa, Pyrus, Crataegus, Filipendula, Prunus, Sanguisorba, Potentilla, Spiraea, Rubus, Fragaria, Poterium, Dasyphora, Acaena* (Rosaceae), *Alnus, Betula* (Betulaceae), *Quercus* (Fagaceae), *Corylus* (Corylaceae), *Trema, Ulmus* (Ulmaceae), *Cornus* (Cornaceae), *Carpinus* (Carpinaceae), *Beta, Salsola, Spinacia, Chenopodium, Atriplex* (Chenopodiaceae), *Humulus, Cannabis* (Cannabaceae), *Passiflora* (Passifloraceae), *Tilia* (Tiliaceae), *Apium* (Apiaceae), *Salix, Populus* (Salicaceae), *Borreria, Coffea* (Rubiaceae), *Mitragyna,* (Naucleaceae), *Rheum, Polygonella, Rumex, Polygonum, Atraphaxis, Muehlenbeckia* (Polygonaceae), *Solanum, Lycopersicon* (Solanaceae), *Chrysanthemum, Carthamus, Lactuca, Ambrosia, Taraxacum, Carduus, Cirsium, Sonchus, Silybum, Onopordum,*

Cnicus, Arctium, Xeranthemum, Cynara, Echinops, Helianthus, Artemisia, Eupatorium, Bellis, Bidens (Asteraceae), *Boswellia* (Burseraceae), *Elaeagnus, Hippophae* (Elaeagnaceae), *Prosopis* (Mimosaceae), *Phaseolus, Glycyrrhiza, Desmodium, Medicago, Trifolium, Melilotus* (Fabaceae), *Cyperus* (Cyperaceae), *Sorghum, Urelytrum, Zea, Oryza, Poa, Triticum, Pennisetum, Panicum, Arundo* (Poaceae), *Citrus, Murraya* (Rutaceae), *Cocos, Elaeis* (Arecaceae), *Sida, Malva, Gossypium* (Malvaceae), *Rhododendron, Calluna, Erica, Philippia, Vaccinium, Kalmia* (Ericaceae), *Empetrum* (Empetraceae), *Acalypha, Euphorbia, Crotonopsis, Croton* (Euphorbiaceae), *Fatoua* (Moraceae), *Impatiens* (Balsaminaceae), *Sagittaria* (Alismataceae), *Mangifera* (Anacardiaceae), *Sinapis, Brassica, Sisymbrium, Cakile, Raphanus, Capsella, Berteroa* (Brassicaceae), *Melastoma* (Melastomaceae), *Piper* (Piperaceae), *Cucumis* (Cucurbitaceae), *Terminalia* (Combretaceae), *Ruellia* (Acanthaceae), *Theobroma* (Sterculiaceae), *Myricaria, Tamarix* (Tamaricaceae), *Linum* (Linaceae), *Chrysobalanus* (Chrysobalanaceae), *Plantago* (Plantaginaceae), *Helianthemum* (Cistaceae), *Lychnis* (Caryophyllaceae), *Convolvulus* (Convol-vulaceae), *Ribes* (Grossulariaceae), *Viburnum* (Caprifoliaceae), *Oenanthe* (Apiaceae), *Musa* (Musaceae), *Verbena, Clerodendrum* (Verbenaceae), *Sesuvium* (Aizoaceae), *Heuchera* (Saxifragaceae), *Fleurya, Urtica* (Urticaceae), *Geranium* (Geraniaceae), *Tabebuia* (Bignoniaceae), *Primula* (Primulaceae), *Dioscorea* (Dioscoreaceae), *Rhizophora* (Rhizophoraceae), *Myrtus* (Myrtaceae), *Laguncularia* (Combretaceae), *Allium* (Alliaceae) and *Gratiola* (Scrophulariaceae). Many other dubious references mention other families as hosts such as Marantaceae in Africa. Frequent mention of the Poaceae as host-plants is perhaps associated with the consumption of pollen by the adults. The previous grouping of secondary host-plants has been undertaken more or less according to host-preference, despite the fact that several of them are extremely doubtful. Basically, the general tendency is towards Onagraceae, Haloragaceae, Lythraceae, Trapaceae, Vitaceae, Rosaceae, Asteraceae, Polygonaceae and Combretaceae. Often, but not always, the food-plants occur along rivers or in swampy areas. *Altica* larvae are free-living or root-feeders according to the species. Nonveiller (1959) has found *Altica oleracea* (L.) in Yugoslavia on 28 different plant species.

46- *Lysathia* Bechyne, 1959.

America. On Onagraceae (*Jussiaea, Ludwigia*), Haloragaceae (*Myriophyllum*) and Lythraceae (*Lagerstroemia*) (all Myrtiflorae). Adults and larvae occur on the top of the stems and on leaves and flowers. Also captured on *Stachytarpheta* (Verbenaceae), *Solanum* (Solanaceae), *Piper* (Piperaceae) and *Cucumis* (Cucurbitaceae) but probably without any trophic relationships.

47- *Nesaecrepida* Blake, 1964.

West Indies. On *Capsicum* (Solanaceae) and *Cucurbita* (Cucurbitaceae).

48- *Macrohaltica* Bechyne, 1959.
New World. Larvae and adults feed openly on leaves of *Gunnera* (Gunneraceae) in Panama (P. Jolivet, pers. obs.).

49- *Hemilactica* Blake, 1937.
On *Micropholis* (Sapotaceae) in Puerto Rico and elsewhere.

50- *Caeporis* Chevrolat, 1837.
Neotropical. On wild and cultivated Brassicaceae (*Lepidium*, *Brassica*, *Raphanus* and *Nasturtium*).

51- *Chlamophora* Chevrolat, 1837.
On *Ilex paraguariensis* (Yerba Mate)(Aquifoliaceae) in Brazil.

52- *Myrcina* Chapuis, 1875.
On *Spathodea* (Bignoniaceae) in East Africa.

53- *Phygasia* Dejean, 1837.
In Asia and Africa on Asclepiadaceae (*Calotropis, Cynanchum, Kanahia, Metaplexis* and *Harmandiella*). This family apears to be the preferred host-plant group. Occasionally also on Convolvulaceae (*Ipomoea*) and accidentally on Crassulaceae (*Crassula*), Rubiaceae (*Paederia, Coffea*), Fabaceae (*Tephrosia*)and Sterculiaceae (*Theobroma*). Also on flowers of *Cyperus* (Cyperaceae) and *Oryza* (Poaceae).

54- *Liprus* Motschulsky, 1860.
Asia. On *Parthenocissus, Vitis* and *Ampelopsis* (Vitaceae). Accidentally and without any trophic relationships on *Equisetum* (Equisetaceae) and *Hosta* (Liliaceae).

55- *Pseudoliprus* Chûjô and Kimoto, 1960.
In Japan on *Ampelopsis, Parthenocissus* and *Vitis* (Vitaceae) and *Actinidia* (Actinidiaceae).

56- *Lypnea* Baly, 1876.
In the Côte d'Ivoire on *Oncoba* (Flacourtiaceae) and in Vietnam on *Adenia* (Passifloraceae). These are closely related plant families.

57- *Batophila* Foudras, 1859.
On *Rubus*, *Fragaria*, *Potentilla*, *Rosa* and *Crataegus* (Rosaceae) in Europe.

58- *Aphthonomorpha* Chen, 1934.
On *Piper* (Piperaceae) in Indonesia and *Sapium* (Euphorbiaceae) in Vietnam.

59- *Manobidia* Chen, 1934.
On *Glochidion* (Euphorbiaceae) in Vietnam.

60- *Chirodica* Germar, 1834.
In the Cape Province, South Africa, on the inflorescences of *Protea repens* (Proteaceae).

61- *Manobia* Jacoby, 1885.
Old World. On *Styphelia* (Epacridaceae), Urticaceae, *Cyathea* (Cyatheaceae), *Olearia* (Asteraceae), *Acalypha* (Euphorbiaceae) and *Metroxylon* (Arecaceae). On *Cyclosorus* (Aspidiaceae) in Japan (Kimoto, 1984).Probably polyphagous, but this needs to be verified.

62- *Arrhenocoela* Foudras, 1859.
On *Erica* and *Calluna* (Ericaceae) in Europe.

63- *Lythraria* Bedel, 1897.
In Europe and Asia on *Lythrum* (Lythraceae), *Lysimachia* (Primulaceae), *Stachys* (Lamiaceae), *Verbena* (Verbenaceae), *Hypericum* (Hypericaceae), etc. The real and preferred host-plants are probably *Lythrum* and *Lysimachia.*

64- *Ochrosis* Foudras, 1859.
Palaearctic. On *Solanum* (Solanaceae), *Euphorbia, Acalypha* (Euphorbiaceae), *Heliotropium* (Boraginaceae) and *Anagallis* (Primulaceae). Perhaps in summer on *Carpinus* (Carpinaceae) but this is doubtful.

65- *Ogloblinia* Csiki, 1939.
In Asia on *Musa* (Musaceae), *Berberis* (Berberidaceae) and *Stephania* and *Cocculus* (Menispermaceae).

66- *Asiorestia* Jacobson, 1925 (= *Crepidora* Auct.)
Cosmopolitan. On Asteraceae and several other families. Collected from *Citrus* (Rutaceae), *Physalis, Atropa, Capsicum, Datura, Solanum* (Solanaceae), *Dimorphanthera, Vaccinium, Gaultheria, Rhododendron* (Ericaceae), *Aconitum* (Ranunculaceae), *Ipomoea* (Convolvulaceae), *Cucurbita, Cucumis* (Cucurbitaceae), *Acalypha* (Euphorbiaceae), *Salix, Populus* (Salicaceae), *Lantana* (Verbenaceae), *Ficus* (Moraceae), *Kleinhovia, Theobroma* (Sterculiaceae), *Limonium* (Plumbaginaceae), *Cirsium, Carduus, Taraxacum, Artemisia, Centaurea, Onopordum, Crepis, Solidago, Leontodon, Senecio, Gnaphalium, Lappa* (Asteraceae), *Prunus, Amygdalus, Crataegus, Potentilla, Malus, Pyrus* (Rosaceae), *Galeopsis, Mentha* (Lamiaceae), *Vicia, Lotus, Medicago, Trifolium, Pisum, Robinia* (Fabaceae), *Urtica* (Urticaceae), *Styphelia* (Epacridaceae), *Symphytum, Borago* (Boraginaceae), *Poa, Bromus, Triticum, Festuca, Secale, Hordeum, Avena, Saccharum* (Poaceae), *Cannabis* (Canna-

baceae), *Juncus* (Juncaceae), *Veratrum, Urgenia* (Liliaceae), *Gossypium, Abutilon* (Malvaceae), Polygonaceae, *Symplocos* (Symplocaceae), *Eurya* (Theaceae), *Hebe* (Scrophulariaceae), *Cyathea* (Cyatheaceae), *Coprosma* (Rubiaceae) and *Ricinus* (Euphorbiaceae). Tree ferns and Liliaceae etc. are not host-plants but Juncaceae and Poaceae are utilized for pollen.

67- *Hornaltica* Barber, 1941.
Nearctic. On *Acalypha* (Euphorbiaceae), *Podophyllum* (Berberidaceae) and *Brassica* (Brassicaceae). The last two records are very doubtful. According to Furth, this genus belongs to the Galerucinae.

68- *Sangariola* Jacobson, 1922.
On various monocots in Japan, i.e. *Lilium* (Liliaceae), *Symplocarpus* (Araceae) and *Smilax* (Smilacaceae).

69- *Arsipoda* Erichson, 1842.
Australia and Indonesia. Polyphagous or artificial genus. Recorded on a large number of families, i.e. *Oryza, Zea* (Poaceae), *Tectona* (Verbenaceae), *Heterodendrum* (Sapindaceae), *Grevillea, Stenocarpus* (Proteaceae), *Styphelia* (Epacridaceae), *Sapium* (Euphorbiaceae), *Pancheria* (Cunoniaceae), *Tristania* (Myrtaceae), *Morus* (Moraceae), *Prunus* (Rosaceae), *Vitis* (Vitaceae) and *Ipomoea* (Convolvulaceae). Records on *Colocasia* (Araceae) are probably accidental.

70- *Crepicnema* Scherer, 1969.
Recorded from Vietnam on *Ipomoea* (Convolvulaceae).

71- *Orestia* Germar, 1845.
On Lamiaceae (*Prasium, Thymus*, etc.). Also captured on *Arundo* (Poaceae) and *Urtica* (Urticaceae) but these do not represent any trophic relationships.

72- *Orthaltica* Crotch, 1873.
On *Rhus* (Anacardiaceae) for all American species.

73- *Microdonacia* Blackburn, 1893.
Australia. Apparently polyphagous. According to Reid (1992), adults feed on *Eucalyptus* (Myrtaceae), *Eucryphia* (Eucryphiaceae), *Grevillea* (Proteaceae), *Pomaderris* (Rhamnaceae) and *Phebalium, Zieria* (Rutaceae). Collected also from *Nothofagus* (Fagaceae) and *Callicoma* (Cunoniaceae) which may or may not be real host-plants (Reid, 1992). Probably a galerucine.

74- *Derocrepis* Weise, 1886.
Cosmopolitan. On *Aesculus* (Hippocastanaceae), *Prunus, Pyrus, Malus* (Rosaceae), *Celtis* (Ulmaceae), *Pisum, Orobus, Vicia, Faba, Lathyrus, Genista,*

Cytisus, Medicago, Phaseolus, Astralagus, Coronilla, Robinia and *Onobrychis* (Fabaceae). On Fabaceae in Europe.

75- *Hippuriphila* Foudras, 1859.
Feeds in Canada, the USA and Europe on *Equisetum* (Equisetaceae), which is a secondary adaptation upon a siliceous and hard plant. Recorded exceptionally on *Rumex* (Polygonaceae) in New York.

76- *Xuthea* Baly, 1865.
In Asia on *Myrtica* and *Fleurya* and in India and the Far East on *Girardinia* (all Urticaceae).

77- *Crepidodera* Chevrolat, 1837 (= *Chalcoides* Foudras, 1859).
Holarctic. Preferentially on *Populus* and *Salix* (Salicaceae). However, also recorded from *Ulmus* (Ulmaceae), *Malus, Pyrus, Prunus, Spiraea, Rubus, Cerasus* (Rosaceae), *Acer* (Aceraceae), *Lonicera* (Caprifoliaceae), *Betula* (Betulaceae) and *Kalmia* (Ericaceae). *Crepidodera* females lay eggs on growing buds where larvae feed until nymphosis.

78- *Trichaltica* Harold, 1876.
On *Fraxinus* and *Ligustrum* (Oleaceae) in the USA. On *Populus* (Salicaceae) in Arkansas.

79- *Epitrix* Foudras, 1859.
Cosmopolitan. Strongly associated with Solanaceae. On *Solanum, Lycopersicon, Nicotiana, Datura, Nicandra, Capsicum, Physalis, Atropa, Hyoscyamus, Lycium, Petunia, Acnistus* and *Withania* (Solanaceae). Also recorded on the following secondary (and accidental) non host/food-plants:- *Ficus* (Moraceae), *Ipomoea* (Convolvulaceae), *Cicer, Vigna, Trifolium, Phaseolus, Glycine, Cajanus, Canavalia, Melilotus, Pisum, Medicago* (Fabaceae), *Prosopis* (Mimosaceae), *Fragaria, Cerasus, Rubus, Amygdalus, Prunus* (Rosaceae), *Cucurbita, Citrullus, Cucumis* (Cucurbitaceae), *Psidium* (Myrtaceae), *Descurainia, Brassica, Sinapis, Raphanus, Armoracia* (Brassicaceae), *Cleome* (Cleomaceae), *Helianthus, Carduus, Taraxacum, Iva, Ambrosia, Leptilon, Lactuca, Arctium, Conyza, Cirsium* (Asteraceae), *Beta, Spinacia, Chenopodium, Kochia* (Chenopodiaceae), *Petroselinum, Daucus, Arracacia* (Apiaceae), *Citrus* (Rutaceae), *Ribes* (Grossulariaceae), *Galium, Borreria, Gardenia* (Rubiaceae), *Althaea, Hibiscus, Gossypium* (Malvaceae), *Lantana* (Verbenaceae), *Corylus* (Corylaceae), *Manihot* (Euphorbiaceae), *Viburnum* (Caprifoliaceae), *Salvia* (Lamiaceae), *Carica* (Caricaceae), *Arbutus, Gaylussacia* (Ericaceae), *Viola* (Violaceae), *Fraxinus, Ligustrum* (Oleaceae), *Oenothera* (Onagraceae), *Phlox* (Polemoniaceae), *Plantago* (Plantaginaceae), *Fagopyrum, Rheum, Rumex* (Polygonaceae), *Smilax* (Smilacaceae), *Hypericum* (Hypericaceae), *Rhus* (Anacardiaceae), *Amaranthus* (Amaranthaceae), *Phytolacca* (Phytolaccaceae),

Allium (Alliaceae), *Avena, Panicum, Oryza, Zea, Sorghum, Triticum, Saccharum* (Poaceae) and *Asparagus* (Liliaceae). *Epitrix* larvae feed on rootlets and subterranean tubers. Although *Epitrix cucumeris* (Harris) mostly feeds on the Solanaceae, it appears to be polyphagous. Larvae are always less selective than the adults because feeding occurs underground.

80- *Alema* Sharp, 1876.
New Zealand. On *Cyathea* (Cyatheaceae), a tree fern.

81- *Analema* Samuelson, 1973.
Probably on Poaceae in the Fiji Islands.

82- *Hemiglyptus* Horn, 1889.
On Brassicaceae (*Brassica, Raphanus*) in the USA.

83- *Glenidion* Clark, 1860.
America. On *Samanea* (Fabaceae).

84- *Minotula* Weise, 1924.
Juan Fernandez Islands. On *Boehmeria* (Urticaceae) and *Solanum* (Solanaceae).

85- *Minota* Kutschera, 1859.
Palaearctic. Common among mosses but has been collected on *Vaccinium* (Ericaceae), although this does not appear to be the true host-plant. It probabiy feeds on *Plantago* (Plantaginaceae) (Hering, 1935). On *Pteridium, Dryopteris* and *Cyrtomium* (Aspidiaceae) in Japan (Kimoto, 1984).

86- *Pseudodera* Baly, 1861.
Asia. On *Smilax* (Smilacaceae) in Japan.

87- *Diphaulaca* Chevrolat, 1837.
Neotropical. The genus appears to be closely associated with Fabaceae (*Desmodium, Phaseolus, Glycine, Cicer, Pisum, Vigna*) but has also been collected on various crops such as *Citrullus, Cucumis* (Cucurbitaceae), *Populus* (Salicaceae), *Solanum* (Solanaceae), *Ricinus* (Euphorbiaceae), *Beta* (Chenopodiaceae) and *Gossypium* (Malvaceae). Probably these represent accidental host records.

88- *Diphaltica* Barber, 1941.
Neotropical. On *Ilex* (Aquifoliaceae) in Brazil. Many Alticinae feed on this plant.

89- *Podagrica* Chevrolat, 1837.
Genus distributed in Africa, Madagascar and the Palaearctic region. The dif-

ferences between *Podagrica* and *Nisotra* are rather subtle. Normally on Malvales, i.e. *Malva, Althaea, Lavatera, Hibiscus, Malope, Kitaibelia, Gossypium, Abutilon, Sida, Cienfuegosia, Urena, Abelmoschus, Malvaviscus, Pavonia, Alcea, Wissadula* (Malvaceae), *Grewia, Triumfetta, Corchorus* (Tiliaceae), *Theobroma, Sterculia* (Sterculiaceae) and *Ceiba, Adansonia, Eriodendron* (Bombacaceae). Collected from various other crops, but often (except for pollen) without any trophic relationships:- *Volkameria* (Clethraceae), *Symphytum* (Boraginaceae), *Marrubium* (Lamiaceae), *Linum* (Linaceae), *Sesamum* (Pedaliaceae), *Basella* (Basellaceae), *Morus* (Moraceae), *Carduus, Arctium, Carthamus, Helianthus* (Asteraceae), *Boehmeria, Urtica* (Urticaceae), *Citrus* (Rutaceae), *Cajanus, Vicia, Dolichos, Indigofera, Vigna, Phaseolus, Arachis, Glycine, Tephrosia, Stylosanthes, Centrosema* (Fabaceae), *Brassica, Raphanus* (Brassicaceae), *Ipomoea* (Convolvulaceae), *Coffea* (Rubiaceae), *Cucurbita* (Cucurbitaceae), *Croton* (Euphorbiaceae) and *Pennisetum, Oryza, Sorghum* (Poaceae). *Podagrica* larvae feed underground on the rootlets of various plants and germinating seeds.

90- *Nisotra* Baly, 1864.

This genus is widely distributed over the Palaeotropics. It feeds on the same plant groups as *Podagrica*, namely the Malvales, i.e. *Malvariscus, Urena, Hibiscus, Abelmoschus, Gossypium, Sida* (Malvaceae), *Ceiba, Eriodendron* (Bombacaceae), *Corchorus, Triumfetta* (Tiliaceae) and *Cola, Theobroma* (Sterculiaceae). It has also been captured on *Morus* (Moraceae), *Clerodendrum* (Verbenaceae), *Thea, Camellia* (Theaceae), *Arachis, Dolichos, Phaseolus, Cajanus* (Fabaceae), *Pyrus* (Rosaceae), *Helianthus, Bidens* (Asteraceae), *Mentha* (Lamiaceae), *Coffea* (Rubiaceae), *Sorghum* (Poaceae) and *Dendrobium* (Orchidaceae). The precise relationships between *Nisotra* and these plants remain to be defined but they are probably not host-plants.

91- *Torodera* Weise, 1902.

On *Oryza* (Poaceae) in Kenya but this is probably not the true host-plant.

92- *Podagricomela* Heikertinger, 1924.

Far East. On Rutaceae (*Citrus, Clausena, Zanthoxylum, Poncirus* and *Aegle*).

93- *Erystus* Jacoby, 1885.

On *Gossypium* (Malvaceae). Larvae feed on roots, while the adults feed on leaves and flowers.

94- *Bimala* Maulik, 1926.

India. On *Daucus* (Apiaceae).

95- *Clitea* Baly, 1877.

South-East Asia. On *Citrus, Aegle* and *Zanthoxylum* (Rutaceae).

The larvae bore into the midribs of the leaves (Zaka-ur-Rab, 1991).

96- *Mantura* Stephens, 1831.

More or less cosmopolitan. Very often on *Rumex* and *Polygonum* (Polygonaceae). Also captured on *Artemisia, Chrysanthemum* (Asteraceae), *Plantago* (Plantaginaceae), *Oxalis* (Oxalidaceae), *Helianthemum* (Cistaceae), *Pyrus, Malus, Crataegus* (Rosaceae) and *Anabasis* (Chenopodiaceae).

97- *Chaetocnema* Stephens, 1831.

Cosmopolitan. The larvae burrow into the plant stems or feed on the rootlets. The genus is divided into two rather distinct subgenera. One, *Tlanoma* Motschulsky 1845, feeds mostly on Polygonales (Polygonaceae) and Centrospermae (Chenopodiaceae, Amaranthaceae), while the subgenus *Chaetocnema* (sensu stricto) feeds on Poaceae, Cyperaceae, Typhaceae, Butomaceae, Juncaceae and Convolvulaceae, but there are some exceptions. The following host-plants have been recorded for *Chaetocnema* (sensu stricto): *Dinebra, Elymus, Lolium, Panicum, Phleum, Pennisetum, Cynodon, Eragrostis, Arrhenatherum. Thysanolaena, Sesleria, Roegneria, Clinelymus, Hordeum, Echinochloa, Digitaria, Setaria, Glyceria, Chrysopogon, Avena, Secale, Phragmites, Agropyrum, Phalaris, Andropogon, Calamagrostis, Chloris, Dactylis, Festuca, Oryza, Sorghum, Zea, Triticum, Agrostis, Alopecurus, Bromus, Poa, Saccharum, Echionochloa, Eleusine* (Poaceae), *Carex, Cyperus, Eleocharis, Scirpus, Bolboschoenus* (Cyperaceae), *Butomus* (Butomaceae), *Typha* (Typhaceae), *Juncus* (Juncaceae), *Ipomoea, Dichandra, Convolvulus* (Convolvulaceae). *Tlanoma* mostly feed on the following plants:- *Canavalia, Medicago, Melilotus, Phaseolus, Arachis, Pisum, Vicia, Glycine, Cicer, Dorycnium, Onobrychis, Gnilandina* (Fabaceae), *Prosopis, Acacia* (Mimosaceae), *Caesalpinia, Erythrophleum* (Caesalpiniaceae), *Nicotiana, Solanum, Lycopersicon* (Solanaceae), *Cannabis, Humulus* (Cannabaceae), *Hibiscus, Gossypium, Althaea* (Malvaceae), *Sinapis, Raphanus, Brassica, Nasturtium, Sisymbrium* (Brassicaceae), *Amaranthus, Achyranthes, Alternanthera* (Amaranthaceae), *Atriplex, Beta, Camphorosma, Chenopodium, Salicornia, Salsola, Spinacia, Suaeda, Kalidium, Halocnemum* (Chenopodiaceae), *Fagopyrum, Polygonum, Rheum, Rumex* (Polygonaceae), *Boehmeria* (Urticaceae), *Lantana* (Verbenaceae), *Carica* (Caricaceae), *Cryptocarya, Persea* (Lauraceae), *Thalia* (Marantaceae), *Rhizophora* (Rhizophoraceae), *Gineria* (Lythraceae), *Galinsoga, Carduus, Senecio,* (Asteraceae), *Conocarpus* (Combretaceae), *Potentilla, Rubus, Fragaria, Duchesnea* (Rosaceae), *Galium*, (Rubiaceae), *Myosotidum* (Boraginaceae), *Cucurbita* (Cucurbitaceae), *Volkameria, Clerodendrum* (Verbenaceae), *Manihot* (Euphorbiaceae), *Mentha* (Lamiaceae), *Psidium* (Myrtaceae), *Azadirachta* (Meliaceae), *Corchorus* (Tiliaceae), *Salix, Populus* (Salicaceae), *Zingiber* (Zingiberaceae) and *Lilium* (Liliaceae).

Chaetocnema basalis Baly is a leaf-miner in India (Zaka-ur-Rab, 1991).

98- *Xenidea* Baly, 1862.
On *Solanum* (Solanaceae). The larvae feed on roots in Queensland.

99- *Disonychodes* Bechyne, 1955.
On *Solanum* (Solanaceae) in Brazil.

100- *Disonycha* Chevrolat, 1837.
New World. Approximately 125 species have been described. Many species feed on Amaranthaceae and have been used in biological control. Amaranthaceae, Chenopodiaceae, Aizoaceae, Portulacaceae and Caryophyllaceae are more or less closely related botanically. The following host-plants have been recorded:- *Cienfuegosia, Gossypium* (Malvaceae), *Spinacia, Beta, Chenopodium, Atriplex, Salsola* (Chenopodiaceae), *Stellaria* (Caryophyllaceae), *Cucurbita, Cucumis* (Cucurbitaceae), *Oryza, Zea, Pennisetum, Saccharum, Panicum* (Poaceae), *Amaranthus, Alternanthera, Iresine, Philoxerus, Chamissoa* (Amaranthaceae), *Ipomoea* (Convolvulaceae), *Jatropha, Euphorbia* (Euphorbiaceae), *Xanthium, Solidago, Chrysothamnus, Tetradymia, Cirsium, Wedelia, Lactuca, Bebbia, Aster, Vernonia* (Asteraceae), *Arachis, Cacara, Pisum, Medicago, Trifolium, Phaseolus, Glycine, Crotalaria, Cajanus* (Fabaceae), *Prosopis, Schrankia* (Mimosaceae), *Cassia* (Caesalpiniaceae), *Passiflora* (Passifloraceae), *Polygonum, Muehlenbeckia* (Polygonaceae), *Phlox* (Polemoniaceae), *Talinum, Portulaca* (Portulacaceae), *Solanum, Capsicum, Lycopersicon* (Solanaceae), *Salix* (Salicaceae), *Psidium* (Myrtaceae), *Condalia, Colubrina* (Rhamnaceae), *Phaulothamnus* (Achatocarpaceae), *Trianthema* (Aizoaceae), *Sagittaria* (Alismataceae), *Thea* (Theaceae), *Heliotropium* (Boraginaceae), *Persea* (Lauraceae), *Quercus* (Fagaceae), *Tabebuia* (Bignoniaceae), *Jasminum* (Oleaceae), *Cydonia* (Rosaceae), *Daucus* (Apiaceae), *Raphanus* (Brassicaceae), *Opuntia* (Cactaceae), *Phacetia* (Hydrophyllaceae), *Pinus* (Pinaceae), *Asparagus* (Liliaceae) and *Ananas* (Bromeliaceae). Any host-plant recorded other than a Centrospermae is doubtful, although there may be secondary host-plants among them.

101- *Phenrica* Bechyne, 1959.
Neotropical. On Amaranthaceae (*Irene, Cyathula* and *Alternanthera*). There are very doubtful records from Fabaceae.

102- *Monomacra* Chevrolat, 1837 (= *Lactica* Erichson, 1847).
New World. On *Passiflora* (Passifloraceae), but doubtful on *Ricinus, Caperonia* (Euphorbiaceae), *Phaseolus* (Fabaceae), *Jussiaea* (Onagraceae), *Zea* (Poaceae), *Phytolacca* (Phytolaccaceae), *Trema* (Ulmaceae), *Solanum* (Solanaceae) and *Turnera* (Turneraceae).

103- *Omophoita* Chevrolat, 1837.
New World. Rather selective according to the species but with a strong ten-

dency towards Lamiaceae, Verbenaceae and Boraginaceae which are closely related families among the Tubiflorae. Several Brazilian species have been remarkably well-studied by Begossi (1984). The eggs are laid in the ground. Unfortunately many food-plant references are only coincidences or represent secondary hosts. On *Clerodendrum*, *Stachytarpheta*, *Aegiphila*, *Lippia*, *Valerianodes*, *Volkameria*, *Phyla*, *Lantana* (Verbenaceae), *Peltodon*, *Hyptis*, *Leonotis*, *Marsypianthes*, *Coleus*, *Ocimum* (Lamiaceae) and *Heliotropium* (Boraginaceae). Reported (accidentally) on *Jussiaea* (Onagraceae), *Crotalaria*, *Glycine*, *Medicago*, *Phaseolus*, *Dipteryx* (Fabaceae), *Vachellia*, *Inga* (Mimosaceae), *Gossypium* (Malvaceae), *Hybanthus* (Violaceae), *Ipomoea* (Convolvulaceae), *Rhizophora* (Rhizophoraceae), *Manihot* (Euphorbiaceae), *Tibouchina* (Melastomataceae), *Tabebuia* (Bignoniaceae), *Cestrum*, *Physalis* (Solanaceae), *Sesamum* (Pedaliaceae), *Eugenia*, *Eucalyptus* (Myrtaceae), *Pyrus*, *Cydonia* (Rosaceae), *Citrus* (Rutaceae), *Pinus* (Pinaceae), *Musa* (Musaceae), *Pennisetum*, *Zea*, *Oryza*, *Saccharum*, *Triticum* (Poaceae), *Allium* (Alliaceae) and *Colocasia* (Araceae). There are also reports on *Theobroma* (Sterculiaceae), *Borreria* (Rubiaceae), *Pluchea*, *Verbesina*, *Vernonia*, *Gochnatia and Eupatorium* (Asteraceae).

104- *Asphaera* Chevrolat, 1842.

Mostly Neotropical. Reported on Lamiaceae (*Peltodon*, *Coleus*, *Hyptis)* and Verbenaceae (*Lantana)*. There are some accidental reports on *Persea* (Lauraceae), *Buddleja* (Buddlejaceae), *Jaegeria* (Asteraceae), *Daucus* (Apiaceae), *Theobroma* (Sterculiaceae) and *Vigna* (Fabaceae).

105- *Pachyonychis* Clark, 1860.

Nearctic. On *Smilax* (Smilacaceae) in the USA.

106- *Kuschelina* Bechyne, 1951.

New World. On *Mentha*, *Teucrium* (Lamiaceae), *Gerardia*, *Dasystoma*, *Aureoluria*, *Pentastemon* (Scrophulariaceae), *Euphorbia* (Euphorbiaceae), *Hypericum* (Hypericaceae), *Polygonum* (Polygonaceae), *Quercus* (Fagaceae) and *Sabal* (Arecaceae). The genus appears to feed normally on Scrophulariaceae and Lamiaceae. Other plants are probably accidental.

107- *Walterianella* Bechyne, 1955.

Neotropical. On Rubiaceae (*Gardenia)* and Bignoniaceae (*Tabebuia*) in Venezuela.

108- *Alagoasa* Bechyne, 1955.

Neotropical. It has been studied by Begossi (1984). On Verbenaceae (*Aegiphila*, *Lantana*, *Duranta*, *Stachytarpheta*, *Clerodendrum*, *Callicarpa*, *Vitex*, *Lippia)* and Lamiaceae (*Coleus*, *Hyptis*, *Peltodon*). Also reported on Asteraceae (*Gochnatia*, *Eupatorium*, *Vernonia*), Bignoniaceae (*Tabebuia*), Pedaliaceae

(*Sesamum*) and Acanthaceae (*Justicia, Thunbergia*). The last two families are certainly not host-plants and the reports on Asteraceae are very doubtful.

109- *Capraita* Bechyne, 1957.
New World. The trophic preferences are not very clear but they tend to be Ericaceae, Lamiaceae, Verbenaceae, Plantaginaceae and Scrophulariaceae. On *Vaccinium* (Ericaceae), *Fagus, Quercus* (Fagaceae), *Juglans* (Juglandaceae), *Plantago* (Plantaginaceae), *Rubus, Amelanchier* (Rosaceae), *Scrophularia, Mimulus, Pentastemon, Verbascum* (Scrophulariaceae), *Viburnum, Symphoricarpus, Triosteum* (Caprifoliaceae), *Evonymus* (Celastraceae), *Campsis* (Bignoniaceae), *Cornus* (Cornaceae), *Teucrium, Stachys, Cunila* (Lamiaceae), *Verbena* (Verbenaceae), *Ilex* (Aquifoliaceae) and *Chionanthus, Fraxinus* (Oleaceae).

110- *Oedionychus* Berthold, 1827.
Southern Europe, North Africa and the Neotropical region. Virtually cosmopolitan with 430 described species. Appears to be polyphagous as a whole, but has tendencies towards Verbenaceae and Ericaceae. On *Gossypium* (Malvaceae), *Citrus* (Rutaceae), *Eupatorium, Verbesina, Salmea* (Asteraceae), *Sesamum* (Pedaliaceae), *Theobroma* (Sterculiaceae), *Crescentia, Tabebuia, Spathodea* (Bignoniaceae), *Heliotropium* (Boraginaceae), *Lantana, Stachytarpheta, Verbena, Duranta, Volkameria, Clerodendrum* (Verbenaceae), *Antirrhinum, Dasystoma, Verbascum* (Scrophulariaceae), *Acalypha, Euphorbia, Croton* (Euphorbiaceae), *Annona* (Annonaceae), *Capparis* (Capparidaceae), *Carica* (Caricaceae), *Gardenia* (Rubiaceae), *Ligustrum, Fraxinus, Chionanthus* (Oleaceae), *Phlox* (Polemoniaceae), *Stellaria* (Caryophyllaceae), *Phaseolus, Lespedeza* (Fabaceae), *Prosopis* (Mimosaceae), *Cassia* (Caesalpiniaceae), *Fagus, Quercus* (Fagaceae), *Juglans* (Juglandaceae), *Plantago* (Plantaginaceae), *Populus* (Salicaceae), *Teucrium, Mentha* (Lamiaceae), *Gaylussacia, Vaccinium, Desmothamnus* (Ericaceae), *Hypericum* (Hypericaceae), *Ilex* (Aquifoliaceae), *Evonymus* (Celastraceae), *Polygonum* (Polygonaceae), *Bambusa, Panicum, Zea* (Poaceae) and *Cattleya* (Orchidaceae).

111- *Philopona* Weise, 1903.
Old World. On *Thunbergia* (Thunbergiaceae) and *Plantago* (Plantaginaceae).

112- *Physonychis* Clark, 1860.
In Tanzania, Africa, on *Thunbergia* (Thunbergiaceae).

113- *Hyphasis* Harold, 1877.
In Asia, on *Clerodendrum* (Verbenaceae) and *Anisomeles* (Lamiaceae). These plant families are closely related.

114- *Hyphasoma* Jacoby, 1903.
On *Anisomeles* (Lamiaceae) and (?) on *Olea* (Oleaceae) in Asia.

115- *Hemipyxis* Dejean, 1837.
Palaeotropical. On *Lamium* (Lamiaceae) and *Premma, Clerodendrum, Callicarpa* (Verbenaceae). Also reported (as secondary or accidental plants) on *Plantago* (Plantaginaceae), *Alnus* (Betulaceae), *Ilex* (Aquifoliaceae), *Mallotus* (Euphorbiaceae), *Verbascum* (Scrophulariaceae), *Buddleja* (Buddlejaceae) and *Psidium* (Myrtaceae).

116- *Zomba* Bryant, 1922
On *Gossypium* (Malvaceae) in Zambia.

117- *Demarchus* Jacoby, 1887.
Asia. On *Loranthus* (Loranthaceae) in Pakistan and Java. Genus adapted to epiphytic and parasitic plants.

118- *Febra* Clark, 1864.
Captured on tree ferns in Fiji and Vanuata (New Hebrides), i.e. *Acrostichum* (Pteridaceae) and *Angiopteris* (Angiopteridaceae). Also on *Nephrolepis* (Oleandraceae). The larva mines the leaves.

119- *Sutrea* Baly, 1876.
On *Theobroma* (Sterculiaceae) in Papua New Guinea and on *Alpinia* (Zingiberaceae) in the Solomon Islands.

120- *Homotyphus* Clark, 1860.
On tubers of *Colocasia* (Araceae) in Guyana.

121- *Distigmoptera* Blake, 1943.
America. On *Trifolium* and *Medicago* (Fabaceae). Also on *Pinus* (Pinaceae), *Azolla* (Azollaceae)(aquatic fern) and Orchidaceae. It appears that this genus is strongly associated with Fabaceae so that the other records are most probably accidental.

122- *Pseudolampsis* Horn, 1889.
On *Azolla* (Azollaceae) in Florida (subaquatic).

123- *Pachyonychus* Crotch, 1873.
On *Smilax* (Smilacaceae) in the USA.

124- *Hypolampsis* Clark, 1860.
Neotropical. On *Buddleja* (Buddlejaceae). (This family is closely related to Loganiaceae and Scrophulariaceae if the selections of several Lepidoptera and Coleoptera are considered). Also accidentally on *Coffea* (Rubiaceae), *Eugenia* (Myrtaceae) and *Areca* flowers (Arecaceae).

125- *Aedmon* Clark, 1860.
West Indies. On various plants such as *Ocotea* (Lauraceae), *Areca* (Arecaceae), *Eugenia* (Myrtaceae), *Piper* (Piperaceae), *Rosa* (Rosaceae), *Palicourea, Coffea* (Rubiaceae) and *Tabebuia* (Bignoniaceae).

126- *Exartematopus* Clark, 1860.
On *Cattleya* (Orchidaceae). It probably feeds on the flowers, leaves and bulbs of this plant. Also found on *Lantana* (Verbenaceae) and *Anthurium* (Araceae).

127- *Allochroma* Clark, 1860.
On *Buddleja* (Buddlejaceae) in Mexico.

128- *Dibolia* Latreille, 1829.
Palaearctic. Also in Africa and America. Divided into three sub-genera. The host-plants are *Teucrium, Stachys, Ballota, Salvia, Mentha, Calamintha, Leonurus, Lamium, Nepeta, Brunella* (Lamiaceae), *Melampyrum, Pentastemon* (Scrophulariaceae), *Cousinia, Echinops, Centaurea* and *Saussurea* (Asteraceae). The larva mines the leaves. Also collected on the following secondary or accidental plants:- *Nasturtium* (Brassicaceae), *Helianthemum* (Cistaceae), *Bupleurum, Eryngium* (Apiaceae), *Agrimonia* (Rosaceae) and *Equisetum* (Equisetaceae). *Eryngium* is also the normal selection of certain species.

129- *Megistops* Boheman, 1859.
Neotropical. On *Tabebuia* (Bignoniaceae), *Clusia* (Clusiaceae), *Scirpus* (Cyperaceae) and *Murraea* (Rubiaceae).

130- *Amphimeloides* Jacoby, 1887.
On *Pipturus* (Urticaceae) in Papua New Guinea.

131- *Acrocrypta* Baly, 1862.
On *Cucumis* (Cucurbitaceae) in Sarawak and on *Alocasia* (Araceae) in Vietnam.

132- *Euphitrea* Baly, 1875.
Indonesia and South-East Asia. On *Ficus* (Moraceae), *Sterculia, Theobroma* (Sterculiaceae), *Eriodendron, Ceiba, Ochroma* (Bombacaceae) and *Boehmeria* (Urticaceae). Mostly on Malvales (Sterculiaceae) and Bombacaceae. In Asia, an insect feeding on *Sterculia* is also adapted to cultivated and introduced *Theobroma.*

133- *Schenklingia* Csiki and Heikertinger, 1940.
Indo-Australian. On a tree fern (petiole) in Micronesia. One species on *Nephrolepis* (Oleandraceae) in Malaysia. On *Lemmaphyllum* and *Colysis* (Polypodiaceae) in Japan (Kato, 1991).

134- *Setsaltica* Samuelson, 1971.
On *Cyathea* (Cyatheaceae) in Papua New Guinea, at an altitude of 3500 m.

135- *Sphaeroderma* Stephens, 1831.
Almost cosmopolitan but absent from South America. Normally feeds on Asteraceae and Ranunculaceae. The larvae mine inside the leaf parenchyma. Nymphosis occurs within the soil inside a brittle cocoon. On *Carduus*, *Silybum*, *Lappa*, *Carduncellus*, *Senecio*, *Wedelia*, *Onopordum*, *Centaurea*, *Carthamnus*, *Cynara*, *Ageratum*, *Bidens*, *Petasites*, *Farfugium*, *Dichrocephala*, *Guizotia*, *Cirsium*, *Serratula*, *Arctium*, *Galactites* (Asteraceae). Also on *Akebia* (Lardizabalaceae, a family related to Ranunculaceae), *Clematis* (Ranunculaceae), *Artocarpus* (Moraceae), *Salix* (Salicaceae), *Vigna* (Fabaceae), *Musa* (Musaceae), *Miscanthus*, *Andropogon*, *Panicum*, *Sasa* (Poaceae), *Coffea*, *Psychotria* (Rubiaceae), *Commelina* (Commelinaceae), *Smilax* (Smilacaceae), *Lilium*, *Scilla* (Liliaceae), *Freycinetia* (Pandanaceae) and various Zingiberaceae.

136- *Bhamoina* Bechyne, 1958.
On *Musa* (Musaceae) in Vietnam.

137- *Chabria* Jacoby, 1887.
On *Melastoma* (Melastomataceae) in Malaysia.

138- *Pentamesa* Harold, 1876.
On *Clematis* (Ranunculaceae) in Central Asia.

139- *Argopus* Fischer, 1824.
Mostly on Ranunculaceae and Asteraceae. The larva is a leaf-miner. On *Clematis*, *Trollius*, *Pulsatilla*, *Ranunculus* (Ranunculaceae), *Quercus* (Fagaceae), *Citrus* (Rutaceae), *Cynara*, *Cirsium* (Asteraceae), *Euphorbia* (Euphorbiaceae) and *Phytolacca* (Phytolaccaceae).

140- *Notomela* Jacoby, 1889.
On *Citrus* and *Zanthoxylum* (Rutaceae) in southern Africa.

141- *Amphimela* Chapuis, 1875.
On *Bersamia* (Velianthaceae) in Kenya, on *Hibiscus* and *Gossypium* (Malvaceae) in the Philippines and on *Citrus* (Rutaceae) in China.

142- *Apteropeda* Stephens, 1839.
Europe and North Africa. On Scrophulariaceae, Plantaginaceae, Lamiaceae and Asteraceae. The larva is a leaf-miner. On *Rhinanthus*, *Veronica*, *Digitalis* (Scrophulariaceae), *Ajuga* (Lamiaceae), *Bellis* (Asteraceae) and *Plantago* (Plantaginaceae). With this data, it is certain that Maulik's (1950) reference concerning *Brassica* (Brassicaceae) as a host, is false.

143- *Mniophila* Stephens, 1831.
Europe. Probably mines *Plantago* (Plantaginaceae) and *Digitalis* (Scrophulariaceae). Often found among grasses.

144- *Mniophilosoma* Wollaston, 1854.
In Madeira, Portugal, found among lichens (?). Host-plant not identified, but most probably *Plantago* (Plantaginaceae).

145- *Kamala* Maulik, 1926.
On Rutaceae (*Clausena*) in Indonesia.

146- *Clavicornaltica* Scherer, 1974.
Endogenous. Polyphagous on roots.

147- *Throscoryssa* Maulik, 1928.
On *Citrus* (Rutaceae) in Asia. The larvae bore into the leaves (Zaka-ur-Rab, 1991).

148- *Argopistes* Motschulsky, 1860.
The cosmopolitan genus *Argopistes* is closely associated everywhere with Oleaceae, except for rare cases of allotrophy on Rutaceae. The larva is a leaf-miner. On *Mayepea, Fraxinus, Osmanthus, Ligustrum, Syringia, Chionanthus, Olea, Forestiera, Linociera, Jasminum* (Oleaceae). Also recorded accidentally on *Citrus* (Rutaceae), *Clerodendrum* (Verbenaceae) and (?) *Casuarina* (Casuarinaceae).

149- *Chilocoristes* Weise, 1895.
Indo-Australian. On *Smilax* (Smilacaceae) in Vietnam and southern China.

150- *Psylliodes* Berthold, 1827.
Cosmopolitan. Mostly on Solanaceae, Fabaceae, Brassicaceae, Chenopodiaceae and Poaceae. In Israel, Furth (1983) provides the following percentages of the total known alticine species on certain host-families:- Brassicaceae (63%), Solanaceae (5%), Poaceae (11%), Chenopodiaceae (5%) and unknown (16%). Mohr (1966) provides the following percentages for middle Europe: Brassicaceae (65%), Solanaceae (13%), Poaceae (9%) and Asteraceae, Scrophulariaceae and Moraceae (4%). These results are comparable to those of Furth. The larvae mine within the stems or the rootlets. Recorded on *Anthemis, Helianthus, Eupatorium, Artemisia, Carduus, Cirsium, Onopordum, Centaurea, Chrysanthemum, Anacyclus, Ormenis, Senecio* (Asteraceae), *Beta, Chenopodium, Atriplex,* (Chenopodiaceae), *Lycopersicon, Solanum, Atropa, Lycium, Hyoscyamus, Nicotiana, Belladonna, Scopolia, Capsicum, Physalis, Datura* (Solanaceae), *Malcolmia, Farsetia, Thlaspi, Sisymbrium, Biscutella, Descurainia, Lepidium, Arabis, Erucastrum, Crambe, Iberis, Cheiranthus,*

Erucaria, Sinapis, Hirschfeldia, Matthiola, Alyssum, Barbarea, Fibigia, Nasturtium, Zilla, Schouwia, Cakile, Erysimum, Alliaria, Lunaria, Isatis, Raphanus, Camelina, Berteroa, Bunias, Brassica, Calepina, Ochthodium, Diplotaxis, Lanaria, Rorippa, Cochlearia, Hesperis, Aubretia, Cardamine, Armoracia, Capsella, Rapistrum (Brassicaceae), *Cannabis, Humulus* (Cannabaceae), *Rumex, Rheum* (Polygonaceae), *Symphoricarpus* (Caprifoliaceae), *Agropyrum, Poa, Triticum, Eremopyrum, Arrhenatherum, Hordeum, Lolium, Scleropoa, Setaria, Barbarea, Erysimum, Psamme, Stipa, Calamagrostis, Ammophila, Bromus, Phragmites* (Poaceae), *Linum* (Linaceae), *Amaranthus* (Amaranthaceae), *Quercus* (Fagaceae), *Pistacia* (Pistaciaceae), *Populus* (Salicaceae), *Urtica* (Urticaceae), *Clerodendrum* (Verbenaceae), *Daucus* (Apiaceae), *Ocimum, Ballota* (Lamiaceae), *Trifolium, Phaseolus, Glycine, Pisum, Vicia, Medicago, Calophaca* (Fabaceae), *Prosopis* (Mimosaceae), *Vitis* (Vitaceae), *Ulmus* (Ulmaceae), *Spergula, Stellaria* (Caryophyllaceae), *Cucurbita, Citrullus* (Cucurbitaceae), *Reseda* (Resedaceae), *Scrophularia* (Scrophulariaceae), *Lysimachia* (Primulaceae), *Lythrum* (Lythraceae) and *Euphorbia, Mercurialis* (Euphorbiaceae).

151- *Nonarthra* Baly, 1862.
Asia, Indo-Australia, Panama. On various plants such as *Mangifera* (Anacardiaceae), *Solanum* (Solanaceae), *Beta* (Chenopodiaceae), *Rosa* (Rosaceae), *Citrus* (Rutaceae), *Erigeron, Saussurea, Carduus* (Asteraceae) and *Argemone* flowers (Papaveraceae). Species are either polyphagous or restricted to Asteraceae.

152- *Omototus* Chevrolat, 1837.
On *Eugenia* (Myrtaceae) and *Inga* (Mimosaceae) in Puerto Rico.

153- *Pachyonychus* Chevrolat, 1837.
Doubtful genus but recorded on *Smilax* (Smilacaceae) in Missouri, USA.

Concluding remarks on Alticinae:

The Alticinae mainly select Brassicaceae but also Malvaceae, Fabaceae, Asteraceae, Solanaceae, Onagraceae, Lythraceae, Polygonaceae, Amaranthaceae, Lamiaceae, Verbenaceae, Boraginaceae, Chenopodiaceae, Euphorbiaceae, Linaceae and several other families. It is very difficult to determine their ancestral food-plant(s) because most of the so-called primitive extant chrysomelid host-plants such as Bryophyta, Equisetaceae, Pteridophyta (at least four or five families, mostly tree-ferns) and Coniferae (two families), appear to be all secondary and represent recent adaptations by highly differentiated species. The selection of Asteraceae by *Sphaeroderma* may be seen as a connection of a most recently evolved plant family with a more specialized genus of beetle, but this is still purely hypothetical.

More than 100 dicotyledon families and 16 monocotyledon families are known as primary (preferred) or secondary host-plants of the Alticinae. However, it must also be pointed out that very often, alticines feed on Poaceae pollen without any other association with those plants. Also, under stress conditions, Alticinae feed on edible leaves of *Quercus* (Fagaceae) and Fabaceae in temperate climates, and on guava, papaw or succulent crops or fruit trees in the tropics. Only the presence of the larvae on or inside the leaves or stems can be a true indication of natural and evolutionary relationships with a plant. Root-feeders, such as Eumolpinae and Alticinae, are generally polyphagous, at least in the larval stages, but larval food cannot yield any information about the adult food choice. Endogenous species such as *Clavicornaltica* are evidently polyphagous. Root-feeding larvae of Eumolpinae will feed on Poaceae roots as well as on the roots of the adult host. It is, however, very interesting to note that, in their great majority, the Alticinae are oligophagous at the genus level (85%), and more often at the species level. Several species are even strictly monophagous, such as *Phyllotreta armoraciae* (Koch) on *Armoracia*. These species must be extremely selective in respect to secondary substances. Very few genera appear to be polyphagous like *Systena* and it is even possible that no single genus is really fundamentally polyphagous. We state this because the trophic relationships of certain genera are poorly known and also because many genera are artificial and actually belong to different monophyletic groups. This explains the problems in interpreting the ancestral food-plants. In addition, many genera are often considered to be polyphagous but their many species are in fact mono- or oligophagous.

The trophic selections of 153 genera out of a total of 520 are known (i.e. about 29%), which is not a poor percentage and provides an adequate sampling for the group. Probably it will be impossible to determine the trophic preferences of the various genera and species feeding in the rain forest canopy, because these insects are killed by fast acting insecticides (pyrethroids) before being collected. For a long time, the trophic selections of these Alticinae will remain a mystery. The rain forests will probably be cut down by the end of the century, most likely before any new investigation technique is developed. Only a few forest reserves will maintain a slight percentage of what is presumed to be the rich canopy fauna. Also, let us not forget that with every 50 to 100 kilometers, the fauna in – which appears to be – uniform forest (such as the Amazon rain forest), changes drastically. A better knowledge of the tropical fauna is definitely needed, especially those of Australia and Madagascar, where many species are already extinct. Concerning host-plant relationships of the Alticinae, Madagascar is still "the land of the unknown", since no data whatsoever have been reported from there, despite the fact that this island possesses one of the richest floras in the world. On the other hand, observations are increasing in Asia and in tropical America and we can reasonably expect that most of the necessary information will be gathered from these regions by the end of the century.

References listed in the bibliography pertaining to Alticinae are as follows:-

Bechyne & Bechyne (1967), Begossi (1984), Bergmann *et al.* (1983), Boppré & Scherer (1981), Burgess (1979, 1981), Chen & Wang (1984, 1987), Chen & Zia (1966), Cordo *et al.* (1984), Doguet (1984), Frost (1973), Furth (1976-1992), Gressitt (1963), Gressitt & Kimoto (1961-1963), Hawkeswood (1988), Hawthorn (1978), Heikertinger (1924-1926), Hering (1935), Inoue (1990), Jacques (1987), Jolivet (1953-1991a), Jolivet & Petitpierre (1981), Jolivet *et al.* (1988), Kalshoven (1951), Kato (1991), Kimoto (1964-1966, 1984), Kostromitin (1973, 1982), Lopatin (1975, 1984), Maes & Tellez Robleto (1988), Martin & Herzog (1982), Maulik (1950), Monros (1959), Nonveiller (1959), Pandit & Pradhan (1991), Parry (1986), Petitpierre (1985), Peyerhimhoff (1911, 1915), Reid (1992), Riley & Enns (1979, 1982), Samuelson (1967-1989), Saunders *et al.* (1983), Scherer (1969-1988), Seeno & Wilcox (1982), Sholes (1987), Shute (1980), Smith (1985), Steinhausen (1982), Virkki (1980, 1984), Virkki & Zambrana (1980, 1983), Wapshere (1982), Zaka-ur-Rab (1991).

SECTION V: CRYPTOSTOMA

2.19. Hispinae

With this subfamily we near the end of our study of the trophic selections of the Chrysomelidae (sensu lato). Hispinae and Cassidinae are very closely related morphologically, are ultra-specialized and are relatively very specialized in their food choices. Within the Cryptostoma, there are no generalists.

There are some 3,000 described hispine species and probably about the same number of cassidines. More species of Hispinae than Cassidinae remain to be described, because the small body size of the first group (e.g. the spiny species) has left many taxa unknown, especially in South-East Asia.

The larvae of the two subfamilies have many characters in common and certain species are even intermediate morphologically between the two groups.

The Hispinae are somewhat important in agriculture, not only because they may be pests of grasses (Poaceae), e.g. rice, corn, sugarcane and bamboo, but also because they feed on ornamental and useful crops such as palms (Arecaceae). The attack upon oil palm *(Elaeis)* by *Coelaenomenodera* is rather serious. Often hispine larvae are leaf-miners, but they can also live between appressed leaves of palm trees and other monocotyledons such as the Zingiberales. Several species change from a mining habit to a free-living one during their life-cycle. The Hispinae are relatively well protected by their mining or hidden habits during the larval stages and by spines in certain species during the adult stage. It is also probable that the mimicry of certain Asian or American species with several other Coleoptera, affords some protection for both partners. Mimetic complexes occur rather frequently in tropical areas.

The biology of certain larvae was studied by Maulik (1919) and since then, many workers have studied, and even raised in the laboratory, leaf-miners of economic importance. A rather complex technique has been developed in the Côte d'Ivoire which has made it easier to rear the mining larvae of the Hispinae than those of the Eumolpinae root-feeders.

Differences between hispine and cassidine larvae have been outlined by several authors, i.e. Boving and Craighead (1931), Chu (1949) and Peterson (1960). The eighth pair of abdominal spiracles are well-developed and are positioned dorsally in hispines but are vestigial in cassidines. The absence of furca on the tergum of the eighth abdominal segment is also peculiar to the Hispinae. Hispine larvae measure mostly between 5 and 10 mm in length, but there are some that are larger. The body is depressed, orthosomatic, with lateral projections on the abdominal segments. The antennae are trisegmented. Four, five or six ocellae are visible, but sometimes they are in total regression and are replaced by a black spot. The larvae are lightly coloured, white, yellow or green, except for the head, the prothorax and the eighth abdominal segment, which are strongly pigmented. Generally, the legs are trisegmented and terminate in a small claw-like tarsungulus and a bifurcate pulvillus. Several larvae (e.g. *Octotoma*) lack segmented legs. The hispine abdomen possesses eight visible segments. The number of larval instars is variable with up to six in *Brontispa*. In the tropics, the number of generations may reach six per year in certain genera such as *Dicladispa*.

Like some Cassidinae, the Hispinae possess stridulatory organs which sometimes exist only in the male (*Spilispa*) or in both sexes. These stridulatory organs are very well developed on the top of the head and articulate with a small triangular flap arising from the anterior margin of the pronotum. From the structure of their stridulating apparatus, it is inferred that these beetles do produce sounds of at least two (or three) degrees of pitch. These sounds appear to be used for communication, warning, etc. Unfortunately this aspect has not been studied well. Certain genera, i.e. *Arescus* and *Alurnus*, appear to have lost the cephalic stridulating organ, but most possess this structure.

Hispine larvae can be divided into four main groups: 1- those which live between the folded leaves of monocotyledons (i.e. Poaceae), between folded folioles of new fronds, at the base of petioles of palms (Arecaceae) and at the bases of crowns; represented by the tribes Cephaloleiini, Botryonopini, Callispini, Leptispini and others; 2- those which burrow into the stems of herbaceous or semi-ligneous plants (tribe Anisoderini); 3- those which mine leaves (tribes Coelaenomenoderini, Hispini, etc.), and 4- free-living larvae which are very few. Generally speaking, many species are leaf-miners, a large number live between young leaves and only a minority are stem-borers. As an exception, *Oediopalpa* (Oediopalpini) displays a behaviour characteristic of Cassidinae in that the larva of this American genus constructs a shield of excreta and exuviae at the rear of its body which is used as a means of defense. In India, *Leptispa pygmaea* Baly has a larva which lives on leaves like the adults. *Brontispa froggatti* Sharp, from Vanuatu (New Hebrides), has a larva which lives with the adult amongst the opened buds of coconut trees (*Cocos nucifera*, Arecaceae)(Maulik, 1919).

Adult Hispinae feed on leaves and consume the mesophyll parenchyma leaving longitudinal white streaks. Many of these adults are nocturnal, while others (e.g. *Dicladispa*) are diurnal. One larva of *Dicladispa armigera* (Olivier) consumes an average of 132.4 mm^2 of rice leaf during its life-history (Budhraja *et al.*, 1979). The adults mostly feed on apical parts of the leaves and then move downwards, producing whitish streaks, after consuming the chlorophyllous parenchyma. One adult eats about 25.4 mm^2 of leaf per day.

Generally, the tribes are rather uniform in larval biology but in certain cases, e.g. among the Gonophorini, the larvae are, according to the species, either miners or free-living larvae between the leaves. Certain tribes also display high specificity to particular host-plant families. For instance, in Asia, Kalshoven (1957) recorded six tribes which utilize only one host family each, i.e. Botryonopini on Arecaceae, Leptispini on Poaceae, Eurispini on Cyperaceae, Coelaenomenoderini on Arecaceae, Oncocephalini on Commelinaceae and Pharangispini on Zingiberaceae. Actually, the selections are more subtle, and now that the Pharangispini has been synonymized with the Coelaenomenoderini, we cannot now speak of such specialization, at least for this group. Among other tribes, the Anisoderini live in Asia on four plant families, the Callispini on five families, the Cryptonychini on three families, the Gonophorini on seven families and the Hispini on nine families, etc. In America, the selected families are related but are more numerous. It is evident that for certain genera such as *Dactylispa*, the host associations are mostly Poaceae, but several species feed on Rubiaceae, Sterculiaceae and Bombacaceae, all three being Dicotyledonae. These selections sometimes closely correlate the plant taxonomy. Uhmann (1958) was able to separate taxonomically and morphologically, species of the genus *Callispa*, by considering their diet which consists of bamboos, palm trees and Orchidaceae.

In certain cases, there are some precise relationships between Hispinae and their host-plant *Cecropia* (Cecropiaceae) in tropical America (Andrade, 1981). For instance, one *Metaxycera* and two *Sceloenopla* live exclusively on various *Cecropia*. All these species are leaf-miners. In Brazil, the adults of *Sceloenopla maculata* (Olivier) chew the underside of the extremity of the leaf lobes of *Cecropia lyratiloba*. The females lay eight eggs which are assembled into a small cylindrical bevel-edged ootheca, measuring 2.5 mm long by 1.5 mm wide, inserted into the mesophyll on the abaxial (ventral) surface of a young leaf. The larvae, which number from one to eight, hatch after 14 days incubation. They are miners but do not cut the median veins of the lobes and after several moults, they reach about 10 mm long. The legs are vestigial and the anterior end is strongly sclerotized. Larval development requires 45 days and the pupal stage, 2 days. Population size is regulated by a chalcid wasp which lays eggs into the oothecae, but the hispines are not chased away by the *Azteca* ants which appear to be indifferent towards them.

The excellent researches of Strong (1977-82) on the Arescini and Cephaloleini from Neotropical Zingiberales, deal with smooth species living between appressed, immature leaves of their host-plants. These species are mono- or oligophagous since we cannot really speak of polyphagy between related families like the Zingiberales. Sometimes, no less than five hispine species can attack different parts of a *Heliconia* plant, but certain species such as *Cephaloleia consanguinea* Baly are strictly monophagous on *Heliconia imbricata.*

In certain cases, several species can congregate in the same leaf-mine (e.g. *Arescus* and *Chelobasis*) in *Heliconia* and *Calathea* (Maulik, 1937). The extraordinary variability in colouration of *Arescus*, *Chelobasis* and other Arescini species, can be linked to a genetic factor, similar to that known among Cassidinae (Vasconcellos-Neto, 1987) but probably has no taxonomic value.

No genus or tribe is common to both the New and Old World Hispinae. A similar situation exists for the palm trees, with the exception of three genera including the coconut palm, *Cocos.* At least thirteen species of Hispinae are known to utilize the foliage of the coconut tree in the Old World. Different species of Alurnini, Hispoleptini, etc., attack the same plant in the New World. Therefore, it is really impossible to prove through a study of Hispinae and their hosts, whether *Cocos nucifera* is American or otherwise. Botanists are in favour of an Asiatic or Polynesian origin of this tree. The adaptation of American Hispinae to the African palm tree, *Elaeis guineensis* (Arecaceae), is progressive in the New World where many local species have adapted to this introduced palm from indigenous species.

The subfamily Hispinae is essentially tropical in origin and has parsimoniously populated the temperate areas. There are no Hispinae in New Zealand and, in the tropical areas, hispines never venture beyond 2000 m altitude, even though their host-plants may occur above this level. Therefore, the group appears to be less cold-tolerant than their host-plants.

Among the Hispinae, the monocots are the preferred host-plants in Asia as well as America, i.e. Zingiberaceae, Arecaceae, Poaceae, Cyperaceae, Musaceae, Orchidaceae, Pandanaceae, etc. The association of a limited number of Hispinae with Dicotyledonae is considered by Kalshoven (1957) as being a relatively recent shift and it is probable that this phenomenon, similar to that which has occurred with Criocerinae, (a switch from mono- to dicotyledons), occurred during their evolution in the Tertiary period (Eocene epoch). This host-plant switch can sometimes be observed in pest beetle species. For instance, the rice pest *Dicladispa armigera* (Olivier), also a pest of *Saccharum* and various other Poaceae, has been observed several times mining the leaves of the teak tree, *Tectona grandis* (Verbenaceae)(Kalshoven, 1957). The larvae probably dropped down from adjacent bamboo plants onto teak leaves. This phenomenon was called xenophagy by Hering (1951).

Many hispine larvae frequently leave their mines and penetrate into a new leaf to start a new mine. This is one of the reasons why it is difficult to rear hispines in the laboratory. At the IRHO at Adiopodoume, Côte d'Ivoire, entomologists introduce the larvae into an incision within a palm leaf. Often the larvae at the end of their development will quit their mines and pupate under other incisions in the leaves. Sometimes, pupation occurs inside the mine itself.

The hypothesis of Kalshoven (1957) on the evolution of Hispinae is interesting. The smooth species of Hispinae are adapted to the terminal leaves of monocotyledons and live inside the tender and nutritive tissue in the "heart" of those plants. No dicotyledon offers such a possibility. Boring into the stems or mining the foliar tissues can be considered as an extension of the habit of living inside appressed leaves. The adults of those miners often have spines on their bodies, which provide them with extra outside protection. The spiny forms appear to be more recent than the smooth forms and they are the ones which mostly attack the dicotyledons. The association with Dicotyledonae appears to be a recent development. The miners of those groups also belong to the most specialized tribes and probably their evolution is still proceeding with the adaptation to Asteraceae.

Finally, some mention must be made concerning the use of Hispinae in the biological control of introduced weeds. There are a few cases recorded in the literature, e.g. *Octotoma scabripennis* Guérin-Méneville and *Uroplata girardi* Pic, which have been recently introduced into South Africa, Ghana and Australia to combat the noxious, introduced weed, *Lantana camara* (Verbenaceae). Other species of *Uroplata* and *Octotoma* naturally attack *Lantana* in tropical America and probably will be eventually imported as well as a control agent.

American Hispinae:

2.19.1. Tribe Oediopalpini

1- *Oediopalpa* Baly, 1858.

Neotropical. Archaic genus. On *Oryza, Panicum, Saccharum, Zea, Poa, Arundo* and *Bambusa* (Poaceae). Recorded also (with little probability) on *Petroselinum* (Apiaceae) in Venezuela. The larvae resemble those of Cassidinae in that they place excreta and exuviae at the extremity of their bodies.

2.19.2. Tribe Cephaloleiini

2- *Homalispa* Baly, 1858.

Neotropical. On *Lasiacis, Olyra* (Poaceae) and *Scleria* (Cyperaceae).

3- *Cladispa* Baly, 1858.

On *Cyrtopodium* (Orchidaceae) in Guyana. The larvae feed on the leaves.

4- *Cephaloleia* Chevrolat, 1837.

On leaves of almost all Central American Zingiberales, except (generally) Cannaceae and Musaceae (Strong, 1977). On *Alpinia, Renealmia* (Zingiberaceae), *Heliconia* (Heliconiaceae), *Calathea, Maranta, Ischnosiphon* (Marantaceae), *Cyclanthus* (Cyclanthaceae), *Costus* (Costaceae) and *Cocos, Diplothemium,*

Elaeis, Geonoma (Arecaceae). The larvae live inside the unrolled leaves and feed on the epidermis. Several species have also been recorded in South America on *Cordyline* (Agavaceae), *Pharus*, *Saccharum*, *Panicum* (Poaceae), *Canna* (Cannaceae), *Musa* (Musaceae), *Ananas* (Bromeliaceae) and Cyperaceae. The adults sometimes pierce the banana fruits.

5- *Stenispa* Baly, 1858.
America. On *Scirpus, Carex, Cyperus* (Cyperaceae), *Spartina, Paspalum* (Poaceae) and Juncaceae. The larvae and adults chew the leaves.

2.19.3. Tribe Arescini

6- *Xenarescus* Weise, 1905.
Neotropical. On *Heliconia* (Heliconiaceae), *Stromanthe* (Marantaceae) and Musaceae.

7- *Arescus* Perty, 1832.
On *Heliconia* (Heliconiaceae) in Brazil.

8- *Chelobasis* Gray, 1832.
Neotropical. On *Heliconia* (Heliconiaceae) and *Ischnosiphon, Calathea* (Marantaceae). The larvae nibble the surface of rolled leaves. Their development is very slow, at least three times the duration of other Chrysomelidae (240 days). This appears to be an adaptation to low-nutrient food (Strong & Wang, 1977).

9- *Nympharescus* Weise, 1905.
Neotropical. On *Heliconia* (Heliconiaceae).

2.19.4. Tribe Alurnini

This tribe is solely Neotropical in distribution. The larvae are free-living, flattened and spineless and live between the appressed leaves of monocotyledons.

10- *Alurnus* Fabricius, 1775.
On *Heliconia* (Heliconiaceae) and *Cocos, Elaeis* (Arecaceae).

11- *Mecistomela* Jacobson, 1899.
Like the following genus on *Cocos*, *Diplothemium*, *Livistona*, *Copernicia, Elaeis, Latania* and *Phoenix* (Arecaceae). The eggs are enormous, measuring 8 mm long by 3.5 mm wide) and develop very slowly. The larvae feed inside unrolled folioles. Pupation occurs within the foliar peduncles. The complete life-cycle lasts 6 to 8 months.

12- *Coraliomela* Jacobson, 1899.
Biology similar to *Mecistomela*, on *Cocos*, *Diplothemium*, *Phoenix, Orbignya, Copernicia, Elaeis* (Arecaceae).

13- *Platyauchenia* Sturm, 1843.
On *Cocos* and *Diplothemium* (Arecaceae). Pupation occurs between the dry leaves of these trees.

2.19.5. Tribe Prosopodontini

14- *Prosopodonta* Baly, 1885.
On *Heliconia* (Heliconiaceae) and Arecaceae in tropical America.

2.19.6. Tribe Sceloenoplini

15- *Sceloenopla* Chevrolat, 1837.
Neotropical. Recorded on *Persea* (Lauraceae), *Chrysophyllum* (Sapotaceae), Rubiaceae, epiphytic Araceae (including *Anthurium* and *Philodendrum*), Cyclanthaceae, *Cocos* (Arecaceae), *Lonchocarpus* (Fabaceae), *Rapanea* (Myrsinaceae) and *Esembeckia* (Rutaceae). Certain species such as *Sceloenopla maculata* (Olivier) and *S. sheppardi* (Baly), are specific on *Cecropia* (Cecropiaceae), while others are specific on *Pourouma* (Cecropiaceae). Another hispine, *Metaxycera subapicalis* Bondar (Chalepini), also occurs on *Cecropia*.

16- *Acentroptera* Baly, 1858.
Neotropical. On *Ananas* (Bromeliaceae).

2.19.7. Tribe Hispoleptini

17- *Hispoleptis* Baly, 1864.
On *Elaeis, Cocos* (Arecaceae) in tropical America.

2.19.8. Tribe Chalepini

18- *Craspedonispa* Weise, 1910.
On *Saccharum* (Poaceae) in Trinidad.

19- *Metaxycera* Baly, 1864.
The larva and adult chew the leaves of *Cecropia* (Cecropiaceae) in Brazil.

20- *Stethispa* Baly, 1864.
Neotropical. Leaf-miner on *Ruprechtia*, *Coccoloba* (Polygonaceae) and *Aristolochia* (Aristolochiaceae).

21- *Charistena* Baly, 1864.
Neotropical. On *Panicum*, *Paspalum* and *Zea* (Poaceae).

22- *Sternostena* Weise, 1910.
On *Paspalum* (Poaceae) in Argentina and Peru.

23- *Anisostena* Weise, 1910.
On *Panicum, Paspalum, Valota, Olyra, Tripsacum* and *Schizachyrium* (Poaceae).

24- *Sumitrosis* Butte, 1969.
North and South America. Polyphagous at least at the genus level. On *Potentilla* (Rosaceae), *Amorpha, Robinia, Desmodium, Strophostyles, Glycine, Amphicarpaea, Phaseolus, Lespedeza, Pueraria, Canavalia, Cajanus, Dolichos, Meibomia* (Fabaceae), *Cassia* (Caesalpiniaceae), *Celastrus* (Celastraceae), *Eupatorium, Aster, Helianthus, Rudbeckia, Solidago, Vernonia, Wedelia* (Asteraceae), *Cyrilla* (Cyrillaceae), *Laportea* (Urticaceae), *Chenopodium* (Chenopodiaceae), *Chusquea, Lasiacis, Bambusa* (Poaceae) and *Quercus* (Fagaceae).

25- *Odontota* Chevrolat, 1837.
Throughout America. The larva is a leaf-miner. Preferred hosts are the Fabaceae and Mimosaceae. Recorded on *Rhododendron* (Ericaceae), *Sophora, Tephrosia, Amphicarpaea, Meibomia, Apios, Robinia, Glycine, Amorpha, Desmodium, Pueraria* (Fabaceae), *Acacia* (Mimosaceae), *Quercus, Fagus* (Fagaceae), *Betula, Alnus* (Betulaceae), *Ulmus* (Ulmaceae), *Malus*, *Crataegus*, *Prunus*, *Pyrus, Amelanchier* (Rosaceae), *Polygonum* (Polygonaceae), *Aster, Eupatorium, Solidago* (Asteraceae), *Tilia* (Tiliaceae), *Carpinus* (Carpinaceae) and *Cornus* (Cornaceae).

26- *Chalepus* Thunberg, 1805.
Throughout America. Polyphagous. The larva is also a leaf-miner, mostly on Fabaceae. Recorded on *Theobroma* (Sterculiaceae), *Valota, Paspalum, Chusquea, Lasiacis, Elymus, Hystrix, Bambusa, Brachiaria, Panicum, Zea, Olyra* (Poaceae), *Vicia, Desmodium, Benthamantha, Crotalaria, Apios, Falcata, Mucuna, Pueraria*, *Dioclea, Phaseolus, Glycine*, *Robinia, Dolichos, Canavalia*, *Calopogonium, Cymbosema, Lathyrus, Meibomia* (Fabaceae), *Cassia, Bauhinia* (Caesalpiniaceae), *Prosopis* (Mimosaceae), *Aloysia, Vitex* (Verbenaceae), *Paullinia* (Sapindaceae), *Philodendrum* (Araceae), *Verbesina*, *Vernonia*, *Eupatorium*, *Cyanus*, *Wedelia* (Asteraceae), *Cordia* (Ehretiaceae), *Quercus* (Fagaceae), *Malus, Pyrus, Cerasus, Crataegus* (Rosaceae), *Ipomoea* (Convolvulaceae), *Commelina* (Commelinaceae), *Acer* (Aceraceae), *Betula* (Betulaceae), *Terminalia* (Combretaceae), *Bromelia* (Bromeliaceae), *Aristolochia* (Aristolochiaceae), *Arrabidea* (Bignoniaceae), *Actinostomon*

(Euphorbiaceae), *Celastrus* (Celastraceae) and Sapindaceae. The fact that several monocotyledons have been recorded, i.e. Bromeliaceae, Poaceae, Commelinaceae, etc., has no special significance for this polyphagous genus.

27- *Temnochalepus* Uhmann, 1935.
Mostly Neotropical. On *Panicum* and *Pharus* (Poaceae) and *Commelina* (Commelinaceae) in Argentina.

28- *Xenochalepus* Weise, 1910.
Throughout America. Appears to be polyphagous but with a clear-cut preference for Fabaceae. On *Theobroma* (Sterculiaceae), *Bambusa, Panicum, Zea, Saccharum, Oryza* (Poaceae), *Lathyrus, Canavalia, Dioclea, Cymbosema, Phaseolus, Glycine, Dolichos, Vigna, Nissolia, Mucuna, Robinia, Wisteria, Faba* (Fabaceae), *Inga* (Mimosaceae), *Bauhinia* (Caesalpiniaceae), *Urera* (Urticaceae), *Schubertia* (Malvaceae), *Prunus* (Rosaceae), *Celtis* (Ulmaceae) and *Ipomoea* (Convolvulaceae).

29- *Oxychalepus* Uhmann, 1936.
Neotropical. Polyphagous, but with a clear-cut tendency towards Fabaceae. The observations on monocotyledons are reliable. On *Mucuna, Canavalia, Dioclea, Cymbosema, Phaseolus* (Fabaceae), *Inga* (Mimosaceae), *Cassia* (Caesalpiniaceae), *Solanum* (Solanaceae), *Heliconia* (Heliconiaceae), *Flagellaria* (Flagellariaceae), *Cocos* (Arecaceae), *Pandanus, Freycinetia* (Pandanaceae) and *Pleomele* (Agavaceae).

30- *Baliosus* Weise, 1905.
America. Polyphagous. Larvae and adults on Bignoniaceae in Brazil. Also on *Ceanothus* (Rhamnaceae), *Hippocratea* (Celastraceae), *Arrabidea* (Bignoniaceae), *Jatropha* (Euphorbiaceae), *Acer* (Aceraceae), *Ulmus* (Ulmaceae), *Tilia* (Tiliaceae), *Robinia, Desmodium, Dioclea, Meibomia* (Fabaceae), *Bauhinia* (Caesalpiniaceae), *Vernonia* (Asteraceae), *Cordia* (Ehretiaceae), *Urtica* (Urticaceae), *Olyra* (Poaceae), *Lippia* (Verbenaceae), *Quercus, Castanea* (Fagaceae), *Salix* (Salicaceae), *Alnus, Betula* (Betulaceae), *Carpinus* (Carpinaceae), *Corylus* (Corylaceae), *Amelanchier, Malus, Prunus, Pyrus, Rubus* (Rosaceae), *Banisteria* (Malpighiaceae), *Citrus* (Rutaceae), *Guazuma* (Sterculiaceae), *Guaiacum* (Zygophyllaceae) and *Basanacantha* (Rubiaceae).

2.19.9. Tribe Uroplatini

31- *Corynispa* Uhmann, 1940.
On *Stigmatophyllum* (Malpighiaceae) in Brazil.

32- *Physocoryna* Guérin-Méneville, 1844.
Neotropical. The larva mines the leaves of *Mucuna, Canavalia, Cymbosema, Dioclea, Phaseolus* (Fabaceae) and *Stigmatophyllum* (Malpighiaceae).

33- *Octotoma* Dejean, 1837.
New World. Appears to prefer Verbenaceae, Lamiaceae, Pedaliaceae and Bignoniaceae, families which are closely related. It is difficult to determine the real host-plants from those which may be accidental or secondary hosts. On *Lantana, Verbena, Clerodendrum, Tectona* (Verbenaceae), *Monarda, Origanum, Mentha, Salvia* (Lamiaceae), *Campsis, Tecoma* (Bignoniaceae), *Lespedeza, Canavalia, Dioclea, Cymbosema* (Fabaceae), *Fraxinus* (Oleaceae), *Sesamum* (Pedaliaceae), *Eupatorium* (Asteraceae), *Quercus* (Fagaceae), *Xanthorrhoea* (Xanthorrhoeaceae) and *Stigmatophyllum* (Malpighiaceae).

34- *Stenopodius* Horn, 1883.
North and Central America. On *Sphaeralcea, Malva* and *Althaea* (Malvaceae).

35- *Brachycoryna* Guérin-Méneville, 1844.
America. On *Abutilon, Sphaeralcea, Sida* (Malvaceae), *Waltheria* (Sterculiaceae) and *Artemisia* (Asteraceae).

36- *Probaenia* Weise, 1904.
Neotropical. On *Verbena* (Verbenaceae), *Inga* (Mimosaceae) and *Verbesina, Rolandra, Vernonia* (Asteraceae).

37- *Uroplata* Chevrolat, 1837.
Neotropical. On *Annona, Rollinia* (Annonaceae), *Vernonia, Baccharis, Wedelia, Rolandra, Eupatorium, Elephantopus, Clibadium, Melanthera* (Asteraceae), *Banisteria, Malpighia* (Malpighiaceae), *Lantana, Verbena, Lippia* (Verbenaceae), *Caesarea* (Vivianaceae), *Arrabidaea, Colea, Pithecoctenium, Bignonia* (Bignoniaceae), *Calopogonium* (Fabaceae), *Aristolochia* (Aristolochiaceae), *Sida, Althaea* (Malvaceae), *Ocotea* (Lauraceae), various Acanthaceae and *Panicum* (Poaceae). The eggs are laid on the leaves and the larvae mine narrow galleries in them. The total development lasts six months.

38- *Glyphuroplata* Uhmann, 1937.
North and Central America. On *Eriochloa, Valota, Panicum* and *Digitaria* (Poaceae).

39- *Heterispa* Chapuis, 1875.
Neotropical. The preferences are towards Malvales but the observations on *Stenotaphrum* and *Panicum* (Poaceae) cannot be rejected. On *Sida, Sphaeralcea, Malvastrum, Althaea* (Malvaceae), *Triumfetta* (Tiliaceae) and *Guazuma* (Sterculiaceae).

40- *Acanthodes* Baly, 1864.
On *Quetzalia* (Celastraceae) in Paraguay.

41- *Cnestispa* Maulik, 1930.
Neotropical. On *Cymbosema*, *Centrosema* and *Desmodium* (Fabaceae).

42- *Octhispa* Chapuis, 1877.
Neotropical. On *Dioclea*, *Machaerium* (Fabaceae), *Inga* (Mimosaceae), *Cassia* (Caesalpiniaceae), *Basanacantha* (Rubiaceae), *Paullinia*, *Serjania*, *Adenocalymna* (Sapindaceae), *Pithecoctenium* (Bignoniaceae), *Cordia* (Ehretiaceae), *Coccoloba* (Polygonaceae) and *Byrsonima*, *Malpighia* (Malpighiaceae).

43- *Platocthispa* Uhmann, 1939.
Neotropical. On *Cassia* (Caesalpiniaceae).

44- *Heptispa* Weise, 1906.
Tropical America. On *Cassia* (Caesalpiniaceae), *Inga* (Mimosaceae), *Machaerium* (Fabaceae) and *Serjania* (Sapindaceae).

45- *Euprionota* Guérin-Méneville, 1844.
Neotropical. On *Vernonia* (Asteraceae) in Costa Rica.

46- *Pentispa* Chapuis, 1875.
New World. On *Colea, Pithecoctenium* (Bignoniaceae), *Paullinia*, *Serjania* (Sapindaceae), *Vernonia, Verbesina, Eupatorium, Elephantopus, Clibadium, Baccharis* (Asteraceae), *Malpighia* (Malpighiaceae) and *Chusquea* (Poaceae).

47- *Carinispa* Uhmann, 1930.
Central America. The larva is a leaf-miner on *Bunchosia* and *Malpighia* (Malpighiaceae) in Costa Rica.

48- *Nonispa* Maulik, 1933.
Neotropical. On *Panicum* and *Paspalum* (Poaceae) in Argentina.

49- *Microrhopala* Chevrolat, 1837.
America. The larva is a leaf-miner on Asteraceae and hibernates in the ground. On *Solidago, Aster, Chrysopsis*, *Brickellia, Encelia, Ambrosia, Helianthus, Franseria, Silphium, Boltonia, Sericocarpus* (Asteraceae). Also on *Salvia* (Lamiaceae).

African, Eurasian and Australian Hispinae:

2.19.10. Tribe Botryonopini

50- *Botryonopa* Blanchard, 1845.
On *Metroxylon* and *Cocos* (Arecaceae) in Java and Sumatra.

2.19.11. Tribe Anisoderini

51- *Estigmena* Hope, 1840.
On *Dendrocalamus, Cephalostachyum* and *Bambusa* (Poaceae) in India and Java. The larva feeds inside an internode, at a density of one larva per internode.

52- *Lasiochila* Weise, 1916.
South-East Asia. Within the internodes of bamboos, *Gigantochloa*, and *Miscanthus* (Poaceae).

53- *Anisodera* Chevrolat, 1837.
On *Imperata, Setaria* (Poaceae), *Zingiber, Amomum* (Zingiberaceae), *Musa* (Musaceae) and *Costus* (Costaceae), etc. Has also been captured on Zamiaceae in Vietnam.

2.19.12. Tribe Aproidini

54- *Aproida* Pascoe, 1863.
Australia. Host-plants are creepers from the rain forest and vine thickets, i.e. *Eustrephus latifolius* (Philesiaceae) and *Drymophila* (Convalariaceae) (Hawkeswood, 1987). Both hosts are monocotyledons. The female lays one egg at a time in an ootheca on a leaf of the host-plant, and as they grow older, the larvae resemble cherry slugs or other sawfly larvae (Hawkeswood, 1987). Pupation occurs on the plant and the pupa is suspended by a thin thread. These peculiar hispines are mostly metallic green and maroon in colour, i.e. *Aproida balyi* Pascoe, and resemble grasshoppers (locusts) or certain rain forest bugs (Hawkeswood, 1987).

2.19.13. Tribe Callispini

55- *Hispodonta* Baly, 1858.
Indo-Australia. On *Musa* (Musaceae), *Alpinia, Zingiber* (Zingiberaceae) and *Pinanga, Heterospathe, Metroxylon* (Arecaceae).

56- *Callispa* Baly, 1858.
Palaeotropical. On *Cocos, Metroxylon, Zalacca, Pinanga, Calamus* (Areca-

ceae), *Musa* (Musaceae), *Bambusa, Dendrocalamus, Phyllostachys, Arundinaria, Linguania, Sinobambusa, Sinocalamus, Thysanolaena, Zea, Oryza, Rottboellia* (Poaceae), *Spathoglottis* (Orchidaceae) and Zingiberaceae.

57- *Amblispa* Baly, 1858.
On Poaceae in India.

2.19.14. Tribe Leptispini

58- *Leptispa* Baly, 1858.
Palaeotropical. On terminal leaves of Poaceae. On *Bambusa, Oryza, Saccharum, Dactylis, Rottboellia, Sinocalamus, Miscanthus, Arundinella, Ischaemum, Paspalum, Vetiveria, Pennisetum, Agrostis* (Poaceae) and *Scirpus* (Cyperaceae).

2.19.15. Tribe Eurispini

59- *Eurispa* Baly, 1858.
On *Gahnia* (Cyperaceae) in Australia.

2.19.16. Tribe Cryptonychini

60- *Gestronella* Weise, 1911.
Madagascar and Mascarene Islands. On *Cocos, Acanthophoenix* and *Dictyosperma* (Arecaceae).

61- *Xiphispa* Chapuis, 1878.
Madagascar. On *Cocos* and *Areca* (Arecaceae).

62- *Cryptonychus* Gyllenhal, 1817.
Africa. On *Cocos, Elaeis* (Arecaceae), *Urelytrum, Loudetia, Oryza* (Poaceae), *Cyperus, Carex* (Cyperaceae), *Costus* (Costaceae), *Amomum, Afromomum* (Zingiberaceae) and *Thalia* (Marantaceae). It has also been recorded on the fern *Ophioglossum* (Ophioglossaceae) and a dicotyledon *Lonchocarpus* (Fabaceae). In the case of Zingiberaceae, the insects chew the terminal buds and the leaf parenchyma.

63- *Gyllenhaleus* Weise, 1903.
Africa. On *Costus* (Costaceae). The beetles chew the terminal buds. Also on *Elaeis* and *Cocos* (Arecaceae).

64- *Plesispa* Chapuis, 1875.
Indo-Australian. *On Areca, Cocos, Heterospathe, Korthalsia, Calamus, Nipa, Metroxylon, Oncosperma, Rhopaloblaste, Archontophoenix, Phoenix, Cyrto-*

stachys, Oreodoxa, Arenga, Daemonorops (Arecaceae), *Saccharum* (Poaceae) and *Flagellaria* (Flagellariaceae).

65- *Ceratispa* Gestro, 1895.
New Guinea and Melanesia. On *Korthalsia, Calamus, Areca, Daemonorops, Heterospathe* and *Pinanga* (Arecaceae).

66- *Palmispa* Gressitt, 1960.
On *Korthalsia* and *Calamus* (Arecaceae) in New Guinea.

67- *Oxycephala* Guérin-Méneville, 1838.
New Guinea. On *Cocos, Calamus, Pinanga* (Arecaceae), *Heliconia* (Heliconiaceae), *Pandanus, Freycinetia* (Pandanaceae) and *Flagellaria* (Flagellariaceae).

68- *Aulostyrax* Maulik, 1929.
Solomon Islands. *On Heterospathe, Bentnickiopsis, Cocos, Pinanga* and *Metroxylon* (Arecaceae).

69- *Brontispa* Sharp, 1903.
Indo-Australian. Larvae feed on the epidermis of the buds but are not leaf-miners. On *Cocos, Archontophoenix, Balaka, Metroxylon, Bentnickiopsis, Latania, Phoenix, Washingtonia, Clinostigma, Calamus, Kentia, Exorrhiza, Arenga, Normanbya, Ptychosperma, Areca, Elaeis, Dictyosperma, Oreodoxa, Acanthophoenix, Mascarena, Chrysalidocarpus, Rhapis, Caryota, Thrinax, Veitchia* (Arecaceae), *Saccharum, Eulalia, Zea* (Poaceae), *Juncus* (Juncaceae) and Cyperaceae.

70- *Octodonta* Chapuis, 1875.
Taiwan and Indo-Australian region. On *Metroxylon, Cocos, Areca, Nipa, Oncosperma, Livistona, Calamus* and *Korthalsia* (Arecaceae). It is known that *Korthalsia* and sometimes other genera such as *Calamus* and *Daemonorops,* are inhabited by ants such as *Camponotus* and *Crematogaster* and these probably provide protection for these hispines.

71- *Nesohispa* Maulik, 1913.
Seychelles. On *Stevensonia* and *Phoenicophorium* (Arecaceae).

72- *Drescheria* Weise, 1922.
On *Bambusa* (Poaceae) in Java.

73- *Caledonispa* Uhmann, 1952.
New Caledonia. Mountainous areas. On *Freycinetia, Pandanus* (Pandanaceae) and *Ptychosperma* (Arecaceae).

74- *Stephanispa* Gressitt, 1960.
New Caledonia. On *Freycinetia* (Pandanaceae).

75- *Callistola* Dejean, 1837.
New Guinea. On *Freycinetia, Pandanus* (Pandanaceae), *Pleomele* (Agavaceae), *Epipremnum, Rhaphidophora* (Araceae), *Flagellaria* (Flagellariaceae) and *Cocos* (Arecaceae).

76- *Isopedispa* Spaeth, 1936.
On *Cocos* (Arecaceae) in New Caledonia.

77- *Teretispa* Gressitt, 1960.
On *Gahnia, Scleria* (Cyperaceae), various Poaceae and native Orchidaceae in New Caledonia.

78- *Calamispa* Gressitt, 1957.
In the Solomon Islands on *Calamus* (Arecaceae). On *Ehrharta* (Poaceae) at the Cape (Uhmann, 1958).

2.19.17. Tribe Coelaenomenoderini

79- *Javeta* Baly, 1858.
On *Areca, Pinanga* and *Metroxylon* (Arecaceae) in Java and Sumatra.

80- *Coelaenomenodera* Blanchard, 1845.
Adults chew the abaxial surface of the leaves producing linear furrows. The larvae are leaf-miners. On *Amomum* (Zingiberaceae), *Cocos, Elaeis, Roystonea* and *Borassus* (Arecaceae) in Africa.

81- *Balyana* Peringuey, 1898.
On *Cocos, Medemia* and *Raphia* (Arecaceae) in Madagascar.

82- *Cyperispa* Gressitt, 1957.
Solomon Islands. On *Hypolytrum*, *Scleria* and *Thoracastachyum* (Cyperaceae).

83- *Pharangispa* Maulik, 1929.
On *Alpinia, Zingiber* (Zingiberaceae) and *Costus* (Costaceae) in the Solomon Islands.

84- *Rhabdotohispa* Maulik, 1913.
On *Roscheria*, *Stevensonia* and *Phoenicophorium* (Arecaceae) in the Seychelles.

85- *Enischispa* Gressitt, 1957.
On *Calamus* and *Daemonorops* (Arecaceae) in New Guinea.

2.19.18. Tribe Promecothecini

86- *Promecotheca* Blanchard, 1853
Asia and Indo-Australian. On *Cocos*, *Metroxylon*, *Elaeis*, *Phoenix*, *Phytelephas*, *Pritchardia*, *Livistona*, *Nipa*, *Caryota*, *Areca*, *Balaka*, *Calamus*, *Ptychosperma* (Arecaceae), *Flagellaria* (Flagellariaceae), *Alpinia* (Zingiberaceae), *Miscanthus*, *Saccharum* (Poaceae), *Heliconia* (Heliconiaceae), *Pandanus*, *Freycinetia* (Pandanaceae) and *Ravenala* (Musaceae).

2.19.19. Tribe Gonophorini

87- *Pistosia* Weise, 1905.
Indo-Australian. On *Areca* and *Phoenix* (Arecaceae).

88- *Wallacispa* Uhmann, 1940.
Indonesia. On *Metroxylon* (Arecaceae).

89- *Downesia* Baly, 1858.
Indo-Australian. On *Bambusa*, *Sinocalamus* and *Saccharum* (Poaceae).

90- *Agonita* Strand, 1942.
Old World. On *Metroxylon*, *Phoenix* (Arecaceae), *Arundina*, *Dendrobium*, *Coelogyne*, *Phalaenopsis*, *Spathoglottis* (Orchidaceae), various Zingiberaceae, *Pandanus* (Pandanaceae), *Loudetia*, *Panicum*, *Hyparrhenia*, *Sporobolus*, *Sorghum*, *Miscanthus*, *Bambusa*, *Rottboellia* (Poaceae), *Cissus* (Vitaceae), *Combretum* (Combretaceae), *Lophira* (Ochnaceae), *Isoberlinia* (Caesalpiniaceae) and *Stereospermum* (Bignoniaceae). Actually most species of this genus appear to feed on monocotyledons. However, one species from Africa, *Agonita fossulata* Guérin-Méneville, has been recorded from those dicotyledons mentioned above, which is probably erroneous.

91- *Klitispa* Uhmann, 1940.
On *Bambusa* (Poaceae) in Java from 250 to 1000 m altitude.

92- *Aspidispa* Baly, 1869.
New Guinea. On *Korthalsia*, *Calamus*, *Daemonorops*, *Pinanga* (Arecaceae), *Flagellaria* (Flagellariaceae), *Dieffenbachia* (Araceae) and *Freycinetia* (Pandanaceae).

93- *Gonophora* Chevrolat, 1837.
Asia and Indo-Australia. On *Dendrobium*, *Phalaenopsis*, *Vanda*, *Arundina*,

Spathoglottis (Orchidaceae), *Alpinia, Curcuma, Amomum, Nicolaia, Elettaria, Zingiber* (Zingiberaceae), *Costus* (Costaceae), *Scleria* (Cyperaceae), *Musa* (Musaceae) and *Donax, Setaria* (Poaceae).

94- *Micrispa* Gestro, 1897.
On *Maranta* (Marantaceae) in Java.

2.19.20. Tribe Oncocephalini

95- *Oncocephala* Agassiz, 1846.
Old World. On Orchidaceae, *Dioscorea* (Dioscoreaceae), *Curcuma* (Zingiberaceae), *Oryza* (Poaceae) and *Ipomoea, Mina* (Convolvulaceae).

96- *Chaeridiona* Baly, 1869.
In Asia on *Curcuma* (Zingiberaceae).

2.19.21. Tribe Hispini

97- *Dactylispa* Weise, 1897.
Old World. The larva mines between the two epidermides. Recorded from a large number of plants, i.e. *Dactyloctenium, Echinochloa, Digitaria, Eleusine, Leptochloa, Leersia, Cynodon, Capillipedium, Oplismenus, Paspalum, Mnesithea, Zizania, Andropogon, Loudetia, Vetiveria, Rottboellia, Imperata, Vossia, Bambusa, Linguania, Sinobambusa, Arundinaria, Oryza, Sporobolus, Setaria, Arthraxon, Pennisetum, Melinis, Saccharum, Triticum, Zea, Panicum, Hymenachne, Urelytrum, Loudetia, Hyparrhenia, Phyllostachys* (Poaceae), *Cinchona, Gardenia, Plectronia, Nertera, Coffea, Canthium* (Rubiaceae), *Filipendula, Malus, Rosa, Rubus, Prunus, Leucosidea* (Rosaceae), *Isodon, Prunella, Plectranthus* (Lamiaceae), *Lobelia* (Campanulaceae), *Croton, Hevea* (Euphorbiaceae), *Phoenix* (Arecaceae), *Callicarpa* (Verbenaceae), *Scleria* (Cyperaceae), *Guaiacum* (Zygophyllaceae), *Ceiba, Durio* (Bombacaceae), *Helicteres, Theobroma, Sterculia* (Sterculiaceae), *Commelina* (Commelinaceae), *Petasites, Artemisia* (Asteraceae), *Semecarpus* (Anacardiaceae), *Jussiaea* (Onagraceae), *Castanea, Quercus, Castanopsis* (Fagaceae), *Carpinus* (Carpinaceae), *Desmodium, Phaseolus, Dalbergia* (Fabaceae), *Piliostigma* (Caesalpiniaceae). Also on certain Malvaceae, Marantaceae and Acanthaceae. The captures from *Raphanus* (Brassicaceae) are very doubtful and are probably accidental. In spite of the clear-cut preference for monocotyledons (mostly Poaceae), it appears that the observations on dicotyledons are real and reliable records.

98- *Asamangulia* Maulik, 1915.
Asia. On *Saccharum, Miscanthus, Oryza* and *Bambusa* (Poaceae).

99- *Rhadinosa* Weise, 1905.
Asia. On *Oryza, Saccharum, Miscanthus, Digitaria, Triticum* and *Zea* (Poaceae).

100- *Hispa* L., 1767.
Palaearctic, Indonesia and Asia Minor. On *Triticum*, *Poa*, *Dactylis*, *Agropyrum, Aegopogon, Digitaria, Sorghum, Saccharum, Zizania*, *Zea* (Poaceae), *Malus* (Rosaceae), *Cistus* (Cistaceae), *Lantana* (Verbenaceae), *Physalis* (Solanaceae), *Robinia* (Fabaceae) and *Zizyphus* (Rhamnaceae).

101- *Dorcathispa* Weise, 1900.
Africa. On *Zea*, *Oryza*, *Sorghum*, *Pennisetum* (Poaceae) and *Pueraria* (Fabaceae). The larva is a leaf-miner.

102- *Hispellinus* Weise, 1897.
Old World. On *Heteropogon, Sorghum, Paspalum, Zea, Saccharum, Zizania, Oryza, Miscanthus, Imperata, Themeda* and *Panicum* (Poaceae).

103- *Acmenychus* Weise, 1905.
On *Phragmites* (Poaceae) in Central Asia.

104- *Polyconia* Weise, 1905.
Africa. On *Oryza, Zea, Pennisetum* and *Sorghum* (Poaceae). The larva is a leaf-miner.

105- *Phidodonta* Weise, 1906.
On *Saccharum, Zea, Andropogon* (Poaceae) in India.

106- *Trichispa* Chapuis, 1875.
Africa. On *Oryza, Setaria, Paspalum, Acroceras* and *Echinochloa* (Poaceae).

107- *Chrysispa* Weise, 1897.
On *Oryza* (Poaceae) in Sierra Leone.

108- *Dicladispa* Gestro, 1897.
Old World. On *Vossia, Oryza, Saccharum, Paspalum, Mnesithea, Eleusine, Dactyloctenium, Echinochloa, Digitaria, Leptochloa, Leersia, Zizania, Vetiveria, Panicum, Cynodon, Capillipedium, Oplismenus* (Poaceae), *Aralia* (Araliaceae), *Grewia* (Tiliaceae), *Malvastrum* (Malvaceae), *Crotalaria, Dalbergia, Rhynchosia* (Fabaceae), *Leucosidea* (Rosaceae), *Petasites* (Asteraceae), *Lobelia* (Campanulaceae), *Costus* (Costaceae) and *Cistus* (Cistaceae).

109- *Thoracispa* Chapuis, 1875.
On flowers of the alpine prairies in Natal (2200 m to 2500m). No host-plant has been formally identified.

110- *Platypria* Guérin-Méneville, 1840.

Old World. On *Quercus* (Fagaceae), *Zizyphus* (Rhamnaceae), *Dolichos, Cajanus, Erythrina, Pueraria, Desmodium, Mucuna, Sesbania, Phaseolus* (Fabaceae), *Uncaria* (Rubiaceae), *Pyrus, Rubus* (Rosaceae) and *Myrica* (Myricaceae). Also on *Saccharum* (Poaceae).

111- *Cassidispa* Gestro, 1899.

Congo and China. On *Dunbaria* and *Galactia* (Fabaceae) in Vietnam.

Concluding remarks on Hispinae:

Maulik (1937) provided a list of 27 plant families utilized as food by the Hispinae. At that time, host-plants had only been recorded for 148 species from a total of 2570 then described. In the present account, 19 monocot and 44 dicot families are recorded, i.e. 63 plant families in total. The observations on Zamiaceae (Cycadales) and on Ophioglossaceae must be confirmed, although they are probable host-plants. If they are determined real hosts, they will bring the total number of known families selected to 65.

About 110 genera (74.4% of the total number of known hispine genera) have host-plants recorded for them. Their preference for Monocotyledonae is very clear-cut; of the 110 genera, 86 (78%) feed on Monocotyledonae, while only 24 (22%) feed on Dicotyledonae. The majority of Hispinae are monophagous or oligophagous (86.1%) while only 13.9% are polyphagous (at least at the genus level). It is evident that the subfamily mainly attacks the Poaceae, Arecaceae and Zingiberales among monocotyledons and the Leguminosae (Fabaceae and Caesalpiniaceae), Malvaceae and Asteraceae among the dicotyledons. Like the related Cassidinae, the Hispinae appear to reject the Brassicaceae and the various records of these beetles on these plants are probably only coincidences and do not represent true trophic relationships. Few species feed on Solanaceae or Convolvulaceae. No Hispinae feed on Chenopodiaceae, a host-plant family commonly utilized by the Cassidinae. Normally hispine adults feed on the larval host-plant but often adults are relatively more polyphagous than their larvae.

Among Hispinae, ring-defence strategy displayed by larvae (Vasconcellos-Neto & Jolivet, 1988, 1989; Jolivet *et al.*, 1990) or maternal protection of the eggs, larvae or pupae by the female beetle (Jolivet, 1988a), is not known. However, the female hispine often inserts an ootheca into the leaf of the host-plant while many larvae are leaf-miners or stem-borers, or even free-living between the appressed monocot leaves. Hispine larvae are so naturally protected that they do not require any chemical, behavioural or mechanical protective systems. Often Hispinae adults are indifferent to co-inhabiting ants and they often also live on *Cecropia* trees (i.e. *Sceloenopla*) beyond the reach of the ants. Several rare cases of totally free-living larvae are known only among three genera and the larvae of one of them (i.e. *Oediopalpa*) protect themselves with the assistance of a shield composed of excreta and exuviae like the true Cassidinae.

This exceptional behaviour casts some doubt on the systematic position of this genus.

In conclusion, the Hispinae, like the Criocerinae, prefer the Monocotyledonae as host-plants. The first family is highly evolved, but the second is primitive, so that the interpretation of this monocot selection remains a mystery. To answer this problem, it is necessary to determine which is the more primitive of the Monocotyledonae and the Dicotyledonae. Probably they evolved simultaneously and the host-preferences of the Hispinae and Criocerinae cannot throw any light of the problem. However, it is probable that for both Hispinae and Criocerinae, the evolutionary shift from Mono- to Dicotyledonae has occurred during more recent geological time.

References listed in the bibliography pertaining to Hispinae are as follows:- Andrade (1981), Boving & Craighead (1931), Budhraja *et al.* (1979), Chu (1949), Gressitt & Kimoto (1961-63), Hawkeswood (1987, 1988), Hering (1951), Jolivet (1988a, 1989a), Jolivet *et al.* (1990), Kalshoven (1957), Kalshoven *et al.* (1981), Kimoto (1964-66), Maulik (1919, 1937, 1945, 1950), McCoy (1984, 1985), Peterson (1960), Siefert & Siefert (1979), Strong (1977-82), Strong & Wang (1977), Uhmann (1958), Williams (1991), Young (1979).

2.20. Cassidinae

The Cassidinae represent the last subfamily of the Chrysomelidae with about 159 described genera, of which about 83 (52.2%) have host-plants recorded for them. This sampling is important and significant. The host-plants of the Cassidinae are rather well known because the beetles have the tendency to adhere to their host-plants. They are poor flyers and drop to the ground only as a last resort. Although the Cassidinae fly clumsily and poorly during the day in sunlight, they are able to disperse well in the forests and to colonize trees and shrubs in the tropics. In the temperate areas, cassidines mostly feed on low herbs. Although all Hispinae are macropterous, the Cassidinae display some cases of apterism (e.g. *Delocrania* and *Elytrogona*) and micropterism (e.g. *Stoiba, Fornicocassis* and the subgenera of *Pilemostoma* and *Mionycha* of *Cassida*). These species evidently do not fly (Jolivet, 1959).

Recently, the Neotropical species have been extensively studied by Buzzi (1975 and many other papers, see Jolivet, 1988d). The recent researches of Vasconcellos-Neto (1987, 1988) on the genetics of Cassidinae completely challenge the present taxonomy of these insects. Many described species appear to be only mutations of various colours and markings of one unique taxon. This is the case with *Chelymorpha cribraria* (Fabricius) and appears also true for many other species such as *Botanochara sedecimpustulata* (Fabricius) and *Stolas discoidea* (L.).

There are fundamental differences between the Cassidinae and the Hispinae, but there are also intermediate genera. In both subfamilies, the antennae of the adults have 11 segments despite some fusion among those of the Hispinae.

The larvae of Cassidinae are generally free-living and exposed on the leaves and often carry the old exuviae and fecal matter on their backs. The shield is carried by the furca, scoli or urogomphi, which are prolongations of the tergum of the 9th abdominal segment. Other species are purported to be leaf-miners.

Most of the Cassidinae are characterized in having rounded or oval body shape with expanded, flattened flanges surrounding the body, including the area above the head in front of the prothorax. The genera *Delocrania* and *Notosacantha* are intermediate between the Hispinae and the Cassidinae since they have elongated bodies with elytral margins not well flattened and the frontal edge of the prothorax flattened with a prominent notch in the middle which exposes the dorsal part of the head. *Platyauchenia* is now considered to belong to Hispinae (Alurnini).

The eggs are sometimes laid isolated (in *Coptocycla*) or covered with excreta (in *Cassida* and *Laccoptera*) but are generally laid in a cluster sometimes called ootheca, composed of lamellae of glandular fluid (in *Aspidomorpha* and *Basipta*). These oothecae, which are attached (cemented) onto leaves or stems, have an upper exit for the emergence of the larvae.

As mentioned previously, the larvae often carry supra-anal extensions (furcae) on the dorsum of the 9th abdominal segment (e.g. in *Aspidomorpha* and *Laccoptera*), while others only have a simple, bifid extension (e.g. *Oocassida*). The structure of the extensions varies in the Neotropical region where certain species and genera may even lack furcae (e.g. *Eurypepla*). The excreta carried by the larvae varies in shape and is often characteristic of the species and genus. Several larvae such as those of *Basilepta* are naked without any protection. Several gregarious larvae of Stolaini obtain maternal protection which is not necessarily associated with the presence of a dorsal shield.

The mature larvae of Cassidinae measure from 6 to 15 mm in length. Several species are large, depressed and ovate in shape and are covered with spines along the lateral edges, while others are more elongated. The general colour is whitish, yellow/orange or green. Their antennae are bi- or trisegmented while their ocellae number 5 to 6. The abdomen, like Hispinae, possess eight normal segments with a 9th and 10th strongly reduced. The final tube (10th segment) deposits the excreta on the furca. Only the 8th segment spiracles are vestigial or absent. There are generally five larval instars, i.e. four moults.

Many papers (e.g. Jolivet, 1988a) have been published on changes in colouration exhibited by adult Cassidinae. The changes during their development are associated with their food but these changes can also have a mimetic basis, i.e. the adult and host-plant gradually change simultaneously from green to yellow during the season. The colours of the Cassidinae are chemical or structural and produce the splendid silvery or golden reflections of living specimens. The cassidines are among the few beetles known to change colour rapidly. Among certain species of *Metriona* and *Deloyala,* the colour change takes less than one minute to return to the initial state of golden yellow, blue-green, violet, orange-brown and vice versa. Certain structural colours can release the sexual instinct (response). Homochromy, warning and sexual attraction are all involved with those changes.

The protection of the larvae is accomplished by the dorsal shield which can be very complicated in structure. The Neotropical Cassidinae of the genera *Hemisphaerota* and *Dorynota* carry a long, dry ball of excrement over their abdomens. Pupation occurs under this "nest" or excremental umbrella, which is very efficient against ants, but also against desiccation and rain. Many larvae manipulate that shield to protect themselves against the attacks of predators (Eisner *et al.*, 1967).

In America, the adults of a number of species are able to cling to their host-plants by the use of a complex system of suckers. Eisner (1972) cited the example of *Hemisphaerota cyanea* (Say) in the south of the USA, but this phenomenon is very common among many species including *Physonota* in Brazil. The terminal pads of the tarsal setae are moistened by an oily secretion which allows a very strong attachment to the leaf surface. Like a tortoise when in danger, the Cassidinae adults withdraw their legs and antennae and clamp downwards so that the body lies flush with the substrate. In this position, their softer body parts and appendages are well protected from ants and other predators. Many adult cassidines are also toxic. Reflex immobilization is rarely employed and only by species adhering superficially to the substrate. Homochromy is sometimes very striking (i.e. green, yellow and brown) and cassidines are rarely the prey of birds, ants or bugs. Ring-defence strategy is a supplementary means of defence among the Neotropical Stolaini, especially *Chelymorpha* (Vasconcellos-Neto & Jolivet, 1988, 1989; Jolivet *et al.*, 1990).

Neotropical (and probably other) Cassidinae produce some sound by rubbing the head against the anterior edge of the prothorax. Is this a warning or alert signal? This system is well-known among the Hispinae.

All Palaearctic Cassidinae live on herbs but many exotic species feed on bushes or small trees such as *Cordia*, *Ipomoea arborescens* and *Tabebuia*. Their larvae feed on the underside of the leaves and the adults feed above, where they are more exposed to predators. The larvae chew the leaf parenchyma except for the upper epidermis and the veins, while the adults eat the branchlets or the limbs by making irregular holes.The few species (70) which are found in the Palaearctic region, mostly feed on Asteraceae, Lamiaceae, Caryophyllaceae, Chenopodiaceae and Convolvulaceae. Everywhere, but mostly in the tropics, the selection of Convolvulaceae predominates. Although there are at least 3,000 described species of Cassidinae, with further research, this number will probably be reduced due to cryptic and polymorphic species presently under more than one name. Most of the known species live in the Neotropical region.

2.20.1. Tribe Notosacanthini

1- *Notosacantha* Chevrolat, 1837.

Old World, except for Europe. On *Areca* (Arecaceae) in Asia but captured (accidentally?) on *Phyllanthus* (Euphorbiaceae) in Vietnam. On *Acacia* in Australia (Hawkeswood, 1989).

2.20.2. Tribe Basiprionotini

2- *Basiprionota* Chevrolat, 1837.
South-East Asia. On *Premma* and *Verbena* (Verbenaceae). Recorded (accidentally?) on *Liquidambar* (Hamamelidaceae) and *Ficus* (Moraceae).

3- *Craspedonta* Chevrolat, 1837.
South-East Asia. On *Gmelina* (Verbenaceae).

2.20.3. Tribe Delocranini

4- *Delocrania* Guérin-Méneville, 1844.
Neotropical. On *Cocos, Attalea, Diplothemium* and *Elaeis* (Arecaceae). Larvae and adults both feed on the abaxial side of the leaves.

2.20.4. Tribe Hemisphaerotini

5- *Spaethiella* Barber and Bridwell, 1940.
Neotropical. On *Cocos, Phoenix, Copernicia, Pritchardia, Arecastrum, Washingtonia, Astrocaryum* and *Elaeis* (Arecaceae). Recently recorded on *Bactris* (Arecaceae) and *Heliconia* (Heliconiaceae) by Windsor *et al.* (1992).

6- *Hemisphaerota* Chevrolat, 1837.
New World. On native and introduced palm trees. The larvae nibble the abaxial side of the palm fronds under a cover of excreta. On *Sabal, Acoelorrhaphe, Arecastrum, Chamaerops, Cocos, Serenoa, Washingtonia, Thrinax* and *Attalea* (Arecaceae). Also recorded (accidentally?) on *Inga* (Mimosaceae) and *Theobroma* (Sterculiaceae).

2.20.5. Tribe Imatidiini

7- *Imatidium* Fabricius, 1801.
Neotropical. On *Astrocaryum, Arecastrum, Bactris, Cocos, Desmoncus* and *Elaeis* (Arecaceae). Also on *Areca, Livistona, Geonoma, Phoenix, Licuala* and *Pritchardia* (Arecaceae). Some of these palm trees have been introduced to Brazil. The larvae and adults nibble the epidermis of the foliar buds, the young fronds, the young sachi and the young fruits. It is a pest of the oil palm (*Elaeis guineensis*). Recently recorded from *Heliconia* (Heliconiaceae) by Windsor *et al.* (1992).

8- *Pseudimatidium* Aslam, 1965.
Neotropical. On *Elaeis* (Arecaceae).

9- *Aslamidium* Borowiec, 1984.
Neotropical. On *Calathea* (Marantaceae) (Windsor *et al.*, 1992).

2.20.6. Tribe Spilophorini

10- *Calyptocephala* Chevrolat, 1837.
Neotropical. On *Arecastrum, Astrocaryum* and *Elaeis* (Arecaceae) in Brazil. On *Chamaedorea* (Arecaceae), *Dioscorea* (Dioscoreaceae) and *Chusquea* (Poaceae) in Panama (Windsor *et al.*, 1992).

2.20.7. Tribe Omocerini

11- *Polychalca* Chevrolat, 1837.
Neotropical. On *Cordia* (Ehretiaceae) and *Hyptis* (Lamiaceae). These are closely related families. The rare, endangered and beautiful *Desmonota variolosa* (Weber) feeds on *Cordia curassanica* (Ehretiaceae) in the Bahia state, Brazil (PJ, pers. obs.).

12- *Omocerus* Chevrolat, 1835.
Neotropical. On *Hyptis* (Lamiaceae).

13- *Discomorpha* Chevrolat, 1837.
Neotropical. On *Cordia* (Ehretiaceae) and *Borago* (Boraginaceae). Other observations on Fabaceae, Rutaceae and Sterculiaceae are errors.

14- *Canistra* Erichson, 1847.
Neotropical. On *Mikania* (Asteraceae) in Paraguay and on *Solanum* (Solanaceae) in Brazil. Accidental on coffee (*Coffea*, Rubiaceae).

15- *Cassidinoma* Hincks, 1950.
Neotropical. On *Cordia* (Ehretiaceae).

2.20.8. Tribe Goniocheniini

16- *Chlamydocassis* Spaeth, 1952.
Neotropical. On *Hyptis* (Lamiaceae).

17- *Polychalma* Barber and Bridwell, 1940.
Neotropical. On *Hyptis* (Lamiaceae) and *Cordia* (Ehretiaceae). Recently recorded in Panama on *Helicteres* (Sterculiaceae) (Windsor *et al.*, 1992).

2.20.9. Tribe Dorynotini

18- *Dorynota* Chevrolat, 1837 (incl. *Akantaka* Maulik).
Neotropical. On *Tecoma, Tabebuia* and *Crescentia* (Bignoniaceae). Other observations are accidental.

19- *Paranota* Monros and Viana, 1949.
Neotropical. On *Tabebuia* and *Tecoma* (Bignoniaceae).

2.20.10. Tribe Eugenysini

20- *Agenysa* Spaeth, 1905.
Neotropical. On *Ipomoea* (Convolvulaceae).

21- *Eugenysa* Chevrolat, 1837.
Neotropical. Recently recorded on *Mikania* (Asteraceae) in Panama (Windsor *et al.*, 1992).

2.20.11. Tribe Stolaini

22- *Stolas* Billberg, 1920.
Neotropical. On *Mikania* and *Neomirandia* (Asteraceae) and *Ipomoea* (Convolvulaceae). Accidentally recorded on *Citrus* (Rutaceae), *Solanum* (Solanaceae), *Cucurbita* (Cucurbitaceae) and *Gossypium* (Malvaceae).

23- *Cyrtonota* Chevrolat, 1837.
Neotropical. On *Ipomoea* and *Convolvulus* (Convolvulaceae).

24- *Mesomphalia* Hope, 1839.
Neotropical. On *Ipomoea* (Convolvulaceae).

25- *Anacassis* Spaeth, 1913.
Neotropical. On *Baccharis, Baccharidastrum* and *Flourensia* (Asteraceae).

26- *Nebraspis* Spaeth, 1913.
Neotropical. On *Ipomoea* (Convolvulaceae) and *Eupatorium* (Asteraceae).

27- *Botanochara* Dejean, 1837.
Neotropical. On *Pharbitis, Merremia, Ipomoea, Convolvulus* and *Calonyctyon* (Convolvulaceae) and *Mikania, Eupatorium* (Asteraceae).

28- *Hilarocassis* Spaeth, 1909.
America. On *Bidens* (Asteraceae). An unidentified species has been recorded recently on *Ipomoea* (Convolvulaceae) in Panama (Windsor *et al.*, 1992).

29- *Zatrephina* Spaeth, 1909.
Neotropical. On *Ipomoea* (Convolvulaceae).

30- *Paraselenis* Spaeth, 1913.
Neotropical. On *Cordia* (Ehretiaceae) and *Convolvulus*, *Ipomoea, Calonyctyon* (Convolvulaceae).

31- *Chelymorpha* Chevrolat, 1837.
New World. On *Ipomoea, Convolvulus, Calystegia, Pharbitis, Merremia* (Convolvulaceae), *Cuscuta* (Cuscutaceae). Recorded accidentally on *Panicum* (Poaceae), *Mandevilla* (Apocynaceae), *Asclepias* (Asclepiadaceae), *Rubus* (Rosaceae) and *Vachellia, Pithecellobium* (Mimosaceae). Only Convolvulaceae are the true host-plants.

32- *Echoma* Chevrolat, 1837.
Neotropical. On *Cordia* (Ehretiaceae) and *Calonyctyon, Convolvulus, Ipomoea* (Convolvulaceae). Several *Echoma* species feed on the flowers of *Mikania* (Asteraceae) in Panama and Brazil (Windsor, Macedo & Toledo, 1994).

33- *Acromis* Chevrolat, 1837.
Neotropical. On *Ipomoea* and *Merremia* (Convolvulaceae).

34- *Omaspides* Chevrolat, 1837.
Neotropical. On *Ipomoea* (Convolvulaceae) and *Cordia* (Ehretiaceae). Accidentally recorded on *Passiflora* (Passifloraceae).

35- *Ogdoecosta* Spaeth, 1909.
Neotropical. On *Ipomoea* (Convolvulaceae)(Windsor *et al.*, 1992).

36- *Terpsis* Spaeth, 1913.
Neotropical. On *Ipomoea* (Convolvulaceae) (Windsor *et al.*, 1992).

2.20.12. Tribe Ischyrosonychini

37- *Eurypedus* Gistel, 1834.
Neotropical. On *Cordia* (Ehretiaceae).

38- *Cistudinella* Champion, 1893.
Neotropical. On *Patagonula* and *Cordia* (Ehretiaceae).

2.20.13. Tribe Physonotini

39- *Physonota* Boheman, 1854.
New World. On *Cordia* (Ehretiaceae), *Franseria, Senecio, Baccharis, Helianthus, Heliopsis, Vernonia* (Asteraceae) and *Monarda, Salvia* (Lamiaceae). Also recorded accidentally on *Gaura* (Onagraceae).

2.20.14. Tribe Aspidomorphini

40- *Aspidomorpha* Hope, 1840.
Old World. On *Ipomoea, Convolvulus, Calonyctyon, Merremia, Calystegia, Argyreia* (Convolvulaceae), *Tectona, Gmelina* (Verbenaceae), *Othonna* (Astera-

ceae) and *Hoslundia*, *Ocimum* (Lamiaceae). Also recorded accidentally on *Elaeis* (Arecaceae) in Sierra Leone and in other areas on *Chenopodium, Beta* (Chenopodiaceae), *Passiflora* (Passifloraceae), *Michelia* (Magnoliaceae), *Citrus* (Rutaceae), *Aleurites*, *Alchornea* (Euphorbiaceae) and *Lycopersicon*, *Solanum*, *Capsicum* (Solanaceae).

41- *Sindia* Weise, 1897.
India. On *Rivea* (Convolvulaceae).

42- *Sindiola* Spaeth, 1903.
Birmania, Vietnam. On *Argyreia* (Convolvulaceae) and *Amaranthus* (Amaranthaceae).

43- *Laccoptera* Boheman, 1855.
Old World. On *Ipomoea, Merremia* and *Calystegia* (Convolvulaceae).

2.20.15. Tribe Cassidini

44- *Macromonycha* Spaeth, 1911.
Caucasus, Turkestan. Recorded doubtfully on Brassicaceae in Central Asia.

45- *Chiridula* Weise, 1889.
Caucasus, Turkestan. On *Artemisia* (Asteraceae).

46- *Eremocassis* Spaeth, 1926.
Transcaspian. On *Lycium* (Solanaceae).

47- *Ischyronota* Weise, 1891.
Palaearctic. On *Anabasis*, *Salsola*, *Kalidium*, *Halocnemum*, *Suaeda, Aellenia* and *Haloxylon* (Chenopodiaceae).

48- *Oxylepus* Desbrochers, 1884.
North Africa and Asia Minor. On *Salsola* and *Suaeda* (Chenopodiaceae) and on various Caryophyllaceae.

49- *Acrocassis* Spaeth, 1924.
Africa. On *Ipomoea* (Convolvulaceae). Recorded also on *Beta* (Chenopodiaceae).

50- *Silana* Spaeth, 1914.
India. On *Zizyphus* (Rhamnaceae) and *Murraya* (Rutaceae).

51- *Glyphocassis* Spaeth, 1914.
India and China. On *Convolvulus* and *Ipomoea* (Convolvulaceae).

52- *Oocassida* Weise, 1897.
Asia and Africa. On *Zizyphus* (Rhamnaceae) and *Ipomoea* (Convolvulaceae).

53- *Cassida* L., 1758.
Cosmopolitan. Considering the genus as a whole (including subgenera) the following host genera have been recorded: *Serratula, Xeranthemum, Erigeron, Echinops, Lactuca, Carlina, Filago, Scorzonera, Cousinia, Onopordum, Carpesium, Pulicaria, Sonchus, Carthamus, Lappa, Arctium, Gaillardia, Helianthus, Solidago, Taraxacum, Silybum, Inula, Tanacetum, Senecio, Dahlia, Crepis, Aster, Saussurea, Tagetes, Leontodon, Chrysanthemum, Cynara, Galactites, Helichrysum, Cirsium, Ormenis, Achillea, Matricaria, Antennaria, Cnicus, Anaphalis, Centaurea, Carduus, Anthemis, Brachylaena, Rhaponticum,* (Asteraceae), *Nelumbo* (Nelumbaceae), *Achyranthes, Amaranthus, Alternanthera, Telanthera, Gomphrena, Celosia* (Amaranthaceae), *Leucas, Salvia, Ballota, Gomphostemma, Ocimum, Galeopsis, Thymus, Mentha, Lycopus, Melissa, Stachys, Nepeta* (Lamiaceae), *Silene, Saponaria, Dianthus, Gypsophila, Isodon, Minuartia, Melandrium, Stellaria, Malachium, Cerastium, Spergula, Sagina, Honckenya, Arenaria, Cucubalus, Petrorhagia, Lychnis* (Caryophyllaceae), *Atriplex, Chenopodium, Salsola, Suaeda, Beta, Spinacia, Salicornia* (Chenopodiaceae), *Ipomoea, Calystegia, Convolvulus, Argyreia* (Convolvulaceae), *Acanthus, Adhatoda* (Acanthaceae), *Verbascum* (Scrophulariaceae) and *Salicaria* (Lythraceae). Overall, *Cassida* feed mostly on Asteraceae, Amaranthaceae, Convolvulaceae, Lamiaceae, Caryophyllaceae and Chenopodiaceae and are mono- or oligophagous at the species level (the genus is polyphagous as a whole). Bordy and Doguet (1987) discovered certain correlative trends between the various subgenera and their host-plants, e.g. *Hypocassida* feeds on Convolvulaceae, *Mionycha* feeds partially on Caryophyllaceae, *Pilemostoma* on *Inula* (Asteraceae) and *Cassida* (sensu stricto) on Asteraceae. Palaearctic species of the subgenera *Odontionycha* and *Lordiconia* may also be differentiated on the basis of their selection of various Lamiaceae. This host-plant distinction appears well-defined at least, for the Palaearctic species, but may be also valid for other species elsewhere. *Cassida* have also been captured on "refuge" plants or accidentally on plants such as *Ornithopus, Trifolium, Medicago, Pisum* (Fabaceae), *Caltha, Clematis* (Ranunculaceae), *Aegle, Citrus* (Rutaceae), *Cestrum, Solanum, Lycium* (Solanaceae), *Gynostemma* (Cucurbitaceae), *Agrimonia, Prunus, Malus, Pyrus, Sorbus, Pourthiaea, Rhaphiolepis* (Rosaceae), *Urtica* (Urticaceae) and *Humulus* (Cannabaceae).

54- *Chiridopsis* Spaeth, 1924.
Africa. On *Ipomoea, Convolvulus, Argyreia* and *Pharbitis* (Convolvulaceae). Recorded (accidentally) on *Tephrosia* (Fabaceae), *Combretum* (Combretaceae), *Alchornea* (Euphorbiaceae) and *Lovoa* (Meliaceae).

55- *Thlaspida* Weise, 1899.
Asia. Mostly on *Callicarpa* (Verbenaceae) and *Ipomoea* (Convolvulaceae). Also on *Ligustrum* and *Fraxinus* (Oleaceae).

56- *Floridocassis* Spaeth, 1952.
USA, Cuba. On Convolvulaceae.

57- *Jonthonota* Spaeth, 1913.
North and Central America. On *Ipomoea* (Convolvulaceae).

58- *Chersinellina* Hincks, 1950.
Central and South America. On *Bonamia* (Convolvulaceae) (Windsor *et al.*, 1992).

59- *Hybosa* Duponchel and Chevrolat, 1842.
Central and South America. On *Phryganocydia* (Bignoniaceae) (Windsor *et al.*, 1992).

60- *Cyclocassis* Spaeth, 1913.
Central and South America. On *Phryganocydia* (Bignoniaceae) (Windsor *et al.*, 1992).

61- *Ischnocodia* Spaeth, 1942.
Central and South America. On *Cordia* (Ehretiaceae) (Windsor *et al.*, 1992).

62- *Saulaspis* Spaeth, 1913.
Neotropical. On *Wedelia* (Asteraceae).

63- *Aporocassida* Spaeth, 1952.
Argentina. On *Wedelia* (Asteraceae).

64- *Agroiconota* Spaeth, 1913.
New World. On *Convolvulus*, *Ipomoea* and *Merremia* (Convolvulaceae). Also recorded on *Erigeron* (Asteraceae) and *Lycopersicon* (Solanaceae). The last record is probably accidental.

65- *Gratiana* Spaeth, 1913.
New World. On *Solanum* and *Lycopersicon* (Solanaceae).

66- *Parorectis* Spaeth, 1901.
On Solanaceae in Mexico.

67- *Orexita* Spaeth, 1911.
Neotropical. On *Phaseolus* (Fabaceae) and *Passiflora* (Passifloraceae).

68- *Coptocycla* Chevrolat, 1837.
New World. Divided into five subgenera. On *Ipomoea*, *Convolvulus* (Convolvulaceae), *Pithecoctenium* (Bignoniaceae), *Ehretia, Cordia* (Ehretiaceae) and *Vernonia* (Asteraceae). Also on *Phaseolus* (Fabaceae) and *Passiflora* (Passifloraceae).

69- *Nuzonia* Spaeth, 1912.
Neotropical. On *Solanum* (Solanaceae).

70- *Plagiometriona* Spaeth, 1899.
New World. On *Solanum, Datura, Physalis, Acnistus, Witheringia, Bassovia* and *Aureliana* (Solanaceae). Feeds occasionally on several other plants such as *Tilia* (Tiliaceae), *Platanus* (Platanaceae), *Quercus* (Fagaceae), *Gossypium* (Malvaceae) etc. This is analogous to the secondary selections of the Alticinae in response to changing climatic/meteorological conditions. Also recorded on *Acer* (Aceraceae).

71- *Metriona* Spaeth, 1912.
Neotropical. On *Ipomoea, Convolvulus, Pharbitis, Calystegia* and *Jacquemontia* (Convolvulaceae) (Woodruff, 1976). Also on *Wedelia* (Asteraceae) and (?) Solanaceae. Recorded (accidentally) on *Elaeis* (Arecaceae), *Zea* (Poaceae), *Camellia* (Theaceae), *Eucalyptus* (Myrtaceae), *Codiaeum, Hura, Sapium* (Euphorbiaceae), *Triplaris* (Polygonaceae), *Asimina* (Annonaceae), *Citrus* (Rutaceae), *Gardenia* (Rubiaceae), *Hibiscus* (Malvaceae), *Ficus* (Moraceae), *Ligustrum* (Oleaceae), *Litchi* (Sapindaceae), *Philodendrum* (Araceae), *Pimenta* (Myrtaceae), *Schinus* (Anacardiaceae), *Prunus, Pyrus* (Rosaceae), *Prosopis* (Mimosaceae), *Vigna, Glycine* (Fabaceae), *Physalis, Capsicum, Solanum, Tochroma* (Solanaceae) and *Lecythis* (Lecythidaceae).

72- *Deloyala* Chevrolat, 1837.
New World. On *Ipomoea, Convolvulus, Merremia, Calystegia* and *Argyreia* (Convolvulaceae). Also on *Capsicum* and *Solanum* (Solanaceae) in Venezuela.

73- *Syngambria* Spaeth, 1911.
Neotropical. On *Pithecoctenium* (Bignoniaceae).

74- *Parachirida* Spaeth, 1942.
South America. On *Jacquemontia* (Convolvulaceae) in Panama (Windsor *et al.*, 1992).

75- *Xenocassis* Spaeth, 1936.
Central America. On *Merremia* (Convolvulaceae) in Panama (Windsor *et al.*, 1992).

76- *Charidotella* Weise, 1896.
America. On *Ipomoea* and *Merremia* (Convolvulaceae).

77- *Metrionella* Spaeth, 1932.
Neotropical. On Convolvulaceae.

78- *Drepanocassis* Spaeth, 1936.
Brazil. On *Hyptis* (Lamiaceae).

79- *Tapinaspis* Spaeth, 1936.
Central America. On *Liabum* (Asteraceae) in Panama (Windsor *et al.*, 1992).

2.20.16. Tribe Charidotini

80- *Charidotis* Boheman, 1854.
Neotropical. On *Lantana* (Verbenaceae), *Crescentia, Bignonia, Jacaranda, Tecoma, Tabebuia, Arrabidaea* and *Pyrostegia* (Bignoniaceae). Recorded also on *Anisomeria* (Rubiaceae), *Patagonula* (Ehretiaceae), *Verbesina* (Asteraceae). Accidental on leaves of *Eucalyptus* (Myrtaceae). Recently recorded on *Anemopaegma, Phryganocydia, Stizophyllum* and *Amphilophium* (Bignoniaceae), *Cordia* (Ehretiaceae), *Liabum* (Asteraceae) and *Citharexylum* (Verbenaceae) in Panama (Windsor *et al.*, 1992).

81- *Cteisella* Weise, 1896.
Neotropical. On *Ipomoea* (Convolvulaceae).

82- *Microctenochira* Spaeth, 1926.
America. On *Merremia* (Convolvulaceae) in Texas. Recently *Phryganocydia* (Bignoniaceae), *Jacquemontia, Ipomoea* (Convolvulaceae) and *Wedelia, Melampodium* (Asteraceae) in Panama (Windsor *et al.*, 1992).

83- *Ctenochira* Chapuis, 1875.
On *Ipomoea* (Convolvulaceae) in Brazil.

Concluding remarks on Cassidinae:
For the Paraguayan species of Cassidinae, Fiebrig (1910) provided the following distribution: 8 species on Bignoniaceae, 4 on Ehretiaceae, 4 on Convolvulaceae, 3 on Asteraceae, 2 on Lamiaceae, 1 on Rubiaceae, 1 on Solanaceae and 1 on Arecaceae. This appears to be an adequate sample for the Neotropical region where the selections of Ehretiaceae and Convolvulaceae prevail. All of the above families, with the exceptions of Asteraceae and Rubiaceae belong to the Tubiflorae. The Arecaceae (Monocotyledonae) are selected by primitive genera of Neotropical Cassidinae. The Cuscutaceae are closely related to the Convolvulaceae. Tempere (1935) emphasized the prefer-

ence of the European Cassidinae for the Centrospermae, a natural order of Dicotyledonae which mainly comprises herbaceous plants. Among the Centrospermae are such families as Amaranthaceae, Caryophyllaceae and Chenopodiaceae. The selection of Convolvulaceae, a family very distant from the above-mentioned ones, is very prevalent world-wide and is also common in Europe and in the whole Palaearctic region. The genus *Cassida* is particularly common on Convolvulaceae. Centrospermae and Tubiflorae remain the dominant orders of plants selected by the Cassidinae, together with several families such as Asteraceae.

A detailed study of the trophic preferences of Cassidinae was presented by Gressitt (1952). Although care must be taken in interpreting the validity of the ecological data provided by Gressitt, most of the host-plant records appear reliable. This American author recorded 14 host-plant families belonging to the Arecaceae, Urticales, Centrospermae, Rosales, Malvales, Contortae, Tubiflorae, Rubiales, Cucurbitales and Campanulatae (Asteraceae). Gressitt tentatively provided a classification based on geographical region, but at that time, the trophic data were rather meagre and incomplete. Gressitt recorded the European *Cassida (Myonica) margaritacea* Schaller, as being relatively polyphagous on 11 plant genera in five families. It should also be noted that there are no reliable records of Cassidinae feeding on grasses (Poaceae), even in the East, in competition (association) with Hispinae. The Cassidinae practically do not feed on flower pollen. The genus *Cassida* feeds on the greatest number of plant families (10) as a genus, but its subgenera are restricted in their selections. Unfortunately, most of the papers on Cassidinae repeat the literature data and provide very few new and reliable observations. It is interesting to note that comparatively few plant families (29) are attacked by the Cassidinae. Many of these families are closely related. Only the hispine-like species feed on palms (Arecaceae).

The economic importance of Cassidinae is not that significant and is rather small. Sporadic attacks may occur in Europe, Asia and North Africa on such plants as sugar-beet, artichoke, mint, camomile etc., and elsewhere on sweet potato. In Argentina, South America, there have been massive attacks on *Tabebuia* (Bignoniaceae) along the roads. The strong attraction of Neotropical Cassidinae towards *Cordia* (Ehretiaceae) has been used as a biological control method against a plant introduced to Mauritius. However, this plan was a failure because *Physonota alutacea* Boheman did not survive on the island. The complex biology of the Stolaini and Physonotini is probably the reason for this failure. The use of *Cassida*, *Aspidomorpha* and *Metriona* has been once considered as a control of introduced *Convolvulus*. Other species have been tested against Caryophyllaceae, Asteraceae, Solanaceae and Amaranthaceae. However, the introduced beetles are not without risk of adapting to local crops and thereby becoming pests themselves.

We have listed here 83 genera from over 159 described in the world. Practically all are oligophagous or monophagous, at least at the level of subgenus or species. The dubious exceptions have to be restudied when the opportunities arise. Like the Alticinae, the Cassidinae appear to occasionally nibble the foliage of plants other than usual hosts, such as Rutaceae (*Citrus*), Brassicaceae, Moraceae, Fabaceae, etc. This does not appear to represent any real trophic relationships. Neutral plants, such as Fabaceae, probably play a transitional or survival role. The primitive species which feed on Arecaceae (i.e. Notasacanthini, Delocranini, Hemisphaerotini and Imatidiini) are generally intermediate between Hispinae and the other Cassidinae. The majority of the genera feed on Convolvulaceae (20%), but also on Verbenaceae, Lamiaceae, Bignoniaceae, Cuscutaceae, Solanaceae, Ehretiaceae, Caryophyllaceae, Chenopodiaceae, Amaranthaceae, Boraginaceae, Asteraceae, Sterculiaceae, Oleaceae, Combretaceae, Nelumbaceae and Lecythidaceae. Much less common are the selections of Cucurbitaceae, Rosaceae, Rubiaceae and plants of nine other families. The genera feeding on Convolvulaceae are sometimes oligophagous and accept related families such as Bignoniaceae. Overall, the predominant selections of the Cassidinae are the Convolvulaceae and related families (i.e. Bignoniaceae, Cuscutaceae, Lamiaceae, Verbenaceae, Ehretiaceae and Boraginaceae) but also other groups such as the Asteraceae and Centrospermae. The original selection from which all others have been derived is probably the Convolvulaceae, except for that of the palmicolous hispine-like species .

References listed in the bibliography pertaining to Cassidinae are as follows:- Araujo & Silva (1968), Bordy & Doguet (1987), Buzzi (1975, 1988, 1994), Costa-Lima (1936, 1955), Eisner (1972), Eisner *et al.* (1967), Fiebrig (1910), Gressitt (1952), Gressitt & Kimoto (1961-63), Hawkeswood (1987, 1988, 1989), Jolivet (1959, 1988a,d), Jolivet *et al.* (1990), Kimoto (1964-66), Knab (1909), Maulik (1919, 1945, 1950), Peterson (1960), Rausher (1983, 1984), Siebert (1975), Takizawa (1978a), Tempere (1935), Vasconcellos-Neto (1987, 1988), Vasconcellos-Neto & Jolivet (1988, 1989), Windsor *et al.* (1992), Woodruff (1976).

3. HOST-PLANTS SELECTED BY LEAF-BEETLES

1. Introduction

In this chapter we have chosen a number of plant families which are selected by the Chrysomelidae. Previously, the selection of some families has been treated in detail as follows:- Asclepiadaceae/Apocynaceae (Jolivet, 1971c, 1982), Araliaceae (Jolivet, 1974), Rutaceae (Jolivet, 1979a) and Malvales (Jolivet, 1972a). The excellent paper of Hsiao (1986) has reviewed the selection of Solanaceae by leaf-beetles. The choice of the plant families here is arbitrary, but it was not our intention to review one by one all of the one hundred or so plant families selected by the Chrysomelidae since this would have produced a large book by itself, and because, for many families, the trophism is secondary, if not accidental. The selection of primitive plants by the Alticinae is often a secondary adaptation, but this is not the case with the Sagrinae, Criocerinae and Aulacoscelinae, where feeding on Cycadales is perhaps, when all records are confirmed, a recollection of old Jurassic or Cretaceous selections when the Bennettitales were abundant and often the only available food.

The plant families or groups are discussed here in alphabetical order within the Angiospermae. Whether or not the Monocotyledonae or the Dicotyledonae are selected, does not necessarily have any phylogenetic significance, e.g. the highly evolved Hispinae feed predominantly on Monocotyledonae which are often regarded as primitive. Botanists now generally believe that both classes of plants originated together and that the monocots are not anterior to the dicots. Selection of the Coniferae among the Gymnospermae represents only a recent adaptation, as the choice of tree ferns *(Cyathea)* by many Oriental Alticinae. There are practically no published records of Chrysomelidae feeding on epiphyllous plants (e.g. algae, lichens) and on fungi. This does not mean that feeding on cellular cryptogams does not occur at all. Probably the subterranean larvae or adults feed on roots but also on fungi (even bacteria) on or in the root nodules of leguminous plants. It is a well-known fact that several Alticinae and Galerucinae (*Cerotoma*) are attracted by the nitrogen-fixing *Rhizobium* nodules. Clytrine larvae are completely polyphagous and feed on all plant or animal material available in the host ant nests, including fungi or algae. The subterranean larvae of *Clavicornaltica* most probably feed on mosses, algae and fungi. Case-bearing larvae of the Camptosoma, mainly Lamprosomatinae and Chlamisinae, feed on algal and lichen deposits on the bark of trees. Morelet (pers. comm.) has observed the special attraction for conidial "lawns" among *Crepidodera* (= *Chalcoides*) *aurea* (Geoffroy) (Alticinae) on *Populus tremula* (Salicaceae). These conidia belong to the fungus *Venturia* (Ascomycetes) and the beetle appears to prefer eating the fungus rather than the healthy foliar parenchyma. This curious phenomenon continues for several years during the vegetative period of the tree and is probably due to the preference for the more tender fungus to the tough parenchyma of the old leaves. According to Morelet, this strange behaviour contributes to the dissemination of the disease on the aspen trees.

Pollen feeders are relatively common among various Chrysomelidae, mostly among Galerucinae, Hispinae, Megascelinae and Alticinae. Many adult Galerucinae (e.g. *Aulacophora* and *Diabrotica*) feed on flowers, generally those of Cucurbitaceae. Samuelson (1989) has shown that many Alticinae, (e.g. *Arsipoda* and *Nonarthra*) from the Pacific, Japan and India have pollen, either from a single plant species or from different ones, densely packed in the gut. Before being digested, this pollen appears to germinate inside the gut of the insects (Arnett, 1969; Haslett, 1983). Actually, there is an exudation of the pollen grain contents through one of the pores in the pollen exine. This may begin as a growing pollen tube and the nectar/sugar inside the gut provides the necessary stimulus.

Strangely enough, no Chrysomelidae have as yet been recorded as feeding on leaves or flowers or even roots of carnivorous plants. However, it is a fact that beetles, including Chrysomelidae, are sometimes captured by pitcher plants and are part of the necrocoenosis. For instance, water beetles (Dytiscidae) are commonly captured by the temporarily flooded *Sarracenia psittacina* (Sarraceniaceae) in the eastern United States. Leaf-beetles are sometimes caught by this pitcher plant.

Chrysomelidae can be collected at very high altitudes in the tropical mountains. Alticinae and Galerucinae are able to survive until the disappearance of the vegetation at the upper levels. *Xingeina* and *Shaira* (both Galerucinae) can be captured on *Potentilla* (Rosaceae) up to 4200-4400 m in the Yunnan, China (Chen *et al.*, 1987). One of us (PJ) has found *Ruwenzoria viridis* Laboissiere (Galerucinae) on *Erica* trees in the Ruwenzori at 3000 m above sea level but there are Alticinae which occur at much higher altitudes. There are also Alticinae on the Venezuelan Tepuys and at very high altitudes in the Andes, South America.

Chrysomelidae are sometimes associated with epiphytic and hemiparasitic plants such as Loranthaceae in the tropics. Others regularly feed on terrestrial or epiphytic, myrmecophilous plants. Chrysomelids even occur in ant gardens. They possess special protective adaptations against the ants with whom they share the food-plant. Some leaf-beetles such as *Coelomera* are protected by reflex bleeding (in the adults) and cycloalexy (in the larvae) as well as by regurgitation. They feed on leaves, food bodies, pearl bodies and probably nectar from the female inflorescence. Others possess trichomes and live inside the stipular thorns of the East African *Acacia*, together with ferocious *Crematogaster* ants. Other chrysomelids, i.e. Hispinae, keep away from the ants, and their larvae mine the leaves and are protected by the leaf tissues. Many Alticinae and Galerucinae live on American and Asian myrmecophilous plants (Jolivet, 1991d).

The leaf-beetles feeding in the canopy of tropical forests are also numerous but their associations with plants remain unknown. Mostly Galerucinae, Alticinae and other good flyers such as Eumolpinae or Megascelinae, colonize the tree tops.

Biological studies on the root-feeding Eumolpinae larvae have been undertaken several times (Ferronatto, 1988). It is well-known how difficult it is to rear these ground-dwelling grubs. Studies on the root-feeding Alticinae, Galerucinae and Synetinae should be very interesting and worth comparing with what is known of the Eumolpinae. As far as is known, some larvae are specialized and among species of *Longitarsus*, one species feeds on the exterior, and another one on the interior, of rootlets. Leaf-beetle larvae are often polyphagous when they are subterranean but their adults can be either polyphagous or very selective. Eventually, the Alticine larvae can be restricted to the roots of their adult host-plants and so be serious pests of crops.

2. *Bryophyta*

There are some classical records of Alticinae which are purported to feed on mosses, e.g. *Mniophila muscorum* Koch, which is commonly found in Europe among *Hypnum striatum,* and *M. ruficollis* Motschulsky, which occurs among mosses in Ceylon (Sri Lanka). *Mniophilosoma laeve* Wollaston from Madeira probably has the same habits and is purported to feed on lichens. The species of the European genus *Minota (Hypnophila)*, the high altitude Singhalese genus *Kamala* and the Asian *Clavicornaltica* are purported to be moss feeders in the adult stage. *Kamala* frequents damp places, but has never been observed feeding. However, this matter is still open to question because Hering (1951) found *Minota* larvae mining leaves of *Plantago* (Plantaginaceae) and *Digitalis* (Scrophulariaceae). *Minota* can even be a fern miner at the larval stage (Kato, 1991). These families are very closely related. *Kamala* has been found on *Clausena* (Rutaceae) in Indonesia. *Clavicornaltica* must be partly, even in the adult stage, a detritus- and root-feeder.

It appears that there are no proven moss or hepatic feeders among the Chrysomelidae, but more observations are needed. Recently (unpublished data) a grasshopper has been found feeding on algae inside a water pool on the summit of the Venezuelan Tepuys. This is most unusual even for a grasshopper so it is probable that there are also unusual methods of feeding among the Alticinae that remain unobserved and thus unrecorded. In addition, Crowson (pers. comm.) has found a galerucine larva in Australia with gut contents clearly showing pieces of moss. The larvae were collected from a mossy, dead, standing tree trunk in southern Queensland.

3. Vascular cryptogams

Equisetales.

Several Alticinae actually feed on *Equisetum*, such as the Holarctic *Hippuriphila*. The mouthparts are probably adapted for feeding on an epidermis rich in silica.

Pteridophyta.

Hendrix (1980) records only five genera and six species of Chrysomelidae feeding on ferns. There are actually many more than this, but most of the fern-feeders belong to the Alticinae, e.g. *Alema* on tree ferns (*Cyathea*, Cyatheaceae) in New Zealand, *Schenklingia* with two species on *Nephrolepis* (Oleandraceae) in Malaysia and Micronesia, *Setsaltica* on *Cyathea* at high altitudes in New Guinea, *Manobia cyatroxae* Samuelson on *Cyathea atrox* and other ferns in New Guinea, *Leptophysa* (*Pseudoepitrix*) *trinitatis* Bryant on *Adiantum tenerum* (Adiantaceae) in Trinidad, *Leptophysa hoffmani* Bryant on a fern in Puerto Rico, *Febra venusta* Clark on *Nephrolepis* (Oleandraceae) and other *Febra* species on *Angiopteris* (Angiopteridaceae) and *Acrostichum* (Pteridaceae). *Febra* is a leaf-miner and the genus contains about 20 species distributed in the eastern tropics. There are actually many more Alticinae adapted to tree ferns in the tropics.

Recently (Kimoto, 1984; Kato, 1991), *Manobia lewisi* Jacoby has been reported feeding on *Cyclosorus acuminatus* (Aspidiaceae); *Minota nigropicea* (Baly) on *Pteridium aquilinum*, *Dryopteris erythrosora* and *Cyrtomium fortunei* (Aspidiaceae); *Schenklingia hiranoi* Takizawa on *Lemmaphyllum microphyllum* and *Loxogramma salicifolia* (Polypodiaceae); *Schenklingia sauteri* (Chen) on *Colysis elliptica* and *Colysis prothifolia* (Polypodiaceae). All are leaf-miners in Japan and Okinawa.

Among Alticinae, some rare *Asiorestia* have been collected on *Cyathea* but this appears to be purely accidental. However, several *Exoceras* (Alticinae) have been found on *Adiantum* (Adiantaceae) in the West Indies and also *Mimastra* (Galerucinae) on *Adiantum* in Thailand. *Galerucella birmanica* Jacoby normally feeds on the water-lilies *Nuphar* and *Nymphaea* (Nymphaeaceae) but has also been captured feeding on *Salvinia* (Salviniaceae). *Distigmoptera* (Alticinae) has also been captured on the aquatic fern *Azolla* (Azollaceae) and *Pseudolampsis guttata* (LeConte) also feeds on *Azolla caroliniana* in Florida, USA.

As noted above, the Alticinae are the dominant group that feed occasionally or normally on Pteridophyta. This phenomenon is more common in the Far East and is certainly a secondary adaptation which, in the Pacific, is due to the scarcity of other suitable food-plants in the mountains.

4. Gymnospermae

Few chrysomelids have been found feeding on Ginkgoaceae, either in the wild (Chekiang) or in places where these trees have been cultivated. At least one galerucine, *Morphosphaera ginkgoae* Gressitt & Kimoto, has been reported feeding on *Ginkgo biloba* in China. However, this is exceptional and since other species of the genus do not feed on *Ginkgo*, we can question this record.

Many Chrysomelidae feed on Gymnospermae. With Cycadales, e.g. *Carpophagus* on *Macrozamia* and *Aulacoscelis* on *Cycas* and *Zamia*, the relationships (if proven in the case of *Carpophagus*) are certainly an indication of antiquity. Two species of *Lilioceris* feed on *Cycas*, i.e. *L. consenta* Lacordaire in Vietnam (Medvedev & Dap, 1983) and *L. clarkii* Baly in New Guinea (Szent-Ivany *et al.*, 1956). The adults and larvae occur in the unfolded fronds of the new leaves where they feed on the leaf tissues.

The very primitive *Cucujopsis* and *Palophagus* have been reared from the male cones of *Araucaria* and/or *Agathis* (Araucariaceae) in Australia (Kuschel & May, 1990). This may represent a modern chrysomelid type which has preserved the attachment to Gymnospermae dating from the origin of the Angiospermae and later to the break-up of Gondwanaland (Crowson, 1991).

There are many reports of various Chrysomelidae feeding on *Pinus, Abies, Larix, Juniperus, Metasequoia, Cunninghamia, Chamaecyparis* etc. In addition, certain *Distigmoptera* (Alticinae) have been found on *Pinus*, certain *Monolepta* and *Arthrotus* (Galerucinae) on *Metasequoia, Sphenoraia* (Galerucinae) on *Pinus*, *Cneorane* (Galerucinae) on *Cunninghamia* and *Luperus* (Galerucinae) on *Pinus*, etc. *Metasequoia glyptostroboides* is the only living species of this genus of Taxodiaceae occurring in China and appears to attract many Chrysomelidae. Apart from *Monolepta* and *Arthrotus* previously mentioned, two galerucines, *Atrachya menetriesi* (Faldermann) and *Paleosepharia liquidambara* Gressitt & Kimoto possibly feed on *Metasequoia* as well. Several Clytrinae and Eumolpinae normally or occasionally feed on pine needles. *Monolepta bifasciata* (Hornstedt) has been observed feeding on *Gnetum* (Gnetales) in Indonesia (Kalshoven, 1951). *Syneta* are polyphagous but several species feed specifically on *Abies*, *Tsuga, Picea, Larix* and *Pinus*.

5. *Angiospermae*

Monocotyledonae.

Are the monocots more primitive than the dicots? This does not appear to be the case. The absence of secondary growth, a feature shared with the more primitive Gymnospermae, is an inferior character, but based on other features, the monocots are advanced, e.g. possession of two integuments by the ovules. According to Cronquist (1981), it appears that monocots are derived from primitive dicots. The origin of the monocots lies in the lower Cretaceous (Aptian), a time when the world was already populated by leaf-beetles which probably fed on the pseudoflowers of the Bennettitales during the Jurassic. The monocots have parallel-veined leaves and it is this feature which determines the direction that the beetles feed.

Highly evolved subfamilies, i.e. Hispinae and Cassidinae, and primitive subfamilies, e.g. Criocerinae and Donaciinae, feed on monocots, but not exclusively. For Cassidinae, monocot feeding is exceptional. Several Criocerinae feed on Solanaceae and many Hispinae also accept dicots. At least 28 families ranging from Pandanaceae to Orchidaceae are selected by the Chrysomelidae but space does not allow a detailed analysis for each of these families. As far as is known, the Hispinae feed on 19 families of monocots and 44 families of dicots. Even if we consider that the selection of Donaciinae is more ecological than botanical, they feed mostly on monocots (at least in the larval stages) and the adults mostly prefer the Nymphaeaceae to any other plants. Only primitive Cassidinae feed on palms (Arecaceae) but many Hispinae do. Monocot feeders are found among most of the subfamilies of leaf-beetles with several notable exceptions including the Zeugophorinae. For instance, *Exosoma* larvae (Galerucinae) feed inside the bulbs of Amaryllidaceae and Liliaceae. Feeding on Poaceae pollen by Alticinae and others such as Megascelinae does not interfere with the normal trophism and merely completes the protein intake of the insects. Orchid feeders mainly occur among the Criocerinae and several genera, e.g. *Petauristes*, are specialized on this diet. The Hispinae on Zingiberales are very selective and their biology has been studied by Strong (1977a-1982) in Central America.

6. *Angiospermae Dicotyledonae*

It has been necessary to select only several of the 120 or more families of dicots upon which leaf-beetles feed because to cover all of them in any detail would certainly require more than one volume just in itself. Also, some plant families are rarely mentioned in the literature, while others are rare, exotic families. Our selection is according to the families that are most often recorded in the literature as host-plants. Despite some omissions, it is hoped that our list will be of interest and use to researchers studying leaf-beetle relationships. A general table arranged family by family or genus by genus was considered but this would have enlarged the book to unmanageable proportions.

We take the opportunity to again emphasize that nothing is known about the food-plants in certain areas of the world such as Madagascar and the Tepuys of the Venezuelan Guyana. If these were known, they would have added many endemic genera or even endemic families of plants to our host-plant lists.

6.1. *Anacardiaceae*

This family contains about 60 genera and 600 species of mainly tropical plants but also some Mediterranean ones. Generally they are trees or shrubs, sometimes edible and non-toxic (e.g. *Mangifera* and *Pistacia*), sometimes with resin which causes allergies or sometimes poisonous to the touch (e.g. *Rhus, Toxicodendron*, *Anacardium* and *Schinus*). Several flea beetles such as *Orthaltica, Blepharida, Podontia* and many others, feed only on Anacardiaceae (Furth, 1982, 1985b). *Serraticollis* (= *Orthaltica*) was previously recorded by Maulik (1945) as well as *Blepharida* and *Longitarsus* on this family. Furth (1982,

1985b) recorded 39 Chrysomelidae from *Rhus tripartita* and other *Rhus* species, including species of Clytrinae, Cryptocephalinae, Chlamisinae and Eumolpinae. Many of these insects are polyphagous. The biology of *Blepharida sacra* (Weise) on the sumac tree *Rhus tripartita* has been studied by Furth (1982). Eighteen flavonoid compounds have been isolated from various *Rhus* and *Schinus* species (Furth & Young, 1988). It appears that flavonoid composition of the different host-plants are correlated with the food-plant preference of the beetle species.

The "venomous" Alticinae from the Kalahari Desert feed, in the case of *Polyclada flexuosa* Baly, on *Sclerocarya caffra* (Anacardiaceae) while two species of *Diamphidia* feed on *Commiphora* (Burseraceae). However, the two families are only weakly related; the Burseraceae family belongs to the Geraniales, while the Anacardiaceae family belongs to the Sapindales.

6.2. Araliaceae

This is a small, mostly tropical family of trees and shrubs. Some are palm-like (e.g. *Schefflera*), some are twiners while others may be ivy root climbers. The Araliaceae possess resin ducts in the stems and branches, are rich in saponins, do not produce alkaloids but do produce ethereal oils and resins, which often gives them a peculiar odour. One famous Oriental and American genus, *Panax*, does not appear to be host for any chrysomelid.

Actually, there are only comparatively few Chrysomelidae which feed on Araliaceae (Jolivet, 1974). The Holarctic *Oomorphus* and the Oriental *Oomorphoides* (Lamprosomatinae) feed on *Hedera, Aralia, Kalopanax, Acanthopanax, Tetrapanax, Dendropanax* and *Schefflera. Oomorphus concolor* Sturm has been recorded on *Hedera* (Araliaceae) and on several Apiaceae (*Astrantia, Aegopodium* and *Chaerophyllum*) in western Europe. In the tropics, the Lamprosomatinae have a completely different selection of *Eugenia, Eucalyptus, Psidium* (Myrtaceae), *Chorisia* (Bombacaceae), *Terminalia* (Combretaceae) and *Vachellia* and *Acacia* (Mimosaceae). *Oomorphus* larvae have a polyphagous diet and feed at night on plant detritus, leaves on or in the ground. One species has been found in Florida feeding on rodent excreta. Outside Lamprosomatinae, the beautiful Chrysomelinae of the genus *Promechus* (= *Aesernia*) feeds on *Schefflera* and *Boerlagiodendron*, palm-like shrubs growing along streams in New Guinea and Queensland. Strangely enough, other leaf-beetles feeding on Araliaceae are not known, but there must be more to be found in the tropics.

6.3. Asclepiadaceae and Apocynaceae (Gentianales)

Mostly the Eumolpinae feed on Asclepiadaceae and the related Apocynaceae (Jolivet, 1971c, 1982, 1987d). It appears that these plants are the basic food choice of the Eumolpinae from which all of the others are derived, including that of Convolvulaceae.

The Asclepiadaceae are erect or twining shrubs or perennial herbs, mostly found in the tropics or subtropics. Their rootstocks are tuberous and frequently eaten by Eumolpinae larvae. They are pollinated mostly by Hymenoptera but Chrysomelidae are not involved. The laticifers produce very toxic cardiotonic glycosides and alkaloids. The related Apocynaceae are mostly tropical and it is only in the tropics or subtropics that they are attacked by chrysomelids, with perhaps some exceptions in Asia.

Several polyphagous eumolpine genera occasionally feed on Gentianales, e.g. *Metachroma* on *Asclepias* and *Rauwolfia, Nodina* on *Secamone, Eumolpus* on *Asclepias* and *Nerium, Phytorus* on *Ochrasia, Colasposoma* on *Cynanchum* and *Colaspis* on *Araujia. Phygasia* (Alticinae) feeds on *Cynanchum, Metaplexis, Calotropis* and *Kanahia* (Asclepiadaceae) in Asia and Africa, but also accepts *Ipomoea* (Convolvulaceae).

Several Eumolpinae are ultra-specialized, such as *Euryope* on *Leptadenia* (Asclepiadaceae) in Africa, *Platycorynus* on *Trachelospermum* (Apocynaceae) and *Tylophora, Antherandra, Calotropis* and *Asclepias* (Asclepiadaceae) in Asia, *Chrysochus* and *Chrysochares* on *Asclepias, Cynanchum, Metaplexis* and *Antitoxicum* (Asclepiadaceae) and *Apocynum, Nerium* and *Trachomitum* (Apocynaceae) in the Holarctic region.

As shown by Dussourd and Eisner (1987), many insects which feed on Asclepiadaceae and other laticiferous plants, cut the leaf veins before feeding distally away from the cuts. Vein-cutting blocks the latex flow to the intended feeding sites and can be regarded as an insect counter-adaptation to the plant's defensive secretions. The vein-cutting behaviour has been described first in *Labidomera clivicollis* (Kirby) a species from the United States which feeds on *Asclepias* and *Cynanchum.*

6.4. Cecropiaceae

The Cecropiaceae, formerly part of the Moraceae or Urticaceae, are dioecious trees, shrubs or lianes, but are sometimes epiphytic with aerial or stilt roots and a rather restricted latex system. Most of them are more or less closely associated with ants, especially the *Cecropia* trees with their trichilia and müllerian bodies. *Cecropia* are pioneer trees in disturbed areas in South America. There is a type of symbiosis between ants and *Cecropia* and the closely related *Pourouma* (Jolivet, 1987d). Among the insects, several Chrysomelidae are specialist feeders on *Cecropia*, such as *Coelomera* (35 species) and *Dircema* (25 species)(both Galerucinae). On *Cecropia* are also found *Cacoscelis marginata* (Fabricius) (Alticinae), *Metaxycera subapicalis* Bondar, *Sceloenopla* spp. (Hispinae) and many other Chrysomelidae, including some that are still undescribed. *Coelomera* spp. and probably *Dircema,* not only feed on leaves, but they also feed on the müllerian bodies of the trichilia, while several species of *Coelomera* (but not all) dig into the prostoma and lay their eggs inside the stem instead of the *Azteca* queen (Andrade, 1984; Jolivet, 1988c). All of the ant-plant Chrysomelidae defend themselves by avoiding contact with the ants, by mining into the leaf (i.e. Hispinae), by being toxic and displaying reflex bleeding or enteric (stomach) discharges (i.e. Galerucinae/Alticinae) or by displaying larval cycloalexy (*Coelomera*) (Jolivet, 1988b). Nothing is known about the Asian or African Cecropiaceae. Their associations with ants appear to be looser but they can also be patrolled by very aggressive ants.

6.5. Chenopodiaceae

Together with the Amaranthaceae and Caryophyllaceae, Chenopodiaceae form part of the Centrospermae. Several Chrysomelidae display this range of feeding selections. Many Cassidinae feed on Chenopodiaceae (e.g. *Oxylepis, Ischyronota, Cassida, Mionycha* and *Cassidulella)*, on Amaranthaceae (e.g. *Cassida* and *Taiwania*) and on Caryophyllaceae (e.g. *Cassida, Odontionycha, Mionycha* and *Cassidulella*). Other leaf-beetles from the Galerucinae and Alticinae feed on Chenopodiaceae and its relatives, but there are not many. For example, *Adoxinia spinipes* Reitter (Eumolpinae), has been frequently collected on *Haloxylon, Aellenia* and *Salsola* (Chenopodiaceae) in Central Asia, but this could be due to the dry and salty soil which does not permit the growth of many plants in these areas (Lopatin, 1984).

6.6. Compositae (Asteraceae)

Many Chrysomelidae feed on Asteraceae, e.g. among the Chrysomelinae are the genera *Chrysolina, Calligrapha, Oreina, Cyrtonus* and *Elytrosphaera*, but they do not feed exclusively on Asteraceae. This selection must be more common in Asia where *Artemisia* is a dominant plant among the fallow land. The selection of Asteraceae appears to be exclusive for some species or is associated with related plant families but it is not certain whether the double selection of Asteraceae and Apiaceae by *Oreina* results from the same species or from the presence of siblings. Experiments must be undertaken on the same individual beetle to determine if it really accepts both plant families. It is most likely that different biotypes of some *Oreina* species exist, or that there are two or more species being confused taxonomically by entomologists. In America, certain species of *Longitarsus* feed on Asteraceae (e.g. *Hymenoclea*). Otherwise, the Chrysomelidae feeding on Asteraceae, e.g. species of Hispinae, Alticinae, Galerucinae, Eumolpinae, Cassidinae and Cryptocephalinae, are numerous. Some polyphagous species like *Systena* (Alticinae), feed on Asteraceae along with many hundreds of other plants.

6.7. Convolvulaceae

There are a number of Cassidinae, apart from the hispinomorphic primitive species which feed on palm trees, which prefer the Convolvulaceae. This appears to be the primitive choice of the subfamily. Tempere (1935) outlined the preference of the European Cassidinae for the Centrospermae, but even in Europe, the tendency towards Convolvulaceae remains latent. In the Neotropics, *Cordia* (Ehretiaceae) and Convolvulaceae are predominant host-plants of Cassidinae, whereas Convolvulaceae alone dominates in the Pacific. In this region, *Merremia, Calystegia, Convolvulus* and *Ipomoea* are frequent choices, together with the Cuscutaceae and Bignoniaceae. Many other chrysomelids feed on Convolvulaceae, either partially like some Eumolpinae (i.e. *Labasa, Colasposoma, Menius* and *Typophorus*), or exclusively like some *Chaetocnema* (Alticinae). Some *Sagra* and many Alticinae and Galerucinae, also feed occasionally on *Ipomoea*.

6. 8. Cruciferae (= Brassicaceae) and related families

The Brassicaceae is a very natural and important cosmopolitan family, with strong development in the north temperate zone. They are mainly herbs with a few undershrubs. This family is related, at least chemically, to the Capparidaceae, Resedaceae, Tropaeolaceae, Cleomaceae, etc. This explains why many Chrysomelidae, e.g. *Phyllotreta* (Alticinae), select the above families. Many chrysomelids are crucifer feeders, e.g. *Colaphellus, Entomoscelis, Phaedon, Microtheca* (Chrysomelinae) and *Phyllotreta, Psylliodes* (Alticinae). It is impossible here to list all of the exclusive or partial crucifer-feeders. Several polyphagous genera, e.g. *Galeruca, Diabrotica, Monoxia* and *Systena*, feed occasionally on various Brassicaceae.

6.9. Cucurbitaceae

In the Old World are to be found many Luperini (Galerucinae) feeding on Cucurbitaceae, e.g. *Aulacophora, Lamprocopa, Leptaulaca, Pseudocophora, Agetocera, Diacantha, Paragetocera, Paridea* and *Asbecesta*, while in the New World there are *Paranapiacaba, Acalymma, Paratriarius, Isotes, Cerotoma* and even the relatively polyphagous *Diabrotica* (all Galerucinae). The Old World genera are oligophagous and are more restricted to Cucurbitaceae. All of these species feed on leaves, pollen and flowers. Their larvae are subterraneous and feed on the roots of various plants. There are other cucurbitaceous feeders among the Chrysomelidae but the great majority occur among the Galerucinae and Alticinae, mostly in the tropics. The specificity on Cucurbitaceae is variable and sometimes the bitter species are rejected by the leaf-beetles; however, almost all Cucurbitaceae genera can be attacked, e.g. *Momordica, Luffa, Citrullus Cucumis, Benincasa, Lagenaria, Trichosanthes, Cucurbita, Echinocystis, Melothrix, Thladiantha, Gynostemma* and *Sechium*.

6. 10. Ehretiaceae and Boraginaceae

In South America, the Cassidinae are frequently found on *Cordia* shrubs (e.g. *Physonota*), but many *Cryptocephalus* and Galerucinae also feed on these plants. Many Alticinae, e.g. *Longitarsus*, feed on Boraginaceae (e.g. *Heliotropium, Symphytum, Anchusa* and *Myosotis*).

With 13 genera and 400 species, the entirely tropical family Ehretiaceae is scarcely distinct from the Boraginaceae. However, the Boraginaceae are not only tropical but are also temperate and especially Mediterranean in distribution. Some recent authors (e.g. Mabberley, 1987) do not separate the two families.

To combat the spread of *Cordia* introduced from the New World, the cassidine *Physonota alutacea* Boheman and the galerucine *Schematiza cordiae* Barber have been introduced into Mauritius from Trinidad but only the second species has survived.

6.11. Euphorbiaceae

The Euphorbiaceae are an enormous, somewhat cosmopolitan family with 300 genera and over 5,000 species. Many species contain toxic latex in special laticiferous vessels. Whether they are trees, shrubs, vines or herbs, the Euphorbiaceae are colonized by many leaf-beetles adapted to their toxicity. Curiously enough, no leaf-beetles appear to feed on *Hevea brasiliensis*, at least in the regions where this tree has been introduced. The milky latex and alkaloids found in many members of this family have not been successful in deterring many Alticinae and other leaf-beetles and insects. *Stignobia* (Eumolpinae) feed on very toxic plants belonging to the Rhamnaceae, Strychnaceae and Urticaceae and on *Glochidion* (Euphorbiaceae) in Samoa. *Cleoporus* (Eumolpinae) feeds on *Phyllanthus* while *Phytorus* feeds on *Mallotus*, but also on *Macaranga*, an ant-tree from South-East Asia. This is not the only leaf-beetle known to feed on *Macaranga*. There is also one species of *Coenobius*, one *Cadmus* (both Cryptocephalinae), two *Demotina*, one *Trichochrysea* (both Eumolpinae) and one *Aphthona* (Alticinae). These beetles have adapted both to the toxicity of this plant and to its fierce ant inhabitants. Several Clytrinae, Eumolpinae, Cryptocephalinae, Galerucinae and Alticinae have been captured on species of *Phyllanthus, Croton, Mercurialis, Mallotus, Ricinus, Caperonia, Trewia, Aleurites, Manihot, Sapium* and *Euphorbia*. Apart from the polyphagous *Systena*, there are Alticinae such as *Aphthona, Hermaeophaga, Psylliodes, Monomacra* (mostly on Passifloraceae, but with some species on *Ricinus* and *Caperonia*), and some *Oedionychus*, which are able to feed on *Euphorbia* and related species.

6.12. Lamiaceae and Verbenaceae

These two families are very closely related and together with the Scrophulariaceae, Bignoniaceae and Boraginaceae (and others) comprise the Tubiflorae. The Lamiaceae have a chief center of diversity in the Mediterranean region but they are also represented in the tropics. They possess epidermal glands which secrete volatile oils with strong scents which attract many insects. The Verbenaceae are herbs, shrubs, lianas and small to large trees, sometimes with an odour approaching that of some Lamiaceae. The genera *Lantana, Clerodendrum* and *Vitex* (Verbenaceae) and *Mentha, Thymus, Teucrium, Rosmarinus, Stachys, Calamintha, Hyptis* and *Ocimum* (Lamiaceae), support many leaf-beetles. Many Chrysomelidae (e.g. *Chrysolina, Apteropeda, Oedionychus, Longitarsus, Mniophila, Cryptocephalus, Phyllobrotica, Arima, Crepidodera, Dibolia* and *Cassida*) feed normally (some *Chrysolina*) or occasionally (*Arima*) on Lamiaceae. Numerous are the flea-beetles which feed on Verbenaceae in

South America, (Begossi, 1984). Species of *Alagoasa, Homophoeta* and *Asphaera* are most commonly found on Verbenaceae but they also feed naturally or experimentally on Lamiaceae. In the Far East, *Phyllocharis* (Chrysomelinae) and *Hoplosomoides* and *Hoplasoma* (both Galerucinae), feed on *Clerodendrum, Vitex* and *Premma* (Verbenaceae) (Jolivet, 1983). *Oidosoma, Pseudomela* (Chrysomelinae), *Argopistes hargreavesi* Bryant, *Cladocera uniformis* Jacoby (Alticinae) and *Xenarthracella* (Galerucinae), feed on *Clerodendrum* in Africa. *Phola* (Chrysomelinae) feeds on *Vitex* in the Far East. In Vietnam, *Xanton* (Eumolpinae) feeds on *Callicarpa* and several Cassidinae feed on *Gmelina* (Verbenaceae). In South America some Alticinae feed on *Clerodendrum, Aegiphila, Phyla* and *Stachytarpheta* (Verbenaceae)(Virkki, 1980). Some appear to be host-specific, i.e. *Oedionychus*.

6.13. Leguminosae (Fabaceae, Mimosaceae, Caesalpiniaceae)

The Leguminosae, containing three families, are the third largest group of flowering plants with over 12,000 species. It is upon the Fabaceae that most of the polyphagous or oligophagous chrysomelids feed. The legumes have a poorly documented evolutionary history. The recent fossil evidence (Crepet & Taylor, 1985) provides data on the timing of the origin of the two derived families, Mimosaceae and Fabaceae. These data strongly suggest the importance of bee pollinators during a major period of angiosperm diversification. Leaf-beetles rarely feed on legume flowers, but many other beetles do. However, Eumolpinae and others feed on legume flowers in Australia (Hawkeswood & Jolivet, 1988). It is impossible here to list all of the leaf-beetles which feed on Leguminosae, but special mention can be made of several genera in Asia and Africa, e.g. *Mesoplatys* and *Phaedonia* (Chrysomelinae) which only feed on Fabaceae (i.e. *Indigofera, Aeschynomene, Sesbania* and *Erythrina*) (Jolivet, 1952a, 1977; Jolivet & Parys, 1977). There are native leaf-beetles which feed on soyabean (*Glycine max)* in Asia, but in the West, several species which are adapted generally to legumes have shifted to cultivated crops, e.g. *Phaedonia inclusa* (Stål), in Indonesia, *Chaetocnema heikertingeri* Ljubiscev in Italy, *Diabrotica undecimpunctata howardi* Barber in Illinois and *Ootheca mutabilis* Sahlberg in Africa. There are also several polyphagous leaf-beetles which feed occasionally on soyabean in the USA, i.e. *Cerotoma, Diabrotica, Acalymma* and *Colaspis*.

Some Chrysomelidae and especially some Galerucinae (i.e. *Craniotectus, Niasia* and *Diaphaenidea*) feed on *Derris* (Fabaceae) in Asia. The plant-like *Lonchocarpus* is insecticidal and used as a fish poison and for natural insecticides. One *Cryptonychus* species (Hispinae) feeds on *Lonchocarpus*.

6.14. Malvales (Malvaceae, Sterculiaceae, Bombacaceae, Tiliaceae)

The Malvales comprise at least 8 families of plants which are more or less closely related. The Malvaceae are herbs, shrubs or trees which are indiscriminately chosen as food by high-flying Alticinae. Their distribution is both tropical and temperate. The Bombacaceae are chiefly trees, often very large, with thick stems (e.g. *Adansonia* and *Bombax*) while the Sterculiaceae are trees and shrubs (e.g. *Cola* and *Theobroma*) and sometimes lianes and are also related to the Euphorbiaceae. The Tiliaceae are trees and shrubs (e.g. *Tilia*, *Triumfetta* and *Corchorus)* and are closely related to the Malvaceae.

The Old World Alticinae of the *Nisotra/Podagrica* complex feed exclusively on the Malvales but the specificity for *Hibiscus*, the most commonly selected host, is not very strong (Jolivet, 1972a). Many *Podagrica* and *Nisotra* feed on cotton (*Gossypium)* and kenaf (*Hibiscus*) in Africa and Asia. Other leaf-beetles feed normally on these plants, e.g. *Cleoporus hibisci* Gressitt and *Rhyparidella hibisci* Gressitt (both Eumolpinae) in Papua New Guinea or occasionally, e.g. *Syagrus* (Eumolpinae) and *Erystus* (Alticinae). Many Eumolpinae are polyphagous, e.g. *Percolaspis*, *Rhapdopterus* etc.) and have adapted very well to cocoa (*Theobroma*, Sterculiaceae) in Brazil (Ferronatto, 1986, 1988).

6.15. Oleaceae

The Oleaceae are shrubs and trees found chiefly in temperate and tropical Asia and Africa. The affinities of the Oleaceae are in dispute. Overall, very few leaf-beetles feed on Oleaceae, but some genera, e.g. *Argopistes*, are strictly *Olea* feeders in Africa and elsewhere. Several polyphagous species feed on Oleaceae occasionally, e.g. *Diabrotica* (Galerucinae), but the genus *Argopistes* (Alticinae) is known only from *Mayepea, Fraxinus, Osmanthus, Ligustrum, Chionanthus, Olea, Jasminum, Syringa, Forestiera* and *Linociera* in Africa, Asia and America (introduced species). Oriental Megalopodinae such as *Temnaspis* and *Colobaspis*, feed on *Fraxinus* (Oleaceae). The female beetles cut the stem apex and lay an egg inside. The larva bores into the stem like the other American species. However, the biology of the Asian Megalopodinae is poorly known while nothing is known about the African species (Yu, 1977; Ahn, 1988, pers. comm.). *Trichaltica* (Alticinae) feeds on *Fraxinus* and *Ligustrum* in the United States.

6.16. Plantaginaceae

This is a cosmopolitan family of annual or perennial herbs, which are allied to the Scrophulariaceae. This relationship is reflected in the leaf-beetle selections. Many Chrysomelidae feed normally or potentially on *Plantago* (Jolivet, 1975), e.g. *Timarcha*, normally feeders upon Rubiaceae but which may switch to *Plantago* and Scrophulariaceae in the Mediterranean region and several *Chrysolina* species (i.e. *C. staphylea* (L.), *C. haemoptera* (L.), *C. limbata* (Fabricius), *C. rossia* (Illiger), *C. carnifex* (Fabricius), *C. banksi* (Fabricius), *C. pyrenaica* (Dufour), *C. crassicornis* (Helliesen), *C. sanguinolenta* (L.) and *C. colasi* (Cobos)), *Mniophila muscorum* Koch, two species of *Longitarsus*, one *Hypnophila*, one *Oedionychus*, one *Phyllotreta* and two *Apteropeda*, feed only on *Plantago* or as a secondary food source. Very often, the selection of Plantaginaceae is linked to either the selection of Rubiaceae/Dipsacaceae or Scrophulariaceae. For instance, *Apteropeda* (Alticinae) feed on Scrophulariaceae, Lamiaceae and Plantaginaceae.

6.17. Polygonaceae

Most of the Polygonaceae are herbs, but some are shrubs or even a few are trees. Mostly temperate in distribution, a few species are tropical. Several genera of Polygonaceae, e.g. *Rumex*, *Rheum*, *Polygonum*, *Fagopyrum* and *Coccoloba*, are selected by a large number of leaf-beetles. *Triplaris,* an ant-plant, is a tree and several Chrysomelidae feed on its leaves. *Gastrophysa* (Chrysomelinae) feeds on *Rumex*, *Rheum* and *Polygonum* (Jolivet, 1951a). *Labidostomis, Smaragdina* and *Cheilotoma* (Clytrinae) occasionally feed on Polygonaceae. Several species of *Galerucella*, *Galerucida*, *Diabrotica* (Galerucinae), *Mantura, Altica, Systena, Chaetocnema, Psylliodes, Disonycha, Oedionychus* (Alticinae), *Colaspis, Nodonota* (Eumolpinae) and *Cryptocephalus* (Cryptocephalinae), feed normally or accidentally on Polygonaceae.

6.18. Ranunculaceae

Most Ranunculaceae are herbaceous perennials, which curiously, are chosen by polyphagous Chrysomelidae which feed on a range of different, non-related plants. Several *Chrysolina* species display these tendencies, e.g. *C. staphylea* (L.), *C. obsoleta* (Brulle) and *C. banksi* (Fabricius). Species of the genus *Argopus* (Alticinae) feed on *Clematis* and *Pulsatilla* (Ranunculaceae) in Europe, Asia and North Africa. Among other Chrysomelidae, some species of *Entomoscelis, Neophaedon, Hydrothassa* (Chrysomelinae), *Galeruca, Luperodes* (Galerucinae), and *Crepidodera, Longitarsus, Psylliodes, Dibolia* (Alticinae), feed on Ranunculaceae, but only a few of them feed exclusively on this family.

6.19. Rosaceae

This cosmopolitan family is composed mostly of perennial trees, shrubs and herbs. The American *Timarcha* (Chrysomelinae) feeds on *Rubus* or *Fragaria. Orsodacne* (Orsodacninae) feeds on flowers of *Spiraea, Pyrus, Crataegus* and others. There are many leaf-beetles which feed on the leaves or flowers of Rosaceae. Some species are actually monophagous on only one or two species of *Rubus*, e.g. *Phaedon fulvescens* (Weise) in Vietnam (Jolivet, 1984). Other Chrysomelidae feeding on Rosaceae occur among the Chrysomelinae (*Calligrapha* and *Paropsides*), Synetinae, Chlamisinae, Clytrinae, Eumolpinae, Galerucinae, Alticinae, Cryptocephalinae and Hispinae. (See Chapter 2 for more details under the subfamilies).

6.20. Rubiaceae

This is one of the largest families of plants. Most are tropical, although a number of genera, including *Galium,* are temperate and also grow on tropical mountains. The family is composed of trees, shrubs, herbs and several are myrmecophilous, e.g. *Myrmecodia, Hydnophytum, Cuviera* and *Duroia*, which also support weevils and leaf-beetles. Relatively few Chrysomelidae feed on leaves of Rubiaceae. However, species of *Rhyparida, Paraivongius, Metachroma, Rhembastus, Typophorus, Colasposoma, Colaspoides* (Eumolpinae) and

Triachus, *Pachybrachis* (Cryptocephalinae) have been recorded from *Randia*, *Coffea*, *Cinchona*, *Uncaria* and *Ixora*, but these beetles are generalists (i.e. polyphagous) so that their choice of Rubiaceae is not specific. Several Galerucinae (i.e. *Diabrotica*, *Dercetis*, *Prosmidia*, *Arima*, *Cerophysa*, *Galerucella*, *Dercetina*, *Sphenoraia* and *Miltina*) occasionally feed on *Rubia*, *Galium*, *Coffea*, *Randia*, *Mitrogyna* and *Morinda* but these beetles are also generalists. In Africa, *Sagra* larvae can be gallicolous on *Coffea* trees, but they usually select leguminous shrubs of the family Fabaceae. Several Cassidinae, e.g. *Aspidomorpha*, *Metriona* and *Notasacantha* (*Hoplionota*), although they have been recorded on Rubiaceae, are not really feeders on this family, and the same is true for several Alticinae, i.e. *Hypolampsis*, *Jamesonia*, *Altica* and *Aphthona*, and Hispinae (*Platypria* and *Dactylispa*). *Systena* (Alticinae) feeds on *Borreria* in America. *Timarcha* readily feeds on Rubiaceae (i.e. *Galium*, *Rubia*, *Crucianella*, *Asperula* and *Sherardia*) but switches in the south of its range to Plantaginaceae, Dipsacaceae and Scrophulariaceae. *Neolochmaea tropica* (Jacoby) and related species of that genus (Galerucinae), eats the leaves of *Borreria verticillata* in Amazonia. *Sermylassa halensis* (L.) (Galerucinae) feeds exclusively on *Galium* in Europe and probably in Asia. Many other Neotropical Galerucinae feed on Rubiaceae. Among other Galerucinae, *Neolochmaea dilatipennis* (Jacoby) lives on *Borreria terminalis* in Miami, United States. This is a tropical species introduced from Panama (White, 1979). One *Dercetina* species has been recorded from *Mitrogyna* in Burma, *Sphenoraia nebulosa* Gyllenhal from *Oldenlandia* in Vietnam (Medvedev & Zaitsev, 1979) and *Miltina dilatata* Chapuis from *Morinda cochinchinensis* in Vietnam.

6.21. Rutaceae

The Rutaceae are mostly shrubs and trees and seldom herbs. Many are xerophytic, tropical or temperate, found extensively in South Africa and Australia. They are aromatic plants, of which the most familiar members are from the genus *Citrus* from South-East Asia. There are two types of Chrysomelidae frequenting the Rutaceae, i.e. the monophagous or oligophagous species and the polyphagous species appearing sporadically and accidentally on the plants where they may eventually nibble the leaves. The food preferences of leaf-beetles on Rutaceae have been previously reviewed by Jolivet (1979a) so only a few examples will be listed here. *Sagra* larvae are found rarely in *Citrus*, as they are, as previously mentioned, mostly legume feeders. Clytrinae are polyphagous and the occurrence of *Euryscopa*, *Labidostomis* and *Physauchenia* on young leaves of Rutaceae is not surprising. The same case exists with Cryptocephalinae (*Cryptocephalus*, *Diachus*), Eumolpinae (*Alethaxius*, *Chrysodina*, *Colasposoma*, *Geloptera*, *Glyptoscelis*, *Iphimeis*, *Maecolaspis*, *Metachroma*, *Nodonota*, *Phytorus*, *Sternocolaspis*), Galerucinae (*Acalymma*, *Cerotoma*, *Cochabamba*, *Derospidea*, *Diabrotica*, *Monolepta*, *Paranapiacaba*), Alticinae (*Altica*, *Aphthona*, *Argopistes*, *Crepidodera*, *Epitrix*, *Hermaeophaga*, *Homophoeta*, *Longitarsus*, *Nonarthra*, *Notomela*, *Oedionychus*, *Systena*) and

Cassidinae (*Cassida, Discomorpha, Metriona*). All these genera may rest accidentally or nibble on the leaves of Rutaceae but the plants are not the real hosts. However, *Amphimela, Cercyonia, Clitea, Podagricomela* and *Throscoryssa* (Alticinae) are oligophagous on *Citrus, Murraya, Clausena, Poncirus, Zanthoxylum, Aegle* and others. Most of these are Asian insects which feed normally on *Citrus*. Some like *Cercyonia* have adapted to cultivated *Citrus* from wild Rutaceae in Africa.

6.22. Salicaceae, Juglandaceae, Betulaceae, Fagaceae, Ulmaceae, Corylaceae and Carpinaceae

All of these plant families are composed mostly of trees, but they also contain shrubs or undershrubs mostly growing in cold areas. The Chrysomelidae feeding upon them are usually strong flyers.

The following genera occur on Salicaceae (e.g. *Populus, Salix*): *Zeugophora* (Zeugophorinae), Cryptocephalinae (*Pachybrachis, Cryptocephalus*), Chrysomelinae (*Chrysomela, Plagiodera, Calligrapha, Gonioctena, Phratora*), Eumolpinae (*Nodonota, Nodostoma*), Clytrinae (*Labidostomis, Smaragdina*), Galerucinae (*Galerucella, Lochmaea, Luperus*) and Alticinae (*Crepidodera, Altica*). Only a few of these (Zeugophorinae and Chrysomelinae) are truly Salicaceae feeders.

There occur several generalists on Juglandaceae such as *Paria* (Eumolpinae) and *Systena* and *Oedionychus* (Alticinae) but there are also some rare specialists (e.g. *Zeugophora*).

Apart from many generalists which feed on Betulaceae, Corylaceae and Carpinaceae, *Agelastica* (Galerucinae) feeds on *Alnus, Betula* and *Corylus* and various Rosaceae trees. *Calligrapha* (Chrysomelinae) feeds on *Alnus, Corylus* and others.

Several *Calligrapha* (Chrysomelinae), *Galerucella* and *Sastra* (Galerucinae) species feed on the leaves of Ulmaceae in America, Europe or the Far East.

There are many Alticinae recorded as accidental feeders on Fagaceae (e.g. *Quercus, Castanea* and *Castanopsis*), but there are also many generalist feeders on this family known from the Clytrinae, Cryptocephalinae and Eumolpinae. Surprisingly enough, there are very few specialists among the Chrysomelinae. However, on the Gondwanian genus *Nothofagus*, several Chrysomelinae have been captured, i.e. *Novocastria nothofagi* Selman (Chrysomelinae) in Australia, and *Mariamela wittmeri* Monros (Megascelinae), *Hornius grandis* (Philippi), *H. sulcifrons* Fairmaire and *Psathyrocerus* spp. (Eumolpinae) in Patagonia (Jolivet, 1987c).

6.23. Scrophulariaceae

The cosmopolitan Scrophulariaceae are mostly comprised of herbs and undershrubs. As noted previously, chrysomelids feeding on Scrophulariaceae such as *Timarcha* and several *Chrysolina* (Chrysomelinae), are associated with selections of *Plantago* (Plantaginaceae) and Rubiaceae. *Prasocuris junci*

(Brahm) (Chrysomelinae) feeds on *Veronica,* but apparently not exclusively. Some species of *Longitarsus* (Alticinae) feed on *Gratiola, Isoplexis, Pedicularis, Rhinanthus, Verbascum* and *Scrophularia.* Other Alticinae such as *Oedionychus, Mniophila, Hypnophila* and *Apteropeda* have also been recorded on Scrophulariaceae, but for many species, this family is only a secondary food source. Species of *Allocharis* (Chrysomelinae) feed on *Veronica* along the rivers in the mountains of New Zealand.

6.24. Solanaceae

The almost cosmopolitan family Solanaceae is composed of 85 genera and 2,800 species of herbs, shrubs, vines or small trees, which commonly produce various alkaloids. The family is best developed in tropical South America, where there are 40 endemic genera, but in Europe and Asia, only two or three genera are represented.

The Chrysomelidae feeding on Solanaceae are rather numerous in tropical and temperate America where many Doryphorina select *Solanum* as a food-plant, e.g. *Leptinotarsa* (pro parte), *Platyphora* and others (Chrysomelinae). These are almost the only Chrysomelinae recorded as feeding on the Solanaceae. Hsiao (1986) has summarized the known information on the "solanophagous" leaf-beetles and a summary of genera known to feed on Solanaceae is provided in Table 5 . The Chrysomelidae select either herbaceous species or small shrubs, but almost never trees. The *Solanum* treelets mainly grow in disturbed habitats in clearings in rain forests. The Convolvulaceae are not too distantly related to the Solanaceae, and even if the gradual evolution from one to the other is not clearly evident, it is possible to observe selections of both families by closely related genera of Cassidinae. Chrysomelidae feeding on Solanaceae are very rare in the Old World but they do occur and not only in the tropics. Some of the Solanaceae feeders are specialized, e.g. *Leptinotarsa* and *Platyphora* (Chrysomelinae), but others which are potentially polyphagous (e.g. Alticinae) chew Solanaceae as secondary hosts without causing much damage. Generally, the Chrysomelidae on Solanaceae are leaf-feeders (Chrysomelinae) and stem borers (Megalopodinae) but at least one genus of Alticinae (*Xenidia*) has a larva feeding on roots. Various species of *Psylliodes* (e.g. *P. difficilis* Baly, *P. affinis* (Paykull), *P. hyoscyami* (L.), *P. angusticollis* Baly, *P. balyi* Jacoby, *P. brettinghami* Baly, *P. dulcamara* Koch, *P. illigeri* Perroud, *P. loriae* Jacoby, *P. luteola* O.F. Müller, *P. plana* Maulik and *P. splendida* Harold) have larvae which mine many genera of Solanaceae. Many other species of *Psylliodes* do not feed on Solanaceae, but instead chew Brassicaceae, Asteraceae and Chenopodiaceae. Several Chrysomelidae, including Alticinae, feed on Solanaceae in South-East Asia, but these are very few.

The shaving activity of the mouth-parts of *Leptinotarsus decemlineata* (Say) removes the spines and hairs from the leaves of *Solanum rostratum* and *S. elaeagnifolium* plants so that the beetles can ingest only the mesodermal tissues. This is why the larvae of the Colorado potato beetle, when offered their natural hosts, initiate feeding on the edge of the leaf (Hsiao, 1986). Siebert (1975) reported that the tortoise beetle, *Gratiana lutescens* (Boheman)(Cassidinae) produced masses of stellate hairs in the container after feeding on leaves of *Solanum elaeagnifolium* in the laboratory and suggested that the trichomes had been nibbled off before feeding proper. Therefore the adaptations of Solanaceae against herbivory are not only chemical but also physical. Glandular trichomes of certain species of *Solanum* can also trap young larvae of insects. It has been recently shown that the larvae of *Platyphora conviva* Stål (Chrysomelinae) in the Atlantic forests of Brazil, cut the hairs on the abaxial side of the leaves of arborescent *Solanum* species and cover their bodies with the brown plant hairs as camouflage (Vasconcellos-Neto & Jolivet, 1988). Among the Chrysomelidae, this type of camouflage was previously known only among some species of Chlamisinae. This exceptional species of *Platyphora* displays cycloalexy or ring-defence behaviour under the leaf (Vasconcellos-Neto & Jolivet, 1988; Jolivet *et al.*, 1990).

As shown by Hsiao (1986), *Leptinotarsa* are very selective in their choice of *Solanum* and only certain species from certain sections of the genus are acceptable to the beetles. Although other *Leptinotarsa* species feed on Asteraceae or Zygophyllaceae, there is no close relationship between these two families with Solanaceae. It is certain now that *Leptinotarsa* has crossed the Ural Mountains and is slowly migrating to the east (Vlasova, 1978). *Leptinotarsa* is unable to adapt to any food, nor to all species of *Solanum,* so it is curious to note that Kulik and Timoshin (1986) found that *Leptinotarsa* can feed, at least occasionally, on leaves of peas, volubilis, corn and several other Poaceae (grasses). However, these do not appear to represent true hosts and the feeding upon these plants does not appear to be stress-related as in Alticinae during periods of drought etc. There appear to be no other reports on this feeding phenomenon displayed by *Leptinotarsa*; however, there have been laboratory reports of the Colorado potato beetle feeding experimentally on *Asclepias* (Asclepiadaceae), *Capsella* (Brassicaceae) and *Lactuca* (Asteraceae) but this is also exceptional and does not occur in nature.

6.25. Tamaricaceae

This family is comprised of desert, steppe and seashore plants, which are often halophilic shrubs and herbs. Very few Chrysomelidae feed on them and then mostly in the Central Asian steppes. Several Eumolpinae (e.g. *Bedelia* and *Chloropteris*), Cryptocephalinae (e.g. *Stylosomus, Cryptocephalus* and *Pachybrachis*) and some Alticinae, feed on *Tamarix*. The genus *Diorhabda* (Galerucinae) has a stronger association with *Tamarix,* and almost exclusively feeds on this genus in Europe, Asia and North Africa (even in the tropical Sahel).

6.26. Umbelliferae (= Apiaceae)

This plant family is cosmopolitan and is composed chiefly of herbs with stout stems and hollow internodes, the latter of which are surprisingly never occupied by ants like other plants with this condition. Many Apiaceae are economically useful but others are poisonous. All are aromatic and attract insects by their smell.

There are a few Chrysomelidae which feed on the Apiaceae, i.e. *Chrysolina, Oreina, Paraphaedon, Prasocuris* (Chrysomelinae), *Dibolia, Syneta* (Alticinae) and *Coptocephala* (Clytrinae). Apiaceae are closely related to Araliaceae and Cornaceae. The close affinity between Araliaceae and Apiaceae is shown by several temperate Lamprosomatinae which feed on both plant families. On the other hand, Apiaceae and Asteraceae are not closely related but several leaf-beetle genera (e.g. *Oreina*) feed on both. In this case, the selection is what we have called "ecological selection".

6.27. Urticaceae

Members of the Urticaceae are herbs or undershrubs (rarely lianas) without latex. Many species are provided with stinging hairs. Few Chrysomelidae feed on them. Some Alticinae feed on *Girardinia, Urtica* and *Fleurya* (Urticaceae) in India and the Far East. *Potaninia* (Chrysomelinae) only feeds on *Boehmeria* and *Maoutia* (Urticaceae) in the Far East. *Hoplostines* (Galerucinae) is restricted to *Urtica* and *Dendrocnide* in Australia. *Psylliodes* has been recorded from *Parietaria* but many other Alticinae records are questionable, e.g. *Aulacophora* (Galerucinae) has been wrongly quoted from *Boehmeria.*

6.28. Vitaceae

The Vitaceae are climbing or rarely erect shrubs, mostly tropical and subtropical in distribution. The type genus *Vitis* occurs throughout the northern hemisphere and even south to Venezuela.

Many Eumolpinae feed on *Vitis* (i.e. *Adoxus, Phytorus, Graphops, Bromius, Fidia, Acrothinium, Glyptoscelis, Colaspis, Rhabdopterus, Scelodonta, Nerissus, Aoria, Paria* and *Pseudocolaspis*). Very often the selection of *Vitis* shrubs is closely associated with the selection of other Vitaceae such as species of *Parthenocissus, Tetrastigma, Ampelopsis, Cissus* and *Cayratia* and with other families of plants such as Onagraceae (*Epilobium, Clarkia, Chamaenerion* and *Fuchsia*) in the case of *Bromius* and other Eumolpinae, or Leeaceae in the case of *Scelodonta* (Eumolpinae). Several Alticinae (i.e. *Altica*) consume *Vitis* leaves, but mostly Galerucinae do (i.e. *Exosoma, Monolepta, Oides, Pseudoliprus, Liprus* and *Gallerucida*). Some are really restricted feeders on *Vitis*, *Ampelopsis* and *Parthenocissus* leaves. Some polyphagous Clytrinae (e.g. *Labidostomis*) are also occasionally found on *Vitis. Altica* feeds on Onagraceae, Vitaceae, Lythraceae, Haloragidaceae and Trapaceae; these families are closely related chemically or taxonomically. *Iscadida* (Chrysomelinae) was recently found feeding on *Rhoicissus* in Transvaal (PJ, unpub. observations).

4. CONCLUDING REMARKS

This study provides a background to the food-plants and general biology of the Chrysomelidae of the world. We can estimate that food-plant records exist for about 30% of the estimated 35,000 described species. This percentage is rather small but we feel that it is sufficient for us to draw certain conclusions concerning the phylogeny of Chrysomelidae based on host-plant preferences as well as various trends and other relationships between species, genera, subfamilies etc. Some regions of the world have been well-explored such as the Holarctic, South-East Asia and tropical America but others are poorly known, e.g. Australia, New Zealand etc. Other areas remain totally unexplored and unstudied in terms of their beetle faunas. For instance, there are 10,000 endemic plant species in Madagascar but nothing is known about their relationships with the Chrysomelidae. It is probable that any data eventually deriving from Madagascar will be in agreement with plant selection trends elsewhere in the world, but since the endemism of Chrysomelidae is very high there, the leaf-beetle food-plants would be very interesting to determine. The Tepuys of southern Venezuela (old sandstone massif) also have a high endemicity of plants and insects. The Chrysomelidae of this region have not yet been formally described but they are mostly Alticinae. Their host-plants would also be very interesting to determine.

Regarding the various subfamilies of Chrysomelidae overall, there are some with very few data recorded for them and even the larvae of such subfamilies as Aulacoscelinae and Megascelinae remain undescribed. The food-plants and biology of most of the important genera such as *Leptinotarsa, Chrysolina, Timarcha, Calligrapha* and *Oreina* (Chrysomelinae) and many Alticinae and Galerucinae, are well-known. However, the food-plants of nocturnal and apterous Entomoscelina (Chrysomelinae) such as *Cyrtonastes* remain virtually unknown.

Many Chrysomelidae are specialists, i.e. mono- or oligophagous. However, it is among the Eumolpinae, Cryptocephalinae and Clytrinae that are found most of the generalists. The Eumolpinae display the basis of host-plant selection in the family, namely that of Convolvulaceae, Asclepiadaceae and Apocynaceae.

Overall, it has been difficult to provide a reliable list of the host-plants because the literature is dispersed among many publications and the data that are presented in them are very variable in terms of their credibility. Some lists are entirely fictitious, and as a consequence, have been ignored. There is also a tendency by applied entomologists to record plants as hosts upon which an insect has only settled for some period of time without feeding or has fed only on the pollen. We have been very careful in accepting or rejecting these data.

The extensive experience of one of us (PJ) in tropical countries has assisted greatly in the screening of the data. All of the personal observations of PJ and the literature data have been recorded over many years on some 20,000 index cards each measuring 21x13.5 cm. P. Jolivet does not claim to have analyzed all of the literature on host-plants but has certainly covered a large majority of it. The archives of the FAO were very useful in adding some unpublished data and some papers printed in confidential journals not usually available to research workers. Even though this book is mainly based on 40 years of observations and data collected by the senior author, there may still be several mistakes or omissions from the work, but all efforts have been made to reduce these to a minimum. Also, it has not been possible to always quote the exact locality from where observations/collections have been undertaken since this is important when the genus in question is cosmopolitan or almost so. To have cited these localities would have enlarged the text enormously. In addition, chapter 3 could have been much larger but only a small number from over two hundred dicot families have been mentioned and discussed. Those dicot families of limited importance have been excluded from discussion because of space limitations and because little is known of their interrelationships.

If there are actually 40-50,000 species of Chrysomelidae in the world, then much field research and collections of leaf-beetles needs to be undertaken before the disappearance of the tropical forests and the destruction of biotopes all over the world by land-clearing, development and the effects of fertilizers and insecticides. We can consider that there are approximately 47 families of monocots and 244 families of dicots but only 26 families of monocots and 95 of dicots have been recorded here in this book as food-plants of Chrysomelidae which means that there are many gaps in knowledge to be filled. Of course, it is probable that several plant families do not in fact support any species of Chrysomelidae.

We hope that this book, the first to attempt a world-wide coverage of the host-plants and general biology of the Chrysomelidae, will encourage and inspire other entomologists all over the globe to undertake more research into these subjects so that one day a more complete and perfect work will be published with the refined data. It is our sincere hope that most of the gaps in our present knowledge, especially those concerning the tropical regions of the world, will start to fill in the near future.

REFERENCES

Alibert, H. (1951). Les Insectes vivant sur les cacaoyers en Afrique Occidentale. *Mem. IFAN.*, 15: 1-171.

Anbiru, T., Shinkaji, N., Amano, H. & Nunokuwa, M. (1987). Deposition sites of egg-masses of the viburnum leaf-beetle, *Pyrrhalta humeralis* (Chen) on the host-tree, *Viburnum awabuki* K. Koch. *Jap. J. Appl. Entomol. Zool.*, 31: 391-394.

Andrade, J.C. de (1981). Biologia de *Cecropia lyratiloba* Miq. var. *nana* Andr. & Car. (Moraceae) na restinga do recreio dos Bandeirantes. Thesis, Rio de Janeiro: 1-71.

Andrade, J.C. de (1984). Desempenho e distribuicao de *Azteca alfari* (Formicidae) em *Cecropia lyratiloba* var. *nana* (Cecropiaceae) na restinga do recreio dos Bandeirantes, Rio de Janeiro. *Atas Soc. Biol., Rio de Janeiro*, 24: 11-15.

Andrews, F.G. & Gilbert, A.J. (1992). *Cadiz hardyi*, a new genus and species of leaf beetle from a western North American sand dune (Coleoptera: Chrysomelidae). *Col. Bull.*, 46: 4-14.

Araujo and Silva (1968). Quarto catalogo dos Insetos que vivem nas plantas do Brasil. Seus parasitos e predatores. Rio de Janeiro, 2: 1-622.

Arnett, R.H. (1969). Determination of host-plants of pollen-feeding beetles through pollen association. *Year book Amer. Phil. Soc.*, 1969: 285-286.

Askevold, I.S. (1987). The genus *Neohaemonia* Szekessy in North America: Systematics, reconstructed phylogeny and geographic history. *Trans. Am. Entomol. Soc.*, 113: 361-430.

Askevold, I.S. (1990). Reconstructed phylogeny and reclassification of the genera of Donaciinae. *Quaest. Entomol.*, 26: 601-664.

Balsbaugh, E.U. (1982). *Colaspis chapalensis* Blake, a new pest of corn in Mexico. *J. Kansas Entomol. Soc.*, 55: 577-580.

Banerjee, B. (1987). Can leaf aspect affect herbivory? A case study with tea. *Ecology*, 68: 839-843.

Battenfield, S.L., Wellso, S.G. & Haynes, D.L. (1982). Bibliography of the cereal leaf beetle, *Oulema melanopus* (L.) (Coleoptera: Chrysomelidae). *Bull. Entomol. Soc. Am.* 28: 291-308.

Bechyne, J. (1980). Evaluacion de los datos sobre los Phytophaga daninos en Venezuela (Coleoptera). Unpub. Ms. Maracay: 1-600.

Bechyne, J. & Bechyne, B.S. (1966). Lista preliminar de los Phytophaga de importancia agricola en Venezuela. Maracaibo: 1-5.

Bechyne, J. & Bechyne, B.S. (1967). Altises neotropicales peu connues ou nouvelles *Ann. Soc. Entomol. Fr.* (N.S.), 3: 1033-1038.

Bechyne, J. & Bechyne, B.S. (1969). La posicion sistematica de *Megascelis* Chevrolat (Col. Phytophaga). *Revta. Fac. Agron. Univ. Centr. Venez.*, 5: 65-75.

Bedel, L. (1891). Faune des Coléoptères du Bassin de la Seine Part V. Phytophaga Chrysomelidae, Paris: 105-406.

Begossi, A. (1984). Habitos alimentares e Coloracao de Advertencia em alguns Alticineos (Coleoptera: Chrysomelidae). *Dissert. M. Biol., Campinas*: 1-231.

Bergmann, E.C. *et al.* (1983). Ocorrencia de *Omophoita sexnotata* Harold, 1876 danificando *Clerodendron philippinum* Schauer em Sao Paulo. *Biologica,* 49: 103-105.

Berti, N. & Rapilly, M. (1976). Faune d'Iran. Liste d'especes et revision du genre *Lilioceris* Reitter (Col. Chrysomelidae). *Ann. Soc. Entomol. Fr.*(N.S.), 12: 31-73.

Bertrand, H. (1965). Contribution à l'étude des premiers états des Coléoptères aquatiques de la région éthiopienne. *Bull. IFAN.*, 27 (A4): 1337-1393.
Blake, D.H. (1950). A revision of the beetles of the genus *Myochrous. Proc. US Nat. Mus.*, 101: 1-64.
Blake, D.H. (1974). The costate species of *Colaspis* in the United States (Coleoptera: Chrysomelidae). *Smith. Contr. Biol.*, 181: 1-24.
Blunk, H. (1954). Handbuch der Pflanzenkranhkeiten. 5.2. Coleoptera, Paul Parey, Berlin: 1-600.
Bokermann, W.C. (1964). Notes on the biology of *Poropleura cuprea* (Klug, 1824). *Anais Acad. Bras. Cienc. Rio de Janeiro,* 36: 165-172.
Boppre, M. & Scherer, G. (1981). A new species of flea beetle (Alticinae) showing male-biased feeding at withered *Heliotropium* plants. *Syst. Entomol.*, 6: 347-354.
Bordy, B. & Doguet, S. (1987). Contribution à la connaissance des Cassidinae de France. Etude de leur spermathèque. *Nouv. Rev. Entomol.*, 4: 161-176.
Borisenko, N. & Burdun, A.M. (1985). Resistance to cereal leaf beetle in pubescent wheat forms. *Selektsiya Semenovostvo USSR*, 6: 19-20.
Borowiec, L. (1983). Zoogeographical study on Donaciinae of the world. *Polskie Pismo Entomol.*, 53: 433-518.
Bosq, J.M. (1942). Segunda lista de Coleopteros de la Republica Argentina dañinos a la Agricultura. *Ing. Agron.*, 4 (18-22): 1-80.
Bourdouxhe, L. & Jolivet, P. (1981). Nouvelles observations sur le complexe mimétique de *Mesoplatys cincta* Olivier (Col. Chrysomelidae) au Sénégal. *Bull. Soc. Linn. Lyon*, 50: 46-48.
Boving, A.G. & Craighead, F.C. (1931). An illustrated synopsis of the principal larval forms of the order Coleoptera. *Entomol. Americana*, 11: 1-351. (Revised edition 1953).
Breden, F. & Wade, M. (1985). The effect of group size and cannabalism rate on larval growth and survivorship in *Plagiodera versicolora* (Col. Chrysomelidae). *Entomogr.*, 3: 455-463.
Breden, F. & Wade, M.J. (1987). An experimental study of the effect of group size on larval growth and survivorship in the imported willow leaf beetle *Plagiodera versicolor. Environ. Entomol.*, 16: 1082-1086.
Briggs, E.M. (1905). The life-history of case bearers. I. *Chlamys plicata. Cold Spring Harb. Mono.*, 4: 3-18.
Broodryk, S.W. (1965). The morphology of *Syagrus rugifrons* Baly. *Sth. Afr. J. Agric. Sci.*, 8: 455-468.
Brown, W.J. (1943). The Canadian species of *Exema* and *Arthrochlamys. Can. Entomol.*, 75: 119-131.
Brown, W.J. (1945). Food-plants and distribution of the species of *Calligrapha* in Canada, with descriptions of new species. *Can. Entomol.*, 77: 117-133.
Brown, W.J. (1952). Some species of Phytophaga (Coleoptera). *Can. Entomol.*, 84: 335-342.
Brown, W.J. (1959). Taxonomic problems with closely related species. *Ann. Rev. Entomol.*, 4: 77-98.
Bryant, G.E. (1923). New injurious Phytophaga from India and Brazil. *Bull. Entomol. Res.*, 13: 261-265.
Bryant, G.E. & Gressitt, J.L. (1957). Chrysomelidae of Fiji. *Pac. Sci.,* 11: 1-191.
Budhraja, K. *et al.* (1979). Feeding behaviour of *Dicladispa armigera. Int. Rice Res. Newsl.*, 4(6): 15-16.
Burgess, L. (1979). A flea beetle that feeds on the common plantain. *Blue Jay,* 37: 86-88.
Burgess, L. (1981). Crucifer-feeding flea beetles occurring in the province of Saskatchewan, Canada. *Col. Bull.*, 35: 307-309.

Burlini, M. (1968). Revision of the Italian species and the majority of the European species of the genus *Pachybrachis* Chevr. *Mem. Soc. Entomol. Ital.*, 47: 11-116.

Buzzi, Z.J. (1975). Contribucao ao conhecimento da biologia de *Anacassis fuscata* (Klug, 1829). *Rev. Bras. Biol.*, 35: 767-774.

Buzzi, Z.J. (1988). Biology of Neotropical Cassidinae. Chap. 32. In: Jolivet, P., Petitpierre, E. & Hsiao, T.H. (Eds.). *Biology of Chrysomelidae*. Kluwer Academic Publishers, Dordrecht: 559-580.

Buzzi, Z.J. (1994). Host Plants of Neotropical Cassidinae. Chap. 13. In: Jolivet, P., Cox, M.L. & Petitpierre, E. (Eds.). *Novel Aspects of the Biology of Chrysomelidae*. Kluwer Academic Publishers, Dordrecht: 205-212.

Cantelo, W.W. & Pholboon, P. (1965). A host list of the insects of Thailand. *Dept. Agric. USOM Bangkok*: 1-149.

Carne, P.B. (1966). Ecological characteristics of the eucalypt-defoliating chrysomelid *Paropsis atomaria* Olivier. *Aust. J. Zool.*, 14: 647-672.

Chapin, J.B. (1979). A review of the Louisiana species of *Colaspis* (Col. Chrysomelidae). *Col. Bull.*, 33: 445-450.

Chapuis, F. (1874). In: Lacordaire, *Histoire Naturelle des Insectes. Genera des Coleopteres*, vol. 10 (1-445), 11 (1-420), Paris.

Chen, S.H. (1934). *Recherches sur les Chrysomelidae de la Chine et du Tonkin*. Thesis, Faculty of Science, University of Paris: 1-105.

Chen, S.H. (1964). Evolution and classification of the chrysomelid beetles. *Act. Entomol. Sinica*, 13: 469-483.

Chen, S.H. (1966). Notes on Chinese Donaciinae. *Acta Entomol. Sinica*, 15: 137-147.

Chen, S.H. (1985). Phylogeny and classification of the Chrysomeloidea. *Entomography*, 3: 465-475.

Chen, S.H. & Jiang, S.Q. (1984). New species of Chinese Galerucinae (Coleoptera: Chrysomelidae). *Entomotaxonomica*, 6: 83-92.

Chen, S.H. & Wang, S.Y. (1962). On the distribution and desert adaptations of the chrysomelid beetles of Sinkiang. *Acta Zool. Sinica*, 14: 337-354.

Chen, S.H. & Wang, S.Y. (1984). Flea-beetles from Hengduan Mountains, Yunnan. Genera *Hespera* and *Yunohespera* (Col., Chrysomelidae). *Acta Entomol. Sinica*, 27: 308-322.

Chen, S.H. & Wang, S. (1987). The alpine genera of flea beetles from Yunnan. *Acta Entomol. Sinica,* 30: 196-200.

Chen, S.H. & Zia, Y. (1966). On the Chinese citrus flea beetles. *Acta Zootax. Sinica*, 3: 67-75.

Chen, S.H. *et al.* (1987). New alpine Galerucinae from Hengduan Mountains of Yunnan and Sichuan. *Sinozoolognia*, 5: 61-71.

Chu, H.F. (1949). *How to know the immature insects*. Brown Co., Dubuque, Iowa: 1-234.

Chujo, M. (1953). A taxonomic study on the Chrysomelidae with special reference to the fauna of Formosa. *Tech. Bull. Kagawa Agric. Coll.*, 5: 19-36, 121-136.

Clark, H. (1864). Descriptions of new Australian Phytophaga. *J. Entomol.*, 2: 247-263.

Cockerell, T.D.A. (1891). Case-making Coleopterous larvae. *Entomol. Month. Mag.*, 27: 190-191.

Coetzee, J.H. & Latsky, L.M. (1986). Faunal list of *Protea repens*. *Acta Hort.*, 185: 241-245.

Cordo, H.A., Deloach, C.J. & Ferrer, M. (1984). Biology and larval host range of the flea beetle *Disonycha argentinensis* on alligatorweed in Argentina. *Ann. Entomol. Soc. Amer.*, 77: 134-141.

Costa, G., Leonardi, M.E. & Petralia, A. (1983). Richerche ecoetologiche sulla fauna delle dune costiere di Portopalo (Siracusa) v. ciclo biologico, ritmi di attivita e comportamento di *Timarcha pimeliodes* Schaf. (Col. Chrysomelidae). *Animalia, Catania*, 10: 113-134.

Costa-Lima, A. da (1936). Terceiro catalogo dos Insectos que vivem nas plantas do Brasil. *Rio de Janeiro*: 1-460.

Costa-Lima, A. da (1955). Insectos do Brasil. 9 tomo. Coleopteros. 3a parte. *Rio de Janeiro* : 1-289.

Cox, M.L. (1981). Notes on the biology of *Orsodacne* Latreille with a subfamily key to the larvae of the British Chrysomelidae (Coleoptera). *Entomol. Gaz.*, 32: 123-135.

Crepet, W.L. & Taylor, D.W. (1985). The diversification of the Leguminosae: First fossil evidence of the Mimosoideae and Papilionoideae. *Science*, 228: 1087-1088.

Cronquist, A. (1981). *An integrated system of classification of flowering plants.* Columbia University Press, New York: 1-1262.

Crowson, R.A. (1946). A revision of the genera of the chrysomelid group Sagrinae (Coleoptera). *Trans. R. Entomol. Soc. Lond.*, 97: 75-115.

Crowson, R.A. (1953). The classification of the families of British Coleoptera. *Entomol. Month. Mag.*, 89: 181-198.

Crowson, R.A. (1967). *The natural classification of the families of Coleoptera.* Classey, London: 1-214.

Crowson, R.A. (1981). *The biology of Coleoptera.* Academic Press, London: 1-802

Crowson, R.A. (1991). The relations of Coleoptera to Cycadales. In: *Advances in Coleopterology*. M. Zunino, X. Belles & M. Blas (eds.). The European Association of Coleopterology, Barcelona: 13-28.

Dap, D.T. (1981). *Coleoptera Chrysomelidae of Vietnam.* Thesis Ac. Sci. USSR: 1-288.

De Little, D.W. (1985). Favourability of fagaceous foliage as food for *Novacastria nothofagi* Selman. *Aust. Entomol. Mag.*, 12: 13-14.

De Wilde, J.W. *et al.* (1969). Effects of host plant age on phytophagous insects. *Entomol. Exp. Appl.*, 12: 714-720.

Dimock, M.B. & Tingey, W.M. (1986). *Solanum neocardenasii*, a new source of potato resistance to the Colorado potato beetle. *J. Econ. Entomol.*, 79: 1269-1275.

Doguet, S. (1984). Contribution à l'étude des espèces d'Afrique du Nord du genre *Phyllotreta* (Coleoptera, Chrysomelidae). *Nouv. Rev. Entomol.*, 1: 243-265.

Dussourd, D.E. & Eisner, T. (1987). Vein-cutting behaviour: insect counterploy to the latex defense of plants. *Science*, 237: 898-901.

Eberhard, W.G. (1981). The natural history of *Doryphora* sp. (Coleoptera, Chrysomelidae) and the function of its sternal horn. *Ann. Entomol. Soc. Am.*, 74: 445-448.

Edwards, J.G. (1953). Species of the genus *Syneta* of the world (Coleoptera: Chrysomelidae). *Wasmann J. Biol.*, 11: 23-82.

Eisner, T. (1972). Chemical Ecology: On arthropods and how they live as chemists. *Verh. Deutsch. Zool. Gesellsch.*, 65: 123-137.

Eisner, T., Tassell, E. van & Carrel, J.E. (1967). Defensive use of a "fecal shield" by a beetle larva. *Science,* 158: 1471-1473.

Erber, D. (1969). Beitrag zur Entwicklungsbiologie mitteleuropäischer Clytrinen und Cryptocephalinen. *Zool. Jb. Syst.*, 96: 453-477.

Erber, D. (1988). Biology of Camptosomata Clytrinae-Cryptocephalinae- Chlamisinae-Lamprosomatinae. Chapter 30 in: P. Jolivet, E. Petitpierre & T.H. Hsiao (eds.) *Biology of Chrysomelidae.* Kluwer Academic Publishers, Dordrecht: 513-552.

Ferronatto, E.M.O. (1986). Observacoes preliminares sobre as plantes hospedeiros das formas immaturas dos principais crisomelideos que occorrem hos cocauais. *Rev. Theobroma*, 16: 107-110.

Ferronatto, E.M.O. (1988). Eumolpinae associated with cacao trees (*Theobroma cacao* L.) in Southeast Bahia. Chap. 31 in: P. Jolivet, E. Petitpierre & T.H. Hsiao (eds.) *Biology of Chrysomelidae.* Kluwer Academic Publishers, Dordrecht: 553-558.

Fiebrig, K. (1910). Cassiden und Cryptocephaliden Paraguays. *Zool. Jahrbr. Suppl.*, 12: 161-264.

Force, D.C. (1966). Reactions of the three-lined potato beetle, *Lema trilineata* (Coleoptera: Chrysomelidae) to its host and certain nonhost plants. *Ann. Entomol. Soc. Am.*, 59: (1112-1119).

Froggatt, W.W. (1907). *Australian Insects*. Sydney: 1-449.

Frost, S.W. (1973). Host and eggs of *Blepharida dorothea* (Coleoptera, Chrysomelidae). *Florida Entomol.*, 56: 120-122.

Furth, D.G. (1976). Alticinae of Israel with special reference to the systematic, zoogeographic and host plant relationships of *Phyllotreta* and *Longitarsus*. Ph.D. Thesis, Cornell University: 1-153.

Furth, D.G. (1979). Zoogeography and host plant ecology of the Alticinae of Israel especially *Phyllotreta* with descriptions of three new species from Israel. *J. Zool.*, 28: 1-37.

Furth, D.G. (1980a). Zoogeography and host plants of *Longitarsus* in Israel with descriptions of six new species. *Israel J. Entomol.*, 13: 79-124.

Furth, D.G. (1980b). Wing polymorphism, host plant ecology and biogeography of *Longitarsus* in Israel. *Israel J. Entomol.*, 13: 125-148.

Furth, D.G. (1980c). *Altica* of Israel. *Israel J. Entomol.* 14: 55-66.

Furth, D.G. (1982). *Blepharida* biology, as demonstrated by the sacred sumac flea beetle (*B. sacra* Weise). *Spixiana Suppl.*, 7: 43-52.

Furth, D.G. (1983). Alticinae of Israel: *Psylliodes*. *Israel J. Entomol.*, 17: 37-58.

Furth, D.G. (1985a). Alticinae of Israel: *Chaetocnema*. *Israel J. Entomol.*, 19: 67-83.

Furth, D.G. (1985b). The natural history of the sumac tree, with an emphasis on the entomofauna. *Trans. Conn. Acad. Arts & Sci.*, 46: 137-234.

Furth, D.G. (1985c). Some flea beetles and their food plants from Kenya. *Col. Bull.*, 39: 259-263.

Furth, D.G. (1992). The New World *Blepharida* group, with a key to genera and description of a new species (Coleoptera: Chrysomelidae), *J. New York Entomol. Soc.*, 100: 399-414.

Furth, D.G. & Young, D.A. (1988). Relationships of herbivore feeding and plant flavonoids (Coleoptera: Chrysomelidae and Anacardiaceae: *Rhus*). *Oecologia*, 74: 496-500.

Gardner, J.C.M. (1934). On some coleopterous larvae from Uganda. *Bull. Entomol. Res.*, 25: 149-153.

Gressitt, J.L. (1942). Plant-beetles from south and west China. I. Sagrinae, Donaciinae, Orsodacninae and Megalopodinae (Coleoptera). *Lingnan Sci. Jour.*, 20: 271-293.

Gressitt, J.L. (1952). The tortoise beetles of China. *Proc. Cal. Acad. Sci.*, 27(17): 433-592.

Gressitt, J.L. (1963). Economic chrysomelid beetles from New Guinea with new species. *Papua New Guin. Agric. J.*, 16: 105-116.

Gressitt, J.L. (1965). Chrysomelid beetles from the Papuan subregion. I. (Sagrinae, Zeugophorinae, Criocerinae). 2. (Clytrinae, Cryptocephalinae, *Chlamisinae*). *Pac. Ins.*, 7: 131-189; 387-449.

Gressitt, J.L. (1966-1969). Chrysomelid beetles from the Papuan subregion. I-VII. *Pac. Ins.*, 8: 777-800; 9: 295-340; 9: 551-562; 11: 1-31.

Gressitt, J.L. & Hart, A.D. (1974). Chrysomelid beetles from the Papuan subregion. 8. *Pac. Ins.*, 16: 261-306.

Gressitt, J.L. & Kimoto, S. (1961-1963). The Chrysomelidae of China and Korea. *Pac. Ins. Mono.*, 1A: 1-299; 1B: 301-1026.

Grillas, P. (1988). *Haemonia appendiculata* Panzer and its impact on *Potamogeton pectinatus* L. and *Myriophyllum spicatum* L. beds in the Camargue, France. *Aquat. Bot.*, 31: 347-353.

Gritsenko, V.V. *et al.* (1986). The main color variants on *Chrysolina varians* L., their heredity and spatial distribution in nature. *Bull. Moskovs. Obshchest. Ispytatelei Prirod. Biol.*, 91(2): 41-50.

Grobbelaar, B. (1993). A revision of the southern African species of *Megalognatha* Baly. *Rep. South Africa Dept. Agric. Memoir*, 86: 1-85

Hare, J.D. & Kennedy, G.G. (1986). Genetic variation in plant-insect associations: survival of *Leptinotarsa decemlineata* populations on *Solanum carolinense*. *Evolution*, 40: 1031-1043.

Hargreaves, H. (1948). List of recorded cotton insects of the world. *Comm. Inst. Entomol.*, 1-50.

Harrison, G.D. (1987). Host-plant discrimination and evolution of feeding preference in the Colorado potato beetle *Leptinotarsa decemlineata. Physiol. Entomol.*, 12: 407-415.

Haslett, J.R. (1983). A photographic account of pollen digestion by adult hoverflies. *Phys. Entomol.*, 8: 167-171.

Hatch, M.H. (1971). *The beetles of the Pacific north-west*. Part V. University of Washington Press, Seattle: 1-662.

Hawkeswood, T.J. (1983). Observations on *Pyrgoides dryops* (Blackburn), a pollen-feeding beetle on *Acacia leiocalyx* (Domin) Pedley at Brisbane, south-east Queensland. *Vict. Nat.*, 100: 156-158.

Hawkeswood, T.J. (1985). Notes on some beetles (Coleoptera) associated with *Xanthorrhoea johnsonii* (Xanthorrhoeaceae) in the Brisbane area, south-east Queensland. *Vict. Nat.*, 102: 162-166.

Hawkeswood, T.J. (1986). New food plants for adults of three species of the Australian genus *Lamprolina* (Coleoptera: Chrysomelidae). *Col. Bull.*, 40: 104.

Hawkeswood, T.J. (1987). *Beetles of Australia*. Angus & Robertson, Sydney: 1-248.

Hawkeswood, T.J. (1988). A survey of the leaf beetles (Coleoptera: Chrysomelidae) from the Townsville district, northern Queensland, Australia. *G. It. Entomol.*, 4: 93-112.

Hawkeswood, T.J. (1989). Studien zu Biologie und Verhalten des australischen Schildkäfers *Hoplionota dorsalis* Waterhouse (Coleoptera: Chrysomelidae: Cassidinae). *Entomol. Zeitschr.*, 99: 346-349.

Hawkeswood, T.J. (1992). Notes on the biology and host plants of the Australian leaf beetle *Lilioceris (Crioceris) nigripes* (Fabricius) (Coleoptera: Chrysomelidae). *Entomol.*, 111: 210-212.

Hawkeswood, T.J. & Jolivet, P. (1988). Notes on the biology and behaviour of the Australian leaf beetle *Cleptor inermis* Lefevre, 1885 (Coleoptera Chrysomelidae Eumolpinae). *Bull. Ann. Soc. Roy. Entomol. Belg.*, 124: 189-194.

Hawthorn, W.R. (1978). Some effects of different *Plantago* species on feeding preference and egg laying in the flea beetle *Dibolia borealis* Chevrolat (Col., Chrysomelidae). *Can. J. Zool.*, 56: 1507-1513.

Heikertinger, F. (1924-1926). Resultate fünfzehnjähriger Untersuchungen über die Nahrungspflanzen einheimischer Halticinen. *Ent. Blatt.*, 20: 214-224; 21: 10-19, 81-92, 119-131, 155-163; 22: 1-9, 49-62.

Hendrix, S.D. (1980). An evolutionary and ecological perspective of the insect fauna of ferns. *Am. Nat.*, 115: 171-196.

Hering, E.M. (1935). Die Blattminen Mittel- und Nordeuropas einschliesslich Englands. Neubrandenburg: 1-631.

Hering, E.M. (1951). *Biology of leaf miners*. W. Junk Publ., 1-420.

Hespenheide, H.A. (1976). Reversed sex-limited mimicry in a beetle. *Evolution*, 29: 780-783.

Hippa, H. & Koponen, S. (1986). Morphological, cytological, ecological and ethological evidence of reproductive isolation between *Galerucella nymphaeae* (L.) and *G. sagittariae* (Gyll.) in Fennoscandia. *Ann. Entomol. Fennici*, 52: 49-62.

Hocking, B. (1970). Insect associations with the swollen thorn acacias. *Trans. Roy. Entomol. Soc. Lond.*, 122: 211-255.

Hoxie, R.P. & Wellso, S.G. (1974). Cereal leaf beetle instars and sex, defined by larval head capsule widths. *Ann. Entomol. Soc. Am.*, 67: 183-186.

Hsiao, T.H. (1974). Chemical influence on feeding behaviour of *Leptinotarsa* beetles. In: Barton-Browne, L. (Ed.) *Experimental Analysis of Insect Behaviour.* Springer Verlag, New York: 237-248.

Hsiao, T.H. (1981). Ecophysiological adaptations among geographic populations of the Colorado potato beetle in North America. In: J.H. Lashomb & R.A. Casagrande (Eds.) *Advances in Potato Pest Management*, Hutchinson Ross Publ., Stroudsburg: 69-85.

Hsiao, T.H. (1986). Specificity of certain chrysomelids for Solanaceae. In: W.G. D'Arcy (Ed.) Solanaceae: *Biology and Systematics.* Columbia University Press, New York: 345-377.

Hsiao, T.H. (1988). Physiological and biochemical studies of insect biotypes, relevance to pest management. *Symp. Ins. Biochem. Physiol.*, 1-9.

Hsiao, T.H. & Fraenkel, G. (1968). Selection and specificity of the Colorado potato beetle for solanaceous and non-solanaceous plants. *Ann. Entomol. Soc. Am.*, 61: 493-502.

Hsiao, T.H. & Hsiao, C. (1983). Chromosomal analysis of *Leptinotarsa* and *Labidomera* species (Col. Chrysomelidae). *Genetica* 60; 139-150 .

Hudson, G.V. (1934). *New Zealand Beetles and their Larvae.* Ferguson & Osborn, Wellington: 1-236.

Inoue, T. (1990). Feeding habits and seasonal development of the flea-beetle, *Argiopistes biplagiatus* Motschulsky in Chiba prefecture, Central Japan. *Jap. J. Appl. Entomol. Zool.*, 34: 153-160.

Jacoby, M. (1908). *The fauna of British India, including Ceylon and Burma. Coleoptera (Chrysomelidae).* London: 1-534.

Jacques, R.L. (1985). The potato beetles of Florida. *Entomol. Circ. Fl.*, 271: 1-2.

Jacques, R.L. (1987). Flea beetles of the genus *Systena* in Florida. *Entomol. Circ. Fl.*, 295: 1-2.

Jacques, R.L. (1988). *The potato beetles.* Brill, New York: 1-144.

Jayanth, K.P. & Nagarkatti, S. (1987). Investigations of the host-specificity and damage potential of *Zygogramma bicolorata* Pallister introduced into India for the biological control of *Parthenium hysterophorus. Entomon.*, 12: 141-145.

Jerez, V. & Ibarra-Vidal, H. (1992). Morfologia y Bionomica de *Hornius grandis* (Phil. & Phil., 1864) (Chrysomelidae, Eumolpinae). *Bol. Soc. Biol. Concepcion*, 63: 93-100.

Jolivet, P. (1948). Les Orsodacninae de la faune Francaise (Col.). *Misc. Ent.*, 45 (3-4): 33-46.

Jolivet, P. (1950). Contribution à l'étude des *Microtheca* Stal. *Bull. Inst. Roy. Sci. Nat. Belg.*, 26 (48): 1-27.

Jolivet, P. (1951a). Contribution à l'étude du genre *Gastrophysa* Chevrolat. *Bull. Inst. Roy. Sci. Nat. Belg.*, 27 (9): 1-11; 27 (13): 1-12; 27 (21): 1-47.

Jolivet, P. (1951b). Contribution à l'étude du genre *Microtheca* (20 note). *Bull. Inst. Roy. Sci. Nat. Belg.*, 27 (38): 1-7.

Jolivet, P. (1952a). Chrysomelidae (Coleoptera: Chrysomeloidea). *Parc Nat. Upemba, Miss. Witte*, 8(3): 23-47.

Jolivet P. (1952b). Quelques données sur la myrmecophile des Clytrides. *Bull. Inst. Roy. Sci. Nat. Belg.*, 28 (8): 1-12.

Jolivet, P. (1952c). Clytridae.(Coleoptera: Chrysomelidae). *Parc. Nat. Albert, Miss. Witte*, 75 (5): 33-43.

Jolivet, P. (1953). Les Chrysomeloidea des Iles Baleares. *Mem. Inst. Roy. Sci. Nat. Belg.*, 2(50): 1-88.

Jolivet, P. (1957). Chrysomelidae: Orsodacninae. In: *Col. Cat. Suppl. Junk Den Haag*, 51 (3): 1-16.

Jolivet, P. (1957-1959). Recherches sur l'aile des Chrysomeloidea. *Mem. Inst. Roy. Sci. Nat. Belg.*, 2(51): 1-180; 2(58): 1-152.

Jolivet, P. (1965). Phytophagie et selection trophique. Vol. V. Van Staelen, 2: 1101-1134.

Jolivet, P. (1966). Plantes-hôtes de quelques Chrysomelides d'Afrique Centrale. *Bull. Ann. Soc. Roy. Entomol. Belg.*, 102: 59-60.

Jolivet, P. (1967). Les Alticides vénéneux de l'Afrique du Sud. *L'Entomol.*, 23 (4): 100-111.

Jolivet, P. (1968a). Les Donaciinae de l'Afrique du Nord. *Bull. Soc. Linn. Lyon*, 37: 307-315.

Jolivet, P. (1968b). De la position systematique du genre *Microdonacia* Blackburn (Coleoptera, Chrysomelidae). *Entomol. Arb. Mus. G. Frey*, 19: 188-192.

Jolivet, P. (1970). Chrysomelidae: Donaciinae. In: *Col. Cat. Suppl.* Junk Den Haag, 51 (2): 1-71.

Jolivet, P. (1971a). Sélection trophique et adaptation écologique chez le genre *Timarcha* Latr. *Proc. 13 Int. Congr. Entomol. Moscow*, 1: 505-506.

Jolivet, P. (1971b). A propos de la plante-hôte de la Chrysomèle des Orchidées *Lema (Petauristes) pectoralis* Baly, 1865 en Thailande, Malaisie et Borneo (Col. Chrysomelidae). *Bull. Soc. Entomol. Fr.*, 76: 248-252.

Jolivet, P. (1971c). Quelques remarques à propos de la plante-hôte des *Platycorynus* Chevrolat en Thailande (Col. Chrysomelidae Eumolpinae). *Bull. Soc. Linn. Lyon*, 40: 189-195.

Jolivet, P. (1972a). Recherches sur *Nisotra gemella* (Erichson, 1834), l'Altise du kenaf (*Hibiscus sabdariffa* L.) en Thailande. Sélection trophique du complexe *Nisotra/Podagrica* (Col. Chrysomelidae Alticinae). *Bull. Soc. Linn. Lyon*, 41: 138-148.

Jolivet, P. (1972b). An interpretation of the host plants selected by species of the genus *Timarcha* (Coleoptera, Chrysomelidae). *Korean J. Entomol.*, 2: 21-26.

Jolivet, P. (1972c). A propos des *Donacia* de la Thailande (Col. Chrysomelidae Donaciinae). *Bull. Mens. Soc. Linn. Lyon*, 41(9): 193-196; 41(10): 197-201.

Jolivet, P. (1974). Les Chrysomèles des Araliacées (Col.) *Bull. Soc. Linn. Lyon*, 43(4): 113-120.

Jolivet, P. (1975). Une excursion entomologique à l'île de Quelpart (Cheju-do) et découverte d'une espèce nouvelle pour la Corée de *Chrysolina* Motsch.(Col. Chrysomelidae). Reflexions sur la plantophagie. *Bull. Soc. Linn. Lyon*, 44: 57-81.

Jolivet, P. (1976). Notes préliminaires sur la Biologie des *Timarcha* du Pacifique Nord Occidental Americain (*Americanotimarcha* Jolivet) (Col. Chrysomelidae). *Cah. Pac.*, 19: 153-165.

Jolivet, P. (1977). Sélection trophique chez les Eupoda (Coleoptera, Chrysomelidae). *Bull. Soc. Linn. Lyon*, 46: 321-336.

Jolivet, P. (1978). Sélection trophique chez les Clytrinae, Cryptocephalinae et Chlamisinae (Camptosoma) et les Lamprosomatinae (Cyclica). *Acta Zool. Pathol. Antwerp.*, 70: 167-200.

Jolivet, P. (1979a). Les Chrysomelidae (Col.) des *Citrus* et apparentes (Rutaceae) en zone tempérée et tropicale. *Bull. Soc. Linn. Lyon*, 48: 197-200, 249-256.

Jolivet, P. (1979b). Réflexions sur l'écologie, l'origine et la distribution des Chrysomelides (Col.) des Iles Mascareignes, océan Indien, avec la description de deux espèces nouvelles. *Bull. Soc. Linn. Lyon*, 48(9-10): 524-528, 606-649.

Jolivet, P. (1982). Les Eumolpinae (Col. Chrysomelidae) des Apocynaceae et des Asclepiadaceae (Gentianales). *Bull. Soc. Linn. Lyon*, 51(7): 214-222.

Jolivet, P. (1983). Un hemimyrmecophyte à Chrysomelides (Col.) du Sud Est Asiatique, *Clerodendrum fragrans* (Vent.) Willd. (Verbenaceae). *Bull. Soc. Linn. Lyon*, 52(8): 242-261.

Jolivet, P. (1984). *Phaedon fulvescens* Weise (Col. Chrysomelidae Chrysomelinae). Un auxiliaire possible dans le controle des *Rubus* aux Tropiques. *Bull. Soc. Linn. Lyon*, 53(7): 235-246.

Jolivet, P. (1985). Un myrmecophyte hors de son pays d'origine: *Clerodendrum fallax*

Lindley, 1844 (Verbenaceae) aux iles du Cap Vert. *Bull. Soc. Linn. Lyon*, 54(5): 122-127.

Jolivet, P. (1986a). *Les Fourmis et les Plantes. Un exemple de Coevolution*. Boubee Ed., Paris: 1-254.

Jolivet, P. (1986b). *Insects and Plants*. E.J. Brill, Flora and Fauna Publ., New York: 1-197.

Jolivet, P. (1987a). Premières données sur la biologie d'*Elytrosphaera xanthopyga* Stal. (Col. Chrysomelidae). *Bull. Soc. Entomol.* Fr., 91(5-6): 163-170.

Jolivet, P. (1987b). Sélection trophique chez les Megascelinae et les Eumolpinae (Cyclica). *Bull. Soc. Linn. Lyon*, 55(6): 199-208, 217-240.

Jolivet, P. (1987c). Aperçu de la sélection trophique chez les Galerucinae étude par genre (Coleoptera Chrysomelidae). *Bull. Annls Soc. Roy. Belge Entomol.*, 123: 283-307.

Jolivet, P. (1987d). Remarques sur la biocénose des *Cecropia* (Cecropiaceae). Biologie des *Coelomera* Chevrolat avec la description d'une nouvelle espèce du Brésil. *Bull. Soc. Linn. Lyon,* 56(8): 255-276.

Jolivet, P. (1988a). Les soins parentaux chez les Chrysomelides. *L'Entomol.*, 44(2): 93-101.

Jolivet, P. (1988b). Une nouvelle stratégie de défense: la stratégie de défense annulaire (cycloalexie) chez quelques larves de Chrysomelides brésiliens. *Bull. Soc. Entomol. Fr.*, 92: 291-299.

Jolivet, P. (1988c). A case of parasitism of an ant-*Cecropia* mutualism in Brazil. *Adv. Myrm.*, 1: 327-335.

Jolivet, P. (1988d). Sélection trophique chez les Cassidinae (Col. Chrysomelidae). *Bull. Soc. Linn. Lyon*, 57: 301-320.

Jolivet, P. (1988e). An apterous beetle genus on Compositae: *Elytrosphaera*. Ann. Int. Symp. Evol. Ecol. Trop. Herb. Campinas, 39.

Jolivet, P. (1988f). Ants, leaf-beetles and plants. *Ann. Int. Symp. Evol. Ecol. Trop. Herb. Campinas*, 40.

Jolivet, P. (1989a). Sélection trophique chez les Hispinae (Cryptostoma). *Bull. Soc. Linn. Lyon*, 58(9): 297-317.

Jolivet, P. (1989b). A propos des *Timarcha* Nord-Américains (Col. Chrysomelidae). *L'Entomol.*, 27-34.

Jolivet, P. (1991a). Sélection trophique chez les Alticinae (Col. Chrysomelidae). *Bull. Soc. Linn. Lyon,* 60(1): 26-40; 60(2): 53-72.

Jolivet, P. (1991b). Le Doryphore (*Leptinotarsa decemlineata* (Say, 1824)) menace l'Asie. *L'Entomol.*, 47: 29-48.

Jolivet, P. (1991c). Distribution et plante-hôtes de *Chrysolina banksi* (Fabricius, 1775) (Coleoptera, Chrysomelinae). *Nouv. Rev. Ent.*, 8: 151-157.

Jolivet, P. (1991d). Ants, plants, and beetles; a triangular relationship. Chapter 26 in: C.R. Huxley & D.F. Cutler, (eds.) *Ant-Plant Interactions*, Oxford University Press, Oxford: 397-406.

Jolivet, P. & Petitpierre, E. (1973). Plantes-hôtes connues des *Timarcha* Latreille. Quelques considérations sur les raisons possibles du trophisme sélectif. *Bull. Soc. Entomol. Fr.*, 78: 9-25.

Jolivet, P. & Petitpierre, E. (1976a). Les plantes-hôtes connues des *Chrysolina*. Essai sur les types de sélection trophique. *Ann. Soc. Entomol. Fr.*, 12: 123-149.

Jolivet, P. & Petitpierre, E. (1976b). Sélection trophique et évolution chromosomique chez les Chrysomelinae. *Acta Zool. Pathol. Antwerp.*, 66: 59-90.

Jolivet, P. & Petitpierre, E. (1981). Biology of Chrysomelidae (Col.). *Bull. Inst. Cat. Hist. Nat.*, 47(4): 105-138.

Jolivet, P., Petitpierre, E. & Daccordi, M. (1986). Les plantes-hôtes des Chrysomelidae (Coleoptera). Quelques nouvelles précisions et additions. *Nouv. Rev. Entomol.*, 3: 341-357.

Jolivet, P., Petitpierre, E. & Hsiao, T.H. (Eds.)(1988). *Biology of Chrysomelidae*. Kluwer Academic Publishers, Dordrecht: 1-613.

Jolivet, P. and Van Parys, E. (1977). Un cas inédit de mimetisme agressif entre un Chrysomelide (*Mesoplatys cincta* Olivier) et un Carabique (*Cyaneodinodes ammon* Fabricius). *Bull. Soc. Linn. Lyon*, 46: 168-180.

Jolivet, P., Vasconcellos-Neto, J. & Weinstein, P. (1990). Cycloalexy: a new concept in the larval defense of insects. *Ins. Mundi*, 4: 133-142.

Jones, D.L. (1972). The pollination of *Prasophyllum alpinum* R.Br. *Vict. Nat.* 89: 260-263.

Juanjie, T. *et al.* (1980). *Economic Insects. Fauna of China 18(1). Chrysomeloidea. 1.* Sueval Press, Beijing: 1-213.

Kalshoven, L.G. (1951). *De Plagen van de Cultuurgewassen in Indonesië.* Amsterdam: 1-556.

Kalshoven, L.G. (1957). An analysis of ethological, ecological and taxonomic data on Oriental Hispinae. *Tidj. fur Entomol.*, 100: 5-24.

Kalshoven, L.G. *et al.* (1981). *The pests of crops in Indonesia.* 2nd Edition. Jakarta: 1-701.

Karren, J.B. (1989). *Chlamisus mimosae* n.sp. from Brazil and imported into Australia and Thailand. *Col. Bull.*, 43: 355-358.

Kasap, H. & Crowson, R.A. (1976). On systematic relations of *Oomorphus concolor* (Sturm) with description of its larva and of an aberrant cryptocephaline larva from Australia. *J. Nat. Hist.*, 10: 99-112.

Kasuya, E. (1985). Size-disassortive mating in the chrysomelid beetle *Chrysolina aurichalcea. Evolution*, 3: 705-707.

Kato, M. (1991). Leaf-mining chrysomelids reared from Pteridophytes. *Jap. J. Entomol.*, 59: 671-674.

Kelly, E.D.G. (1915). On *Myochrous denticollis. US Dept. Agric. Bull.*, 221: 4.

Kimoto, S. (1964-1966). The Chrysomelidae of Japan and the Ryukyu Islands. *J. Fac. Agric. Kyushu Univ.*, 13: 1-671.

Kimoto, S. (1984). Chrysomelidae. in: M. Hayashi, K. Morimoto & S. Kimoto (eds.), *The Coleoptera of Japan in Color*, 4: 147-224, Hoikusha Publ. Co., Osaka.

Knab, F. (1909). Nuptial colors in the Chrysomelidae. *Proc. Entomol. Soc. Wash.*, 11: 151-153.

Kogan, M. & Goeden, R. (1970a). The systematic status of *Lema trilineata daturaphila*, new name, with notes on the morphology of chemoreceptors of adults (Coleoptera: Chrysomelidae). *Ann. Entomol. Soc. Am.*, 63: 529-537.

Kogan, M. & Goeden, R. (1970b). The biology of *Lema trilineata daturaphila* (Coleoptera: Chrysomelidae) with notes on efficiency of food utilization by larvae. *Ann. Entomol. Soc. Am.*, 63: 537-546.

Kogan, M. & Goeden, R. (1970c). The host-plant range of *Lema trilineata daturaphila* (Coleoptera: Chrysomelidae). *Ann. Entomol. Soc. Am.*, 63: 1175-1180.

Komiya, Y. (1986). Notes on Taiwanese Megalopodinae with description of a new species. *Elytra*, 14: 1-9.

Kostromitin, V.B. (1973). Food specialization of *Phyllotreta vittula. Zool. Zh.*, 52: 1415-1417.

Kostromitin, V.B. (1982). Habitat, distribution and trophic relations of flea beetles of the genus *Phyllotreta* Chvr. in the Tselinograd district. *Ent. Oborz.*, 61: 81-83.

Kostromitin, V.B.& Muraveva, N.N. (1982). Criocerinae. *Zashch. Rasteni*, 10: 58-59.

Kovalev, D.V. & Medvedev, L.N. (1983). Theoretical principles for the introduction of ambrosia leaf beetles of the genus *Zygogramma* Chevr. into the USSR for the biological control of Ambrosia. *Entomol. Rev. Wash.*, 62: 1-19 .

Kovtun, I.V. (1966). The biological characteristics of the development of the colorado beetle in the western regions of the Ukranian SSR. *Zakhyst Roslyn,* 3: 3-7.

Kreslavsky, A.G. *et al.* (1976). Intrapopulation ecological differentiation in the chrysomelid beetle *Chrysochloa cacaliae*. *Zool. Zh.*, 55: 1163-1171.

Kreslavsky, A.G. *et al.* (1981). Genetic exchange and isolating mechanisms in sympatric races of *Lochmaea caprae*. *Zool. Zh.*, 60: 62-68.

Krysan, J.L. & Miller, T.A. (1986). *Methods for the study of pest Diabrotica.* Springer Verlag, New York: 1-259.

Kulik, L.V. & Timoshin, A.A. (1986). Unusual tastes of the Colorado potato beetle. *Proroda Mask.*, 10: 98.

Kurcheva, G.F. (1958). Soil-dwelling larvae of Eumolpinae (Coleoptera, Chrysomelidae) the commonest subfamily of leaf-beetles in the European part of the USSR. *Acta Soc. Entomol. Czechoslov.*, 55: 383-393.

Kurcheva, G.F. (1967). Leaf beetle larvae of the subfamily Eumolpinae and of *Syneta betulae* F. (Coleoptera: Chrysomelidae). *Entomol. Rev. Wash.*, 46: 132-137.

Kurcheva, G.F. (1975). Larvae of *Basilepta fulvipes* Motsch. (Coleoptera: Chrysomelidae) from southern Primorge. *Trud. Biol. Pech. Inst. Vlad.*, 28: 107-113.

Kuschel, G. & May, B.M. (1990). Palophaginae, a new subfamily for leaf-beetles, feeding as adult and larva on araucarian pollen in Australia (Coleoptera: Megalopodidae). *Invert. Taxon.*, 3: 697-719.

Kuwayama, S. (1954). A contribution to our knowledge on the food-plants of the rice leaf-beetle. *Oyo-Kontyu,* 10: 59-62.

Laboissiere, V. (1934). Galerucinae de la faune francaise. *Ann. Soc. Entomol., Fr.*, 103: 1-108.

Larsson, S. & Ohmart, C.P. (1988). Leaf age and larval performance of the leaf beetle *Paropsis atomaria*. *Ecol. Entomol.*, 13: 19-24.

Lazell, J.D., Keirans, J.E. & Samuelson, G.A. (1991). The Sulawesi black racer, *Coluber (Ptyas) dipsas*, and a remarkable ectoparasitic aggregation. *Pac. Sci.*, 45: 355-361.

Lee, J.E. (1990). Description of the first-instar larva of *Syneta adamsi* Baly from Japan with notes on the systematic position of Synetinae (Chrysomelidae). *Esakia,* 29: 77-81.

Le Pelley, R.H. (1959). *Agricultural Insects of East Africa.* Nairobi: 1-307.

LeSage, L. (1982). The immature stages of *Exema canadensis* Pierce (Coleoptera: Chrysomelidae). *Col. Bull.*, 36: 318-327.

LeSage, L. (1984a). Immature stages of Canadian *Neochlamisus* Karren (Coleoptera: Chrysomelidae). *Can. Ent.*, 116: 383-409.

LeSage, L. (1984b). Egg, larva and pupa of *Lexiphanes saponatus* (Coleoptera: Chrysomelidae: Cryptocephalinae). *Can. Ent.*, 116: 537-548.

LeSage, L. (1985). The eggs and larvae of *Pachybrachis peccans* and *P. bivittatus*, with a key to the known immature stages of the Nearctic genera of Cryptocephalinae (Coleoptera: Chrysomelidae). *Can. Ent.*, 117: 203-220.

LeSage, L. (1986). The eggs and larvae of *Cryptocephalus quadruplex* Newman and *C. venustus* Fabricius, with a key to the known immature stages of the Nearctic genera of cryptocephaline leaf-beetles (Coleoptera: Chrysomelidae). *Can. Ent.*, 118: 97-111.

Lopatin, I.K. (1975). The chrysomelid beetles (Coleoptera, Chrysomelidae) of the Mongolian People's Republic. *Ins. Mong.*, 3: 191-234.

Lopatin, I.K. (1984). *Leaf beetles (Chrysomelidae) of Central Asia and Kazakhstan.* Oxonian Press, New Delhi: 1-416.

Mabberley, D. (1987). *The Plant Book.* Cambridge Univ. Press: 1-706.

Maes, J.M. & Staines, C.L. (1991). Catalogo de les Chrysomelidae (Coleoptera) de Nicaragua. *Rev. Nicar. Entomol.*, 18: 1-53.

Maes, J.M. & Tellez Robleto, J. (1988). Catalogo de los Insectos y Artropodos terrestris asociados a los principales plantas en Nicaragua. *Rev. Nicar. Entomol.*, 5: 1-96.

Mafra-Neto, A. & Jolivet, P. (1994). Entomophagy in Chrysomelidae: adult *Aristobrotica angulicollis* (Erichson) feeding on adult meloids (Coleoptera). Chap. 9. In: Jolivet,

P., Cox, M.L. & Petitpierre, E. (Eds.). Novel Aspects of the Biology of Chrysomelidae. Kluwer Academic Publishers, Dordrecht: 171-178.

Mann, J.S. & Crowson, R.A. (1981). The systematic positions of *Orsodacne* Latr. and *Syneta* Lacord. (Coleoptera: Chrysomelidae), in relation to characters of larvae, internal anatomy and tarsal vestiture. *J. Nat. Hist.*, 15: 727-749.

Martin, W.D. & Herzog, G.A. (1987). Life history studies of the tobacco flea beetle, *Epitrix hirtipennis* (Melsheimer), *J. Entomol. Sci.*, 22: 237-244.

Martorell, L.F. (1976). Annotated food plant catalog of the insects of Puerto Rico. *Agric. Exp. Station Puerto Rico*: 1-303.

Maulik, S. (1919). *The fauna of British India, including Ceylon and Burma. Coleoptera. Chrysomelidae (Hispinae and Cassidinae).* London: 1-439.

Maulik, S. (1926). *The fauna of British India, including Ceylon and Burma. Coleoptera. Chrysomelidae (Chrysomelinae and Halticinae).* London: 1-442.

Maulik, S. (1937). Distributional correlation between hispine beetles and their host plants. *Proc. Zool. Soc. Lond.*, A2: 129-159.

Maulik, S. (1941). Biology and morphology of the Sagrinae. *Ann. Mag. Nat. Hist.*, 2: 235-254.

Maulik, S. (1945). *An inquiry into the relationships between the chrysomelid beetles and their food-plants.* Ms., London: 1-281.

Maulik, S. (1947). Willow as host-plant of *Crioceris merdigera. Nature*, 160: 609-610.

Maulik, S. (1950). An alphabetical list of names of plant genera attacked by beetles of the family Chrysomelidae. Ms., London.

Mayet,V. (1907) Métamorphoses du *Malacosoma lusitanicum* (Col.). *Bull. Soc. ent. Fr.*: 115-117.

McCoy, E.D. (1984). Colonization by herbivores of *Heliconia* spp. plants. *Biotropica*, 16: 10-13.

McCoy, E.D. (1985). Interactions among leaf-top herbivores of *Heliconia imbricata* (Zingiberales, Heliconiaceae). *Biotropica*, 17: 326-329.

McFadyen, P.J. (1987). Host specificity and biology of *Lioplacis elliptica* introduced into Australia for the biological control of *Baccharis halimifolia* (Compositae). *Entomophaga*, 32: 19-21.

Medeiros, L. (1991). *Aspectos da interaçao entre especies de Chrysomelinae (Col. Chrys.) e plantas da familia Solanaceae na Serra do Japi, Jurdiai, S.P.* Thesis, Unicamp, Campinas 1-86.

Medvedev, L.N. (1962). Systematics and biology of the subfamily Clytrinae (Coleoptera, Chrysomelidae). *Zool. Zh.*, 41: 1334-1344.

Medvedev, L.N. (1968). Leaf-beetles of the Kara Tau Jurassic. In: *Symposium Jurassic Insects of Kara Tau.* Nauka Publ. House, USSR, Moscow: 155-165.

Medvedev, L.N. (1970). A contribution to the knowledge of African and Oriental Clytrinae (Coleoptera, Chrysomelidae). *Entomol. Arb. Mus. Frey*, 21: 184-195.

Medvedev, L.N. & Dap, T.D. (1982). Relations trophiques des chrysomelides du Vietnam. *Animal Kingdom Vietnam*, Moscow: 84-97.

Medvedev, L.N. & Dap, T.D. (1983). Plantes-hôtes des Chrysomelides du Vietnam. *Acad. USSR Fauna Vietnam*: 145-148.

Medvedev, L.N. & Pavlov, S.I. (1987). Mating behaviour of the leaf-beetles. *Ent. Oboz.*, 66: 746-753.

Medvedev, L.N. & Roginskaya, E.Ja. (1988). Catalog of the host-plants of the leaf-beetles of the USSR. *Acad. Sci. USSR Moscow*: 1-190.

Medvedev, L.N. & Zaitsev, Y.M. (1979). Larvae of leafbeetles (Col., Chrysomelidae) of Vietnam. *Zool. Zh.*, 58: 1157-1167.

Messina, F.J. (1982). Comparative biology of the Goldenrod Leaf Beetles, *Trirhabda virgata* LeConte and *T. borealis* Blake (Coleoptera: Chrysomelidae). *Col. Bull.*, 36: 255-269.

Mikheev, A.V. & Kreslavsky, A.G. (1980). Interrelations of the willow and birch races of *Lochmaea capreae* L. (Coleoptera, Chrysomelidae) with their host-plants. *Zool. Zh.*, 59: 705-714.
Mikheev, A.V. & Kreslavsky, A.G. (1981). The variation of morphological features in the willow and birch races of *Lochmaea capreae* (Coleoptera, Chrysomelidae). *Zool. Zh.*, 60: 200-204.
Mikheev, A.V. & Kreslavksy, A.G. (1984). Relations between host plants and structure of the willow race of *Lochmaea capreae* L. (Coleoptera, Chrysomelidae). *Zool. Zh.*, 63: 209-217.
Mikheev, A,V. & Kreslavsky, A.G. (1986). Ecological and genetic principles of race differentiation in *Lochmaea capreae* L. (Coleoptera, Chrysomelidae). *Zool. Zh.*, 65: 1826-1834.
Mohr, K.H. (1966). Familie: Chrysomelidae. In: Freude, H., Harde, K.W. & Lohse, G.A. (Eds.). *Die Käfer Mitteleuropas*, 9: 95-280.
Moldenke, A.R. (1970). *A revision of the Clytrinae of North America north of the Isthmus of Panama*. Stanford Univ. Press: 1-310.
Moldenke, A.R. (1971). Host-plant relations of phytophagous beetles in Mexico. *Pan-Pac. Entomol.*, 47: 105-116.
Monros, F. (1949). Descripcion de las metamorphosis de *Lamprosoma chorisiae* monao y consideraciones taxonomicias sobre Lamprosominae. *Acta Zool. Lill.*, 7: 449-466.
Monros, F. (1951). Nuevo genero de Megascelinae de la Patagonia. *Rev. Chil. Entomol.*, 1: 171-173.
Monros, F. (1952a). Revision de las especies argentinos de Chlamisinae (Coleoptera, Chrysomelidae). *Acta Zool. Lilloana* 10: 489-672.
Monros, F. (1952b). Notas sobre algunas Eumolpinae neotropicales (Coleoptera: Chrysomelidae). *Rev. Chil. Entomol.*, 2: 187-196.
Monros, F. (1954a). *Megalopus jacobyi*, nueva plaga de Solanaceae en el Noroeste Argentino, con notas sobre biologia y taxonomica de Megalopinae. *Rev. Agron. Nor. Arg.*, 1: 167-179.
Monros, F. (1954b). Revision of the chrysomelid subfamily Aulacoscelinae. *Bull. Mus. Comp. Zool.*, 112: 321-359.
Monros, F. (1955). Biologia y descripcion de la larva de *Atalasis sagroides*. *Agron. Nor. Argentina*, 1: 275-281.
Monros, F. (1956). Sur le genre *Megamerus* MacLeay. *Rev. Fr. Entomol.*, 23(2): 104-115.
Monros, F. (1959). Los generos de Chrysomelidae. *Opera Lilloana*: 3: 1-337.
Morris, M.G. (1970). *Phyllodecta polaris* Schneider (Coleoptera, Chrysomelidae) new to British Isles from Wester Ross and Inverness-shire, Scotland. *Entomol. Month. Mag.*, 106: 48-53.
Mummery, R.S. & Valadon, L.R.G. (1974). Carotenoids of the lily beetle (*Lilioceris lilii*) and of its food-plant (*Lilium hansonii*). *J. Ins. Physiol.*, 20: 429-433.
Neal, J.W. (1989). Bionomics of immature stages of ethology of *Neochlamisus plateni* on American sycamore (Coleoptera: Chrysomelidae). *Ann. Entomol. Soc. Am.*, 82: 64-72.
Niemelae, P. & Tuomi, J. (1987). Does the leaf morphology of some plants mimic caterpillar damage? *Oikos*, 50: 256-257.
Nishida, R. *et al.* (1986). Isolation of feeding stimulants of Brazilian leaf beetles (*Diabrotica speciosa* and *Cerotoma arcuata*) from the root of *Ceratosanthes hilariana. Agric. Biol. Chem.*, 50: 2831-2836.
Nonveiller, G. (1959). Un cas d'apparition de *Haltica oleracea* L. sur le fraisier avec de nouvelles données sur ses plantes-hôtes. *Zastita Bilja*, 55: 87-94.
Olalquiaga, F.G. (1980) Aspectos fitosanitarios de la Isla de Pascua. *Rev. Chil. Entomol.*, 10: 101-102.
Omer-Cooper, J. & Miles, P. (1951). On *Lema trilineata*, a beetle closely resembling the tobacco slug attacking the Cape gooseberry. *Sth. Afr. J. Sci.*, 67: 330-333.

Ostmark, H.E. (1975). Banana pests in the genus *Colaspis*, including description of a new species. *Fl. Entomol.*, 58: 1-8.
Otto, C. (1985). Effects of temporal and spatial variations for good availability on life-cycle and palatability of a chrysomelid beetle *Donacia cinerea*. *Aquat. Ins.*, 7: 19-28.
Paik, W.H. (1967). Insect pests of Korea. In: *Major Insect Pests of the rice plant*. John Hopkins Univ. Press: 1-657.
Pallister, J.C. (1953). The leaf beetles of North Central Mexico collected on the David Rockefeller Mexican expedition (Coleoptera, Chrysomelidae). *Am. Mus. Novit.*, 1623: 1-95.
Palmer, W.A. & Goeden, R.D. (1991). Host range of *Ophraella communa* LeSage (Coleoptera: Chrysomelidae). *Col. Bull.*, 45: 115-120.
Pandit, N.C. & Pradhan, S.K. (1991). Bionomics of the sunhemp flea beetle, *Longitarsus belgaumensis* Jacoby (Col., Chrysomelidae). *Environ. Entomol.*, 9: 983-985.
Parry, R.H. (1986). The systematics and biology of the flea beetle genus *Crepidodera* Chevrolat in America North of Mexico (Col.: Chrysomelidae). *Insecta Mundi*, 1: 156-196.
Passoa, S. (1983). Lista de los Insectos asociados con los granos basicos y otros cultivos selectos en Honduras. *Ceiba*, 25: 1-98.
Paterson, N.F. (1930). The bionomics and morphology of the early stages of *Paraphaedon tumidulus* Germ. *Proc. Zool. Soc. Lond.*, 41: 627-676.
Paterson, N.F. (1931). The bionomics and comparative morphology of the early stages of certain Chrysomelidae. *Proc. Zool. Soc. Lond.*, 42: 879-949.
Paterson, N.F. (1931-1932). Embryological development of *Euryope terminalis*. 1: Early development. 2: Organogeny. *Sth. Afr. J. Sci.*, 28: 344-371; 29: 414-418.
Paterson, N.F. (1936). Observations on the embryology of *Corynodes pusis* (Coleoptera, Chrysomelidae). *Q.J. Micros. Sci.*, 78: 91-132.
Pavlov, S.I. (1984), On the mechanism of defense reaction in the *Donacia* species. *Zool. Zh.*, 63: 1192-1195.
Peterson, A. (1960). Larvae of insects. Columbus, Ohio, 2: 1-416.
Petitpierre, E. (1976). Further cytotaxonomical and evolutionary studies on the genus *Timarcha* Latr. (Coleoptera, Chrysomelidae). *Genet. Iber.*, 28: 58-72.
Petitpierre, E. (1985). Notas faunisticas y ecologicas sobre Chrysomelidae de Mallorca y Catalunya. *Boll. Soc. Hist. Nat. Balears*, 29: 31-36.
Peyerhimhoff, P. de (1911). Notes sur la biologie de quelques Coléoptères phytophages du Nord africain (1 serie). *Ann. Soc. Entomol. Fr.*, 80: 283-314.
Peyerhimhoff, P. de (1915). Notes sur la biologie de quelques Coléoptères phytophages du Nord africain (2 série). *Ann. Soc. Entomol. Fr.*, 84: 19-61.
Pillai, K.S. and Rajamma, P. (1987). New records of the incidence of spotted beetle, *Oides affinis* Jacoby on elephant foot yam and the scale insect *Aspidella hartii* Cockerell. *Entomon.*, 12: 113-114.
Powell, E.F. (1941). Relationships within the family Chrysomelidae as indicated by the male genitalia of certain species. *Am. Mid. Nat.*, 25: 148-195.
Rausher, M.D. (1983). Conditioning and genetic variation as causes of individual variation in the oviposition behavior of the tortoise beetle, *Deloyala guttata*. *Anim. Behav.*, 31: 743-747.
Rausher, M.D. (1984). Tradeoffs in performance on different hosts: Evidence from within and between site variation in the beetle *Deloyala guttata*. *Evolution*, 38: 582-595.
Reid, C.A.M. (1988). *Diachus auratus* (F.)(Col. Chrysomelidae): a recent immigrant to the South-West Pacific region on legumes. *Gen. Appl. Entomol.*, 20: 5-8.
Reid, C.A.M. (1989). The Australian species of the tribe Zeugophorini (Coleoptera: Chrysomelinae: Megalopodinae). *Gen. Appl. Entomol.*, 21: 39-47.
Reid, C.A.M. (1991). Immature stages of the genera *Johannica* Blackburn, *Lamprolina*

Baly and *Chalcolampra* Blanchard (Coleoptera, Chrysomelidae, Chrysomelinae). *J. Nat. Hist.*, 25: 341-357.
Reid, C.A.M. (1992). The leaf-beetle genus *Microdonacia* Blackburn (Coleoptera: Chrysomelidae: Galerucinae): revision and systematic placement. *Syst. Entomol.*, 17: 359-387.
Reid, C.A.M. (1993). Description of the *constricta* species-group of the genus *Chalcolampra* Blanchard (Coleoptera: Chrysomelidae: Chrysomelinae). *J. Aust. Ent. Soc.*, 32: 253-263.
Riley, E.G. & Enns, W.R. (1979). An annotated checklist of Missouri leaf-beetles. *Trans. Missouri Acad. Sci.*, 13: 53-83.
Riley, E.G. & Enns, W.R. (1982). Supplement to an annotated checklist of Missouri leaf-beetles: New State records and host-plant associations. *Entomol. News*, 93: 32-36.
Root, R.B. & Messina, F.J. (1983). Defensive adaptations and natural enemies of a case-bearing beetle, *Exema canadensis* (Coleoptera: Chrysomelidae). *Psyche*, 90: 67-80.
Salt, G. (1928). A study of *Colaspis hypochlora* Lef. (Col. Chrysomelidae). *Bull. Entomol. Res.*, 19: 295-308.
Samedov, N.G. & Mirzoeva, N.B. (1982). Ecological-faunistic and zoogeographical grouping of leaf-beetles of the Lesser Caucasus in Azerbaijan. *Entomol. Obozr.*, 61: 795-800.
Samsinak, K. (1956). Pripad phoresie u larve rodu *Clytra* Laich *Vestn. Cesko-sl. Zool. Spolec.*, 20: 375-376.
Samuelson, G.A. (1967). Alticinae of the Solomon Islands. *Pac. Ins.*, 9: 139-174.
Samuelson, G.A. (1973). Alticinae of Oceania. *Pac. Ins. Mono.*, 30: 1-165.
Samuelson, G.A. (1984). Plant associated Alticinae from the Bismarck Range, New Guinea. *Esakia*, 21: 31-47.
Samuelson, G.A. (1985). Description of a new species of *Alagoasa* from southern Brasil associated with *Lantana* (Verbenaceae). *Rev. Bras. Entomol.*, 29(3-4): 579-585.
Samuelson, G.A. (1989). Pollen feeding in Alticinae (Chrysomelidae). *Entomogr.*, 6: 407-411.
Santos, H.R. (1980). Distribuicao a danos do *Agathomerus sellatus* (Germar, 1824) am cultura de tomateiro. *Rev. Setor Ciencias Agrarias*, 2: 157-162.
Santos, H.R. (1981a). Redescricao do *Agathomerus sellatus* (Germar, 1824) praga do tomateiro (*Lycopersicum esculentum* Mill.). *Rev. Setor Ciencias Agrarias*, 3: 107-116.
Santos, H.R. (1981b). Biologia de *Agathomerus sellatus* (Germar, 1824) broca de tomateiro. *Rev. Bras. Entomol.*, 25(2): 165-170.
Saulnier, C. (1947). L'osier est-il une nouvelle plant-hôte de *Crioceris merdigera* L. ? *Misc. Entomol.*, 43: 43.
Saunders, J.L., King, B.S. & Vargas, C.L. (1983). Plagas de cultivos en America Central. *Turrialba*: 1-90.
Schaefer, P.W. (1987). Trench-feeding behavior of *Aulacophora femoralis* Motschulsky on a cucurbit in Central China. *Col. Bull.*, 41: 136.
Scherer, G. (1969). Die Alticinae des Indischen Subkontinentes (Coleoptera, Chrysomelidae). *Pac. Ins. Mono.*, 22: 1-251.
Scherer, G. (1974). *Clavicornaltica*, a new genus from Ceylon (Col., Chrysomelidae). *Rev. Suisse Zool.*, 81: 57-68.
Scherer, G. (1979). *Clavicornaltica* recorded also from the Philippine Islands (Col., Chrysomelidae). *Rev. Suisse Zool.*, 86: 713-714.
Scherer, G. (1982). Origins of the Alticinae (Coleoptera, Chrysomelidae). *Spixiana (Suppl.)*, 7: 7-9.
Scherer, G. (1988). The origins of the Alticinae. Chapter 8 in: P. Jolivet, E. Petitpierre & T.H. Hsiao (Eds.) *Biology of Chrysomelidae.* Kluwer Academic Publishers, Dordrecht: 115-130.

Schmitt, M. (1988). The Criocerinae: Biology, Phylogeny and Evolution. Chap. 28 in: P. Jolivet, E. Petitpierre & T.H. Hsiao (Eds.) *Biology of Chrysomelidae*. Kluwer Academic Publishers, Dordrecht: 475-495.

Seeno, T.N. & Wilcox, J.A. (1982). *Leaf beetle genera (Coleoptera: Chrysomelidae)*. Entomography Publ., Sacramento: 1-221.

Seifert, R.P. & Seifert, H.F. (1979). Utilization of *Heliconia* (Musaceae) by the beetle *Xenarescus monoceros* (Olivier) in a Venezuelan forest. *Biotropica*, 11: 51-59.

Selman, B.J. (1962). Remarkable new chrysomeloids found in the nests of arboreal ants in Tanganyika (Coleoptera: Clytridae and Cryptocephalidae). *Ann. Mag. Nat. Hist.* 13(5): 295-299.

Selman, B.J. (1963). Coléoptères Chrysomelides récoltés par M.J. Mateu dans l'Ennedi et au Tchad. *Bull. IFAN,* 25(A4): 1148-1162.

Selman, B.J. (1972). Eumolpinae. In: *Expl. Parc Nat. Garamba*, Bruxelles, 55: 1-95.

Sharp, D. & Muir, F. (1912). The comparative anatomy of the male genital tube in Coleoptera. *Trans. Entomol. Soc. Lond.*, 1912: 558-565.

Sholes, O.D.V. (1987). Host plants and seasonal abundance of adult *Capraita subvittata* (Col., Chrysomelidae). *Proc. Entomol. Soc. Wash.*, 89: 818-820.

Shute, S.L. (1980). Wing polymorphism in British species of *Longitarsus* beetles. *Syst. Entomol.*, 5: 437-448.

Shute, S.L. (1983). Key to the genera of galerucine beetles of New Guinea with a review of *Sastra* and related new taxa. (Coleoptera, Chrysomelidae). *Bull. Br. Mus. Nat. Hist. (Entomol.)*, 46: 205-266.

Siddappaji, C. & Gowda, T.K.S. (1980). Rhizobial nodules eating insect-Rivellia sp.- a new pest of pulse crops in India. *Curr. Res.*, 9: 122-123.

Siebert, M.W. (1975). Candidates for the biological control of *Solanum elagnifolium* Cav. (Solanaceae) in South Africa. 1. Laboratory studies on the biology of *Gratiana lutescens* (Boheman) and *Gratiana pallidula* (Boheman) (Col. Chryso. Cassidinae). *J. Entomol. Soc. Sth. Afr.*, 38: 297-304.

Smith, E.H. (1985). Revision of the genus *Phyllotreta* Chevrolat of America north of Mexico. Part 1. The maculate species. *Field. Zool.*, 1364: 1-168.

Staines, C.L. (1986). An unusual food plant for adult *Cerotoma trifurcata*. *Proc. Entomol. Soc. Wash.*, 88: 392.

Staines, C.L. (1991). A host-plant for adult *Spintherophyta globosa* (Olivier) (Coleoptera: Chrysomelidae). *Col. Bull.*, 45: 200.

Stehr, F.W. (1991). *Immature Insects* 2: 1-974.

Steinhausen, W. (1982). Status of Palaearctic Alticinae larvae research. *Spixiana (Suppl)*, 7: 39-42.

Strong, D.R. (1977a). Insect species richness: hispine beetles of *Heliconia latispatha. Ecology*, 58: 573-582.

Strong, D.R. (1977b). Rolled-leaf hispine beetles and their Zingiberales host plants of Middle America. *Biotropica,* 9: 156-169.

Strong, D.R. (1981). The possibility of insect communities without competition; hispine beetles on *Heliconia.* In: Denno & Dingle (Eds.) *Insect Life History Patterns.* Springer Verlag, New York: 183-194.

Strong, D.R. (1982a). Harmonious coexistence of hispine beetles on *Heliconia* in experimental and natural communities. *Ecology*, 63: 1039-1049.

Strong, D.R. (1982b). Potential interspecific competition and host specificity: hispine beetles on *Heliconia. Ecol. Entomol.*, 7: 217-220.

Strong, D.R. and Wang, M.D. (1977). Evolution of insect life-histories and hostplant chemistry: hispine beetles on *Heliconia. Evolution*, 31: 854-862.

Swain, T. (Ed) (1963). *Chemical Plant Taxonomy.* Academic Press, London: 1-543.

Szel, G. & Adam, L. (1989). Beetles damaging reeds in the lake Balaton. *Fol. Entomol. Hung.*, 50: 184-185.

Szent-Ivany, J.J.H., Womersley, J.S. & Ardley, J.H. (1956). Some insects of *Cycas* in New Guinea. *Papua New Guinea Agr. J.*, 11: 53-56.

Takhtajan, A. (1969). *Flowering Plants: Origin and Dispersal.* Oliver & Boyd, Edinburgh: 1-310.

Takizawa, H. (1970). Descriptions of five new species of the genus *Chrysolina* Motsch. (Col., Chrysomelidae) in Japan. *Kontyu*, 38(2): 117-125.

Takizawa, H. (1971a). On the larvae of the genus *Chrysolina* Motsch. (Col., Chrysomelidae) in Japan. *Entomol. Rev. Jap.*, 23(2): 102-109.

Takizawa, H. (1971b). Note on the pupae of the genus *Chrysolina* Motsch. (Col., Chrysomelidae) in Japan. *Entomol. Rev. Jap.* 23(2): 110-111.

Takizawa, H. (1972). Descriptions of larvae of glanduliferous group of Galerucinae in Japan. *Insecta Matsumurana Suppl.*, 10: 1-14.

Takizawa, H. (1976). Larvae of the genus *Gonioctena* Chevrolat. Descriptions of Japanese species and implications of larval characters for the phylogeny (Col., Chrysomelidae). *Kontyu*, 44(4): 444-468.

Takizawa, H. (1977). Larva of *Agrosteomela indica* (Hope)from Taiwan(Col. Chrys.). *Akitu, N.S.*, 10: 1-4.

Takizawa, H. (1978a). Notes on Taiwanese chrysomelid larvae. *Entomol. Rev. Jap.*, 31(1-2): 75-84.

Takizawa, H. (1978b). Descriptions of two larvae of the genus *Chrysolina* Motschulsky in Japan (Col., Chrysomelidae). *Entomol. Rev. Jap.*, 31(1-2): 111-117.

Tandon, P.L., Krishnaiah, K. & Prasad, V.G. (1975). *Sagra femorata* Drury (Sagrinae, Chrysomelidae, Coleoptera), a new pest of *Dolichos lablab* L. *Curr. Sci.*, 44(21): 786-787.

Tayade, S. (1978). Bionomics of *Sagra femorata* Drury (Col., Chrysomelidae, Sagrinae). *Mysore J. Agric. Sci.*, 12(4): 582-587.

Tella, R. (1952). Contribucos para a conhecimento do *Agathomerus sellatus* Germar. *R. Agric. Piracicaba*, 27: 373-376.

Tempere, G. (1927). Régime alimentaire anormal de *Crioceris lilii* Scop. *Pro. Verb. Soc. Linn. Bordeaux*, 78: 131-133.

Tempere, G. (1935). Les Phanerogames Centrospermees et l'instinct botanique de quelques Coléoptères. *Proc. Verb. Soc. Linn. Bordeaux*, 1-5.

Tempere, G. (1946). L'instinct botanique des Insectes Phytophages. *L'Entomol.*, 2: 219-224.

Tempere, G. (1967). Un critère méconnu des systématiciens phanerogamistes: l'instinct des Insectes Phytophages. *Trav. Biol. Vet. P. Dangeard Le Botaniste:* 473-482.

Thomas, C.D. (1987). Behavioural determination of diet breadth in insect herbivores: the effect of leaf age on choice of host species by beetles feeding on *Passiflora* vines. *Oikos*, 48: 211-216.

Tieberghien, G. (1971). Etudes systématiques sur les Chrysomelides du genre *Timarcha. Bull. Soc. Entomol. Fr.*, 78: 185-192.

Tieberghien, G. (1972). Ecologie et Ethologie de diverses espèces de *Timarcha* Latr. (Col. Chrysomelidae). *Bull. Soc. Hist. Nat. Toulouse*, 108(1-2): 238-338.

Tillyard, R.J. (1926). *Insects of Australia and New Zealand.* Sydney: 1-560.

Uhmann, E. (1958). Hispinae aus Südafrika III. Teil. *J. Entomol. Soc. Sth. Afr.*, 21: 214-226.

Van den Berg, M.A. (1982). Coleoptera attacking *Acacia dealbata* Link., *Acacia decurrens Willd.*, *Acacia mearnsii* De Willd. and *Acacia melanoxylon* R.Br. in Australia. *Phytophylactica* 14: 51-55.

Vasconcellos-Neto, J. (1987). Genetica ecologica de *Chelymorpha cribraria* F. 1775 (Cassidinae, Chrysomelidae). Tese de Doutorado, Unicamp, Campinas: 1-254.

Vasconcellos-Neto, J. (1988). Genetics of *Chelymorpha cribraria*, Cassidinae: colour patterns and their ecological meanings. In: P. Jolivet, E. Petitpierre & T.H. Hsiao

(Eds.) *Biology of Chrysomelidae*. Kluwer Academic Publishers, Dordrecht: 217-232.
Vasconcellos-Neto, J. & Jolivet, P. (1988). Une nouvelle stratégie de défense: la stratégie de défense annulaire (cycloalexie) chez quelques larves de Chrysomelides brésiliens (Coleoptera). *Bull. Soc. Entomol. Fr.*, 92: 291-299.
Vasconcellos-Neto, J. & Jolivet, P. (1989). Ring defense strategy (Cycloalexy) among Brazilian chrysomelid larvae (Coleoptera). *Entomogr.*, 6: 347-354.
Vinogradova, E.B. (1988). Peculiarities of the reproduction and pattern of the imaginal diapause in the leaf-beetle *Zygogramma suturalis* F. in Stauropol province. *Rev. Entomol. USSR*, 47: 468-479.
Virkki, N. (1980). Flea beetles, especially Oedionychina, of a Puerto Rican marshland in 1969-72. *J. Agric. Univ. Puerto Rico,* 64: 63-92.
Virkki, N. (1984). Additional observations on the life-history of the Oedionychina flea beetles: *Alagoasa januaria* Bechyne. *J. Agric. Univ. Puerto Rico*, 68: 107-109.
Virkki, N. & Zambrana, I.(1980). Demes of a Puerto Rican flea beetle *Alagoasa bicolor* (L.) differing in mean body size and foodplant association. *J. Agric. Univ. Puerto Rico*, 64: 264-274.
Virkki, N. & Zambrana, I. (1983). Life history of *Alagoasa bicolor* (L.) in indoor rearing conditions. *Entomol. Arb. Mus. G. Frey*, 31-32: 131-155.
Vlasova, J.A. (1978). A prognosis of the distribution area of the Colorado potato beetle in the Asiatic part. *Zasheita Rast.*, 6: 44-45.
Wade, M.J. and Breden, F. (1986). Life-history of natural populations of the imported willow leaf beetle, *Plagiodera versicolora. Ann. Entomol. Soc. Am.*, 79: 73-79.
Walker, J.J. (1906). Antipodean field notes. III. A sketch of the entomology of Sydney, N.S.W. *Entomol. Mon. Mag.*, 17: 22-27, 50-55.
Wallace, J.B. (1970). The defensive function of a case on a chrysomelid larva. *J. Georgia Entomol. Soc.*, 5(1): 19-24.
Wapshere, A.J. (1982). Life-histories and host specialities of the *Echium* flea beetles *Longitarsus echii* and *L. aenus* (Coleoptera: Chrysomelidae). *Entomophaga*, 27: 173-181.
Webb, G.A. (1986). Some insect pollinators of *Kunzea ambigua* (Sm.) Druce (Myrtaceae) near Sydney, New South Wales. *Vict. Nat.*, 103: 12-15.
Weise, J. (1916). *Col. Cat. (Chrysomelidae: Chrysomelinae)*. 68(12): 1-255.
Wellso, S.G. & Hoxie, R.P. (1988). Biology of *Oulema*. In: P. Jolivet, E. Petitpierre & T.H. Hsiao (Eds.), *Biology of Chrysomelidae*. Kluwer Academic Publishers, Dordrecht: 497-511.
White, R.E. (1979). A Neotropical leaf-beetle established in the United States (Chrysomelidae). *Ann. Entomol. Soc. Am.*, 72: 269-270.
Wilcox, J.A. (1979). Leaf beetle host plants in northeastern North America. *World Nat. Hist. Publ.*, 1-30.
Williams, C.E. (1991). New England aster, *Aster novae-angliae*: a new host record for *Microrhopala xerene* (Coleoptera: Chrysomelidae). *Proc. Entomol. Soc. Wash.*, 93: 790.
Willis, J.C. & Airy Shaw, H.K. (1973). *Dictionary of the flowering plants and ferns.* Cambridge Univ. Press, 8th Ed., 1-1245.
Windsor, D.M., Riley, E.G. & Stockwell, H.P. (1992). An introduction to the biology and systematics of Panamanian tortoise beetles (Coleoptera: Chrysomelidae: Cassidinae). Chap. 25 in: D. Quintero & A. Diello (Eds.) *Insects of Panama and Mesoamericae*. Oxford Sci. Publ.: 372-391.
Wood, G.W. (1966). Life history and control of a case bearer, *Chlamisus cribripennis* (Coleoptera, Chrysomelidae), on blueberry. *J. Econ. Entomol.*, 59: 823-825.
Wood, G.W. (1978). Studies on *Neochlamisus cribripennis* (LeConte) with particular reference to its reproductive biology. *Diss. Abstrs. Int.*, 39(3): 1097.

Woodruff, R.E. (1965). The leaf beetle genus *Chrysomela* in Florida (Col. Chrys.) *DAI Entomol. Circ.* 40: 1-2.

Woodruff, R.E. (1976). The tortoise beetles of Florida. IV. *Metriona bicolor* (Fab.). *DAI Entomol. Circ.*, 164: 1-2.

Wyniger, R. (1962). Pests of crops in warm climates and their control. *Acta Tropica, Basel*: 1-555.

Xue, J.J. (1986). Notes on the life-history of *Entomoscelis orientalis* Motschulsky. *Acta Pyt. Sinica*, 13: 102.

Yaroshenko, V.A. (1985). Ecological-faunistical characteristics of leaf beetles of the northern Caucasus. *Vestnik Zool.*, 4: 52-55.

Young, A.M. (1979). Notes on a population outbreak of the beetle *Uroplata* sp. on the tree *Bunchosia pilosa* (Malpighiaceae) in Costa Rica. *J. New York Entomol. Soc.*, 87: 289-298.

Yu, P. (1977). On *Temnaspis nankinea* (Pic) a new pest of *Fraxinus chinensis* Roxb. *Acta Entomol. Sinica*, 20: 482-484.

Yu, P. (1988). A preliminary study of *Syneta adamsi* Baly. *Scientia Silvae Sinicae*, 24: 235-238.

Zaitsev, Y.M. (1982). Larva of *Sagra femoralis* from Vietnam. *Zool. Zh.*, 61: 458-460.

Zaitsev, Y.M. (1983). Preimaginal stages and biology of *Agasta formosa* Hope and the generic position within the system. In: L.N. Medvedev, *Fauna Ecology Animal Vietnam*, Moscow: 119-122.

Zaitsev, Y.M. (1986). Larvae of *Lycaria wastermanni* and taxonomic position of the genus (Col., Chrysomelidae). *Zool. Journ. Moscow*, 65(9): 1424-1427.

Zaitsev, Y.M. (1988). Morphoecological characteristics of *Orsodacne cerasi* L. *Vestnik Zool.*, 3: 80-82.

Zaka-ur-Rab, M. (1991). Leaf mining Coleoptera of the Indian subcontinent. *J. Entomol. Res.*, 15: 20-30.

Zapata, S. (1970). Biologia de *Plagiodera erythroptera* Blanchard (Coleoptera, Chrysomelidae). *Publ. Cont. Estud. Univ. Chile*, 10: 57-63.

Zulueta, A. de (1925). La herencia ligada al sexo en el Coleoptero *Phytodecta variabilis. Eos,* 1: 203-231.

Zulueta, A. de (1929). La mutacion "jaspeado" del Coleoptero Phytodecta variabilis (su aparicion y herencia). *Mem. Real Soc. Esp. Hist. Nat.*, 15: 819-824.

ADDENDA

The following list adds data not available or overlooked at the time when the first drafts were completed and submitted for publication.

Subfamily Cryptocephalinae

Diandichus Chapuis, 1874.

On flowers of *Acacia* (Mimosaceae) in Western Australia (Van den Berg, 1980).

Reference: Van den Berg, M.A. (1980). Natural enemies of *Acacia cyclops* A. Cunn. ex G. Don and *Acacia saligna* (Labill.) Wendl. in Western Australia. II. Coleoptera. *Phytophylactica*, 12: 169-171.

Semelvillea Reid, 1991.

On *Acacia* (Mimosaceae), *Nothofagus* (Fagaceae) and *Eucalyptus* (Myrtaceae) in south-eastern Australia (Reid, 1991).

Reference: Reid, C.A.M. (1991). A new genus of *Cryptocephalinae* from Australia (Coleoptera: Chrysomelidae). *Entomol. Scand.*, 22: 139-157.

Subfamily Eumolpinae

Eboo Reid, 1993.

One species, *Eboo virudula* (Erichson) has been recorded on *Eucalyptus* leaves (Myrtaceae) in Australia (Reid, 1993).

Reference: Reid, C.A.M. (1993). *Eboo*, nom. nov.: Redescription of type species (Coleoptera: Chrysomelidae: Eumolpinae). *Col. Bull.*, 47: 61-67.

Spilopyra Baly, 1860.

On *Guoia semiglauca* and *Cupaniopsis anacardioides* (Sapindaceae) in New South Wales, Australia (Hawkeswood, 1991, 1992).

References: Hawkeswood, T.J. (1991). Notes on *Spilopyra sumptuosa* Baly (Coleoptera: Chrysomelidae), a spectacular leaf beetle from the rainforests of Queensland and New South Wales. *Vict. Entomol.*, 21: 125-126.

Hawkeswood, T.J. (1992). Further notes on *Spilopyra sumptuosa* Baly (Coleoptera: Chrysomelidae). *Vict. Entomol.*, 22: 49-50.

Subfamily Galerucinae

Syphaxia Baly, 1886.

On *Cecropia* (Cecropiaceae) in Peru (D. Davidson, pers. comm.) (Jolivet, 1993).

Subfamily Alticinae

Procalus Clark, 1865.

On Anacardiaceae in Chile, i.e. *Lithrea caustica, Schinus montanus, S. polygamus* and *S. latifolius* (Jerez, 1992).

Reference: Jerez, V. (1992). Revision taxonomica del genero *Procalus* Clark, 1865 (Chrysomelidae, Alticinae). *Gayana, Zool.*, 56(3-4): 109-125.

Table 1. Classification of the Chrysomelidae (Modified from Jolivet, 1959)

Section 1. Eupoda	*Section 3. Cyclica*
1. Sagrinae	13. Lamprosomatinae
2. Aulacoscelinae	14. Megascelinae
3. Palophaginae	15. Eumolpinae
4. Orsodacninae	
5. Zeugophorinae	*Section 4. Trichostoma*
6. Donaciinae	16. Chrysomelinae
7. Megalopodinae	17. Galerucinae
8. Criocerinae	18. Alticinae
9. Synetinae	
Section 2. Camptosoma	*Section 5. Cryptostoma*
10. Clytrinae	19. Hispinae
11. Cryptocephalinae	20. Cassidinae
12. Chlamisinae	

Note: The Lamprosomatinae belong to the Camptosoma on the basis of their biology and larval structure, but to Cyclica on the basis of their wing venation and other characters

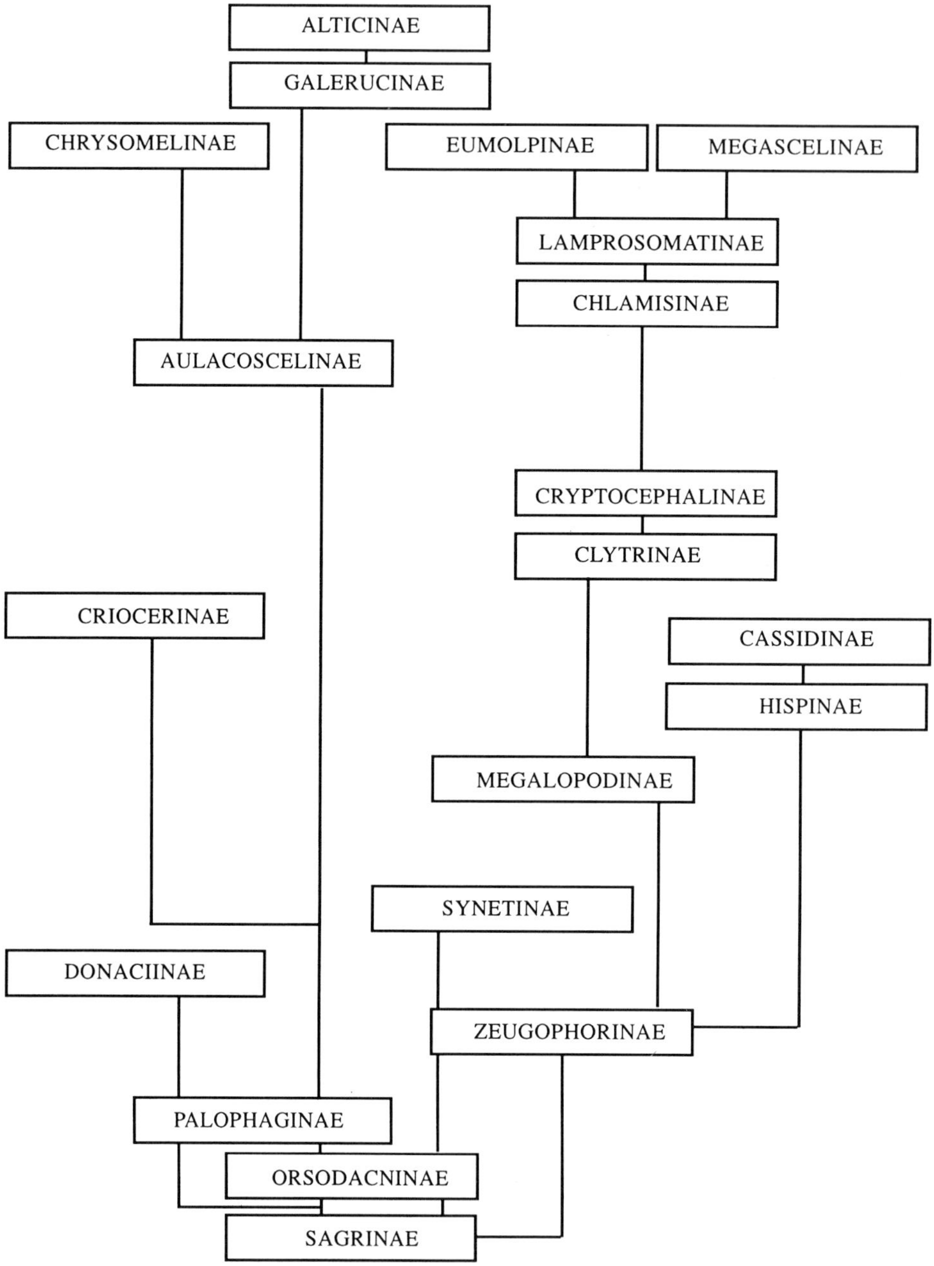

Table 2. Relationships among the 20 subfamilies of Chrysomelidae (Modified from Jolivet, 1959). Suzuki's cladogram (1993, in print) recognizes only ten subfamilies with 20 divisions.

Table 3. Trophic selection of *Chlamisus* species (Modified from Jolivet, 1978).

Species	Host-plant	Family	Country
C. alni Brown	*Alnus inacana americana*	Betulaceae	Canada
C. bebbianae Brown	*Salix bebbiana, S. discolor*	Salicaceae	Canada
C. boliviana Jacoby	*Psidium guajava*	Myrtaceae	Bolivia
C. brittocostai Bokermann	*Byrsonima intermedia, B. variabilis*	Malpighiaceae	Brazil
C. chamaedaphnes Brown	*Chamaedaphne calyculata*	Ericaceae	Canada
	Potentilla fruticosa	Rosaceae	Canada
C. comptoniae Brown	*Comptonia peregrina*	Myricaceae	Canada
C. consimilis Chujo	*Quercus dentata, Q. mongolica*	Fagaceae	Japan
C. coryli Brown	*Corylus americana*	Corylaceae	Canada
C. cribripennis LeConte	*Vaccinium angustifolium*	Ericaceae	Canada
C. discipennis Jacoby	*Waltheria americana*	Sterculiaceae	Brazil
C. echinatus Klug	*Croton horridulus*	Euphorbiaceae	Brazil
C. elongatus Kollar	*Byrsonima intermedia*	Malpighiaceae	Brazil
C. erodus Monros	Unidentified species	Myrtaceae	Brazil
C. eubati Brown	*Rubus frondosus, Potentilla fruticosa*	Rosaceae	Canada
C. fragariae Brown	*Fragaria ananassa*	Rosaceae	Canada
C. geniculatus Jacoby	*Rubus sieboldi*	Rosaceae	Japan
C. gibbicollis Lacordaire	*Waltheria americana*	Sterculiaceae	Brazil
C. gibbosa Fabricius	*Acer pseudoplatanus*	Aceraceae	North America
C. globosus Kollar	*Tetrapteris phlomoides*	Malpighiaceae	Brazil
C. handroi Bokermann	*Banisteria laevigata*	Malpighiaceae	Brazil
C. heiroglyphicus Kollar	*Banisteria campestris, Byrsonima verbascifolia*	Malpighiaceae	Brazil
C. hircinus Lacordaire	*Banisteria laevigata*	Malpighiaceae	Brazil
C. hispidulus Klug	*Acacia caven, Acacia* spp.	Mimosaceae	Argentina
C. interjectus (Baly)	*Cirsium* sp.	Asteraceae	Japan

Table 3. Trophic selection of *Chlamisus* species (Modified from Jolivet, 1978) — *continued*

Species	Host-plant	Family	Country
C. jutahyensis Bowditch	*Mascagnia cordifolia*	Malpighiaceae	Brazil
C. knochii (Kollar)	*Psidium guajava, P. subfruticosum, Campomanesia obversa, C. salviaefolia*	Myrtaceae	Brazil, Bolivia
C. langsdorfii Kollar	*Bauhinia rufa*	Caesalpiniaceae	Brazil
C. laticollis Chujo	*Rhododendron* sp.	Ericaceae	Japan
C. lewisii Baly	*Camellia japonica*	Theaceae	Japan
C. maculipes Chevrolat	*Ipomoea pedicellaris*	Convolvulaceae	Mexico
C. melochiae Monros	*Sphaeralcea* sp.	Malvaceae	Argentina, Brazil
	Melochia sp., *Waltheria americana*	Sterculiaceae	
C. mimicus Monros	*Tibouchina* sp.	Melastomataceae	Brazil
C. minax Lacordaire	*Banisteria campestris, B. stellaris*	Malpighiaceae	Brazil
C. nattereri Kollar	*Byrsonima variabilis, B. intermedia, B. verbascifolia*	Malpighiaceae	Brazil
	Psidium araca	Myrtaceae	
C. ornaticollis Lacordaire	*Tibouchina* spp.	Melastomataceae	Brazil
C. pilaga Monros	*Serjania* sp.	Sapindaceae	Bolivia
C. platani Brown	*Platanus occidentalis*	Platanaceae	USA
C. plicatus Fabricius	*Rubus* spp., *R. villosus*	Rosaceae	USA
	Gaylussacia baccata, Vaccinium angustifolium, V. myrtilloides	Ericaceae	
C. puncticollis Germar	*Serjania* sp., *Cardiospermum grandifolium*	Sapindaceae	Paraguay
C. rhododendri Gressitt	*Rhododendron* sp.	Ericaceae	New Guinea
C. schotti Kollar	Unidentified species	Malpighiaceae	Brazil
C. schachti Kollar	*Tibouchina* sp.	Melastomataceae	Brazil
C. scortillus Lacordaire	*Banisteria campestris, B. crotonifolia, B. laevigata*	Malpighiaceae	Brazil

Table 3. Trophic selection of *Chlamisus* species (Modified from Jolivet, 1978) — *continued*

Species	Host-plant	Family	Country
C. sidae Monros	*Sida rhombifolia*	Malvaceae	Argentina
C. spilotus Baly	*Castanea crenata, Quercus* spp.	Fagaceae	Japan
	Corylus heterophyllus	Corylaceae	
	Prunus donarium	Rosaceae	
	Rhododendron sp.	Ericaceae	
C. subspinosus Klug	*Byrsonima verbascifolia*	Malpighiaceae	Brazil
C. sulcatus Kollar	*Heteropteris seringiifolia*	Malpighiaceae	Brazil
C. takasei Bokermann	*Tibouchina multiceps*	Melastomataceae	Brazil
C. tarsatus Klug	*Banisteria argyrophylla, B. laevigata*	Malpighiaceae	Brazil
C. tectus Brown	*Corylus americana*	Corylaceae	Canada
	Alnus rugosa	Betulaceae	
C. tuberosa Koch	*Betula alba*	Betulaceae	USA
	Quercus sp.	Fagaceae	
	Platanus occidentalis	Platanaceae	
	Amorpha canescens	Fabaceae	
C. tucumanus Monros	*Croton* sp.	Euphorbiaceae	Argentina
C. umbrosa Klug	*Byrsonima verbascifolia*	Malpighiaceae	Brazil
C. vagans Lacordaire	*Byrsonima variabilis, B. intermedia, B. verbascifolia*	Malpighiaceae	Brazil

Table 4. Trophic selection of *Exema* species (Modified from Jolivet, 1978).

Species	Host-Plant	Family	Country
E. byersi Karren	*Echinacea angustifolia, Gutierrezia dracunculoides*	Asteraceae	USA
	Fraxinus pensylvanica	Oleaceae	
	Asclepias tuberosa	Asclepiadaceae	
E. canadensis Pierce	*Solidago altissima, S. neglecta, Ambrosia* sp., *Vernonia* sp., *Haplopappus phyllocephalus, Erigeron quercifolius*	Asteraceae	USA Canada
E. conspersa Mann	*Aster* spp., *Artemisia califor-nica, Hymenoclea monogyna, Ambrosia* spp.	Asteraceae	USA
E. deserti Pierce	*Ambrosia dumosa, A. chenopod-iifolia, A. eriocentra*	Asteraceae	USA
E. dispar Lacordaire	*Ambrosia* spp., *Helianthus* spp., *Eupatorium* spp.	Asteraceae	USA
E. elliptica Karren	*Iva fructescens, Baccharis halimifolia*	Asteraceae	USA
E. gibber Fabricius	*Aster* sp., *Baccharis halimifolia*	Asteraceae	USA
	Quercus sp.	Fagaceae	
	Myrica cerifera	Myricaceae	
	Crataegus sp., *Rubus* sp.	Rosaceae	
	Carya illinoensis	Juglandaceae	
	Litchi chinensis	Sapindaceae	
E. morio Kollar	Unidentified species of bamboo (Bambuseae)	Poaceae	Argentina
E. neglecta Blatchley	*Baccharis halimifolia, Solidago* spp., *Chondrophora virgata, Eupatorium* sp.	Asteraceae	USA
	Strophostyles helvala	Fabaceae	
	Arundinaria sp. (Bambuseae)	Poaceae	
E. serjaniae Monros	*Serjania* sp.	Sapindaceae	Argentina
E. variopicta Monros	*Lantana hypoleuca, Lippia salviaefolia*	Verbenaceae	Brazil

Table 5. A preliminary compilation of Solanaceae genera used as hosts by Chrysomelidae (Modified from Hsiao, 1986).

Chrysomelidae subfamily and genus	Solanaceae genus														
	Atropa	*Capsicum*	*Chamaesaracha*	*Datura*	*Hyoscyamus*	*Lycium*	*Lycopersicon*	*Nicandra*	*Nicotiana*	*Petunia*	*Physalis*	*Salpichroa*	*Solandra*	*Solanum*	*Others*
Criocerinae															
Lema	·		·		·	·	·	·	·		·	·	·	·	·
Oulema									·						
Megalopodinae															
Agathomerus		·					·		·					·	
Bothromegalopus														·	
Colobaspis														·	
Megalopus							·							·	
Clytrinae															
Smaragdina														·	
Cryptocephalinae															
Cryptocephalus		·		·			·		·					·	
Eumolpinae															
Colaspis														·	
Maecolaspis														·	
Chrysomelinae															
Leptinotarsa					·						·			·	
Platyphora													·		
Galerucinae															
Acalymma														·	
Dercetina														·	
Diabrotica				·										·	
Triarius														·	
Alticinae															
Altica														·	
Chaetocnema														·	
Disonycha														·	
Disonychodes														·	
Epitrix		·		·			·		·	·				·	
Glyptina															
Haltica									·						
Lactina															·
Leptophysa														·	
Licyllus														·	
Nisotra														·	
Ochrosis														·	
Phyllotreta									·						
Psylliodes	·				·	·	·		·					·	
Systena									·					·	
Xenidea														·	
Cassidinae															
Aspidomorpha														·	
Eremocassis						·									
Gratiana														·	
Nuzonia														·	
Parorectis														·	
Plagiometriona		·									·			·	

Table 6. List of the host-plants of the genus *Timarcha*
(After Jolivet & Petitpierre, 1973; Tieberghien, 1972; Costa *et al.*, 1983).(* = observations by P.H. Jolivet).

Timarcha species	Host-plants	Plant family
Genus *Timarcha*		
Subgenus *Metallotimarcha* Motschulsky		
T. metallica Laicharting	*Galium* spp.	Rubiaceae
Subgenus *Americanotimarcha* Jolivet		
T. cerdo Stal	*Fragaria vesca, F. chiloensis, Rubus parviflorus, R. vitifolius, R. spectabilis*	Rosaceae
T. intricata Halderman	As for *T. cerdo*	
Subgenus *Timarcha* (sensu stricto) and subgenus *Timarchostoma* Motschulsky		
T. affinis Laboissiere	*Galium mollugo, Rubia peregrina*	Rubiaceae
T. balearica Gory	*Rubia peregrina*	Rubiaceae
T. catalaunensis Fairmaire	*Scabiosa maritima*	Dipsacaceae
	Asperula cynanchica, Galium mollugo	Rubiaceae
T. cyanescens Fairmaire & Allard	*Plantago lanceolata, P. maritima, P. media, P. coronopus*	Plantaginaceae
	Asperula sp., *Galium arenarium, G. hercynicum, G. verum, G. uliginosum, G. mollugo, Galium* sp.	Rubiaceae
	Centaurea sp.	Asteraceae
T. cyrenaica Peyerhimhoff	*Plantago albicans*	Plantaginaceae
T. daillei Laboissiere	*Galium hercynicum*	Rubiaceae
	Bellis perennis	Asteraceae
T. espagnoli Bechyne	*Plantago coronopus*	Plantaginaceae
T. fallax Perez	*Scabiosa maritima*	Dipsacaceae
	Galium mollugo	Rubiaceae

Table 6. List of the host-plants of the genus *Timarcha*
(After Jolivet & Petitpierre, 1973; Tieberghien, 1972; Costa *et al.*, 1983).(* = observations by P.H. Jolivet).(*cont.*)

Timarcha species	Host-plants	Plant family
T. generosa khalouensis Jolivet	*Galium* sp.	Rubiaceae
T. generosa mesatlantica Kocher	*Asperula hirsuta*	Rubiaceae
	Plantago mauritanica	Plantaginaceae
T. geniculata Germar	*Plantago alpina*, *Plantago* spp.	Plantaginaceae
	Galium spp.	Rubiaceae
T. g. goettingensis L.	*Galium mollugo, G. verum, G. cruciata, G. aparinae, Asperula odorata*	Rubiaceae
T. g. splendorifera Bechyne	*Galium uliginosum*	Rubiaceae
T. goettingensis gallica Fairmaire	*Scabiosa maritima*	Dipsacaceae
	*Rubia peregrina**	Rubiaceae
T. goettingensis normanna Reiche	*Galium verum* var. *littorale*	Rubiaceae
T. intermedia Herrich-Schäffer	*Alyssum maritimum**, *Carrichlera annua*	Brassicaceae
T. interstitialis Fairmaire	*Crucianella maritima*, *Galium verum*	Rubiaceae
T. laevigata L. (and subspecies)	*Plantago albicans*	Plantaginaceae
T. lugens Rosenhauer	*Plantago nivalis*	Plantaginaceae
	Veronica fruticosa	Scrophulariaceae
	Alyssum spinosum	Brassicaceae
T. lusitanica Fabricius	*Galium* sp.	Rubiaceae
T. mairei humberti Peyerhimhoff	*Asperula* sp.	Rubiaceae
T. marginicollis Rosenhauer	*Galium* sp.	Rubiaceae
T. maritima Perris	*Galium arenarium*, *G. mollugo**	Rubiaceae
	Plantago lanceolata, *P. maritima*	Plantaginaceae
T. maroccana Weise	*Asperula hirsuta*	Rubiaceae
T. monserratensis Bechyne	*Galium cruciata*, *G. maritimum*, *G. mollugo*	Rubiaceae
T. monticola Dufour	*Plantago montana*	Plantaginaceae
	Galium pusillum	Rubiaceae

Table 6. List of the host-plants of the genus *Timarcha* (After Jolivet & Petitpierre, 1973; Tieberghien, 1972; Costa *et al.*, 1983).(* = observations by P.H. Jolivet).(*cont.*)

Timarcha species	Host-plants	Plant family
T. sp. aff. *T. monticola* Dufour	*Galium verum*	Rubiaceae
T. nicaeensis Villa	*Galium mollugo*	Rubiaceae
T. obsoleta Laboissiere	*Plantago lanceolata*	Plantaginaceae
	Galium pumilum, G. verum	Rubiaceae
T. perezi Fairmaire	*Galium* sp.	Rubiaceae
	Plantago lanceolata, Plantago spp. (in vitro)	Plantaginaceae
T. pimelioides Herrich-Schäffer	*Plantago coronopus*	Plantaginaceae
	Crucianella maritima, Rubia peregrina	Rubiaceae
	Launaea resedifolia	Asteraceae
T. punctella Marseul	*Plantago albicans*	Plantaginaceae
T. recticollis Fairmaire	*Asperula hirta*	Rubiaceae
T. rugosa rugata Kocher	*Asperula* sp.	Rubiaceae
	Plantago albicans	Plantaginaceae
T. sardea Villa	*Rubia peregrina*	Rubiaceae
T. scabripennis Fairmaire	*Galium* sp., *Sherardia arvensis*	Rubiaceae
T. sinuatocollis Fairmaire	*Galium pumilum, G. verum*	Rubiaceae
T. strangulata (sensu stricto) Fairmaire	*Galium anisophyllum*	Rubiaceae
T. strangulata sculptipennis Bechyne	*Asperula hirta*	Rubiaceae
	Iberis sp.	Brassicaceae
	Plantago alpina	Plantaginaceae
T. sublaevis Fairmaire	*Galium* sp., *Rubia* sp	Rubiaceae
T. tangeriana Bechyne	*Plantago* sp.	Plantaginaceae
	Galium sp.	Rubiaceae
T. tenebricosa tenebricosa Fabricius	*Galium aparine, G. cruciata, G. mollugo, G. parisiense, G. verum*	Rubiaceae
T. tenebricosa transgrediens Bechyne	*Asperula cynanchica, Galium parisiense*	Rubiaceae

Table 6. List of the host-plants of the genus *Timarcha*
(After Jolivet & Petitpierre, 1973; Tieberghien, 1972; Costa *et al.*, 1983).(* = observations by P.H. Jolivet). (*cont.*)

Timarcha species	Host-plants	Plant family
T. theryi Kocher	*Veronica rosea*	Scrophulariaceae
T. tingitana Fairmaire	*Plantago* spp.	Plantaginaceae
T. ventricosa atlantica Bechyne	*Plantago lagopus*	Plantaginaceae
T. ventricosa aurulenta Kocher	*Crucianella maritima*	Rubiaceae
T. ventricosa mima Bechyne	*Plantago lagopus*	Plantaginaceae
T. vermiculata Fairmaire	*Galium* sp.	Rubiaceae

GLOSSARY

Allotrophy: a drastic food change among animals in response to certain factors, often occurring under stress conditions as a survival strategy (cf. xenophagy).
Aposematic: warning colouration or markings which signal to a predator that an organism is toxic, dangerous or distasteful.
Apterous: wingless.

Batesian mimicry: resemblance of one animal (the mimic) to another (the model) to the benefit of the mimic, as when the model is dangerous or unpalatable, described by the English naturalist H.W. Bates.
Biocoenosis: community of organisms inhabiting a particular biotope.
Brachypterous/brachyptery: wing muscle reduction.

Cycloalexy: ring-defence strategy.
Cyphosomatic: having a bent or crooked body (cf. orthosomatic).

Detriticolous: living in detritus.

Epigeous/epigeal, epigean: (of insects) living near the ground.

Gallicolous: living in plant galls.

Homochromy (crypsis): of one colour; the phenomenon of being camouflaged to resemble part of the environment.
Hypertelia/hypertely: excessive imitation of colour or pattern, being of problematical utility.

Macropterous: with unusually large wings.

Necrocoenosis: an assemblage of dead organisms.
Nymphosis: the process of changing into a nymph or pupa.

Ootheca: egg-case.
Orophilous: thriving in subalpine or mountainous regions.
Orthosomatic: (of larval insects) having a straight body (cf. cyphosomatic).

Peckhamian (aggressive) mimicry: a type of mimicry complex where the predator (mimick) resembles the prey (model) in form, colour, shape, pattern etc., so that the predator can more easily approach and capture its prey, e.g. the larvae of *Lebistina* (a genus of Carabidae, a predacious group of beetles) have adapted morphological and colour similarities to the larvae of *Diamphidia* (Alticinae) upon which they prey.

Phoresia/phoresy: the carrying of one organism by another, without parasitism, as in e.g. ants and beetle-larvae.
Phyllophagous: feeding on leaves.

Radicolous: living in or on roots.

Subligneous: a condition intermediate between herbaceous and woody stems.

Symphily: the sitation where one species of insect lives as a guest (symphile) in the nest of a social insect which feeds and protects it in return for its secretions which are used as food, as do certain beetles in the nests of ants and termites.

Xenophagy: (of insects) an (abnormal) change in diet from mono- to dicotyledons (e.g. the rice pest beetle, *Dicladispa armigera* (Olivier), a pest of Poaceae, has been observed mining the leaves of the teak tree, *Tectona grandis*. The larvae probably dropped down from adjacent bamboo plants onto teak leaves. Hering coined the phrase xenophagy in 1951). (cf. allotrophy).

Plates 1-17

Plate 1. – 1: *Cecropia adenopus* (Cecropiaceae) showing foliage which has been eaten by *Coelomera* (Galerucinae) and various caterpillars. Viçosa, MG, Brazil; 2: Eggs of *Coelomeria lanio* Dalman (Galerucinae) laid on the underside of the folioles; 3: First instar larvae of *C. lanio*; 4: Second instar larvae of *C. lanio*; 5: Third instar larvae of *C. lanio*; 6: Adults of *C. lanio* in copula; (The eggs of some species of *Coelomera* can be deposited inside the stems; the second instar larvae may show cycloalexy – see fig. 4) (Photos: P. Jolivet).

Plate 2. – 1: *Platycorynus peregrinus* (Herbst) (Eumolpinae) on a stem of *Calotropis gigantea* (Asclepiadaceae), a toxic plant with latex- Pham Ram, Vietnam; 2: *Phyllocharis undulata* (L.) larvae on a leaf of *Clerodendrum fragrans* (Verbenaceae)- Cuc Phuong Forest, Vietnam; 3 & 4: Adults of *P. undulata* on leaves of *C. fragrans*- Cuc Phuong Forest, Vietnam; 5: *Phaedon fulvescens* (Weise), male and female on *Rubus moluccanus* (Rosaceae)- Tam Dao Forest, Vietnam; 6: *Rubus moluccanus* (Rosaceae) the plant eaten by *Phaedon* (Chrysomelinae) (Photos: P. Jolivet).

Plate 3.–1-4: Galled stems of *Centrosema* (Fabaceae) containing cocoons and adults of *Sagra* sp. (Sagrinae) ready to hatch, at different magnifications- Konedobu, Central Province, Papua New Guinea, November 1974. (Photos: T.L. Fenner).

Plate 4.- 1: *Elytrosphaera xanthopyga* Stål (Chrysomelinae), a species belonging to a middle or high altitude-inhabiting genus near *Leptinotarsa* but apterous in South America; 2: Larva of *E. xanthopyga;* 3: Pupa of *E. xanthopyga*; 4: Host-plant of *E. xanthopyga*, *Adenostemma brasilanum* (Asteraceae) in the primary Atlantic forest of Brazil (Photos: P. Jolivet).

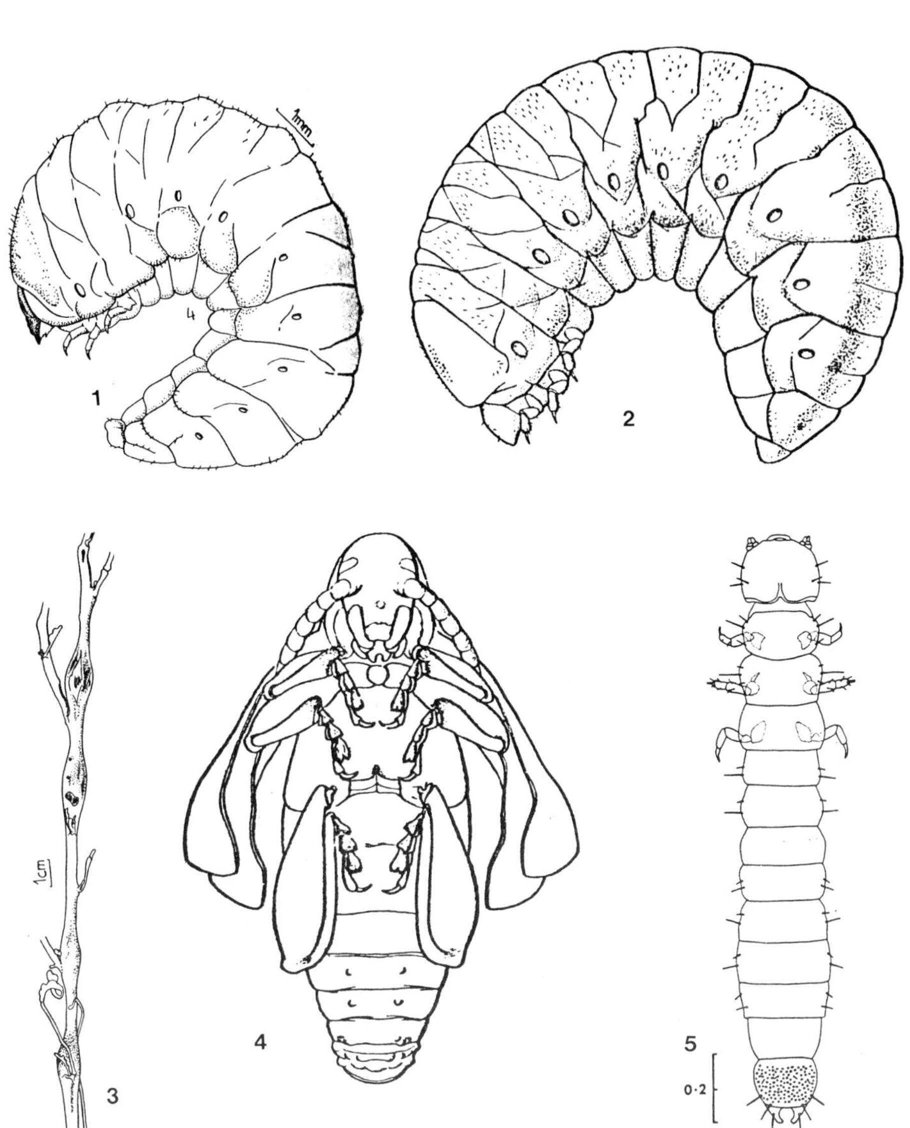

Plate 5. – 1: Mature larva of *Atalasis sagroides* Lacordaire (Sagrinae) from Argentina; 2: Larva of *Sagra femorata* Drury (Sagrinae) from India; 3: Stem of *Sphaeralcea* (Malvaceae) with galls produced by the larva of *Atalasia sagroides*; 4: Ventral view of the pupa of *S. femorata* Drury; 5: Larva of *Orsodacne cerasi* (L.) (Orsodacninae), first instar, from England (1, 3 after Monros, 1955; 2, 4 after Maulik, 1941; 5 after Cox, 1981).

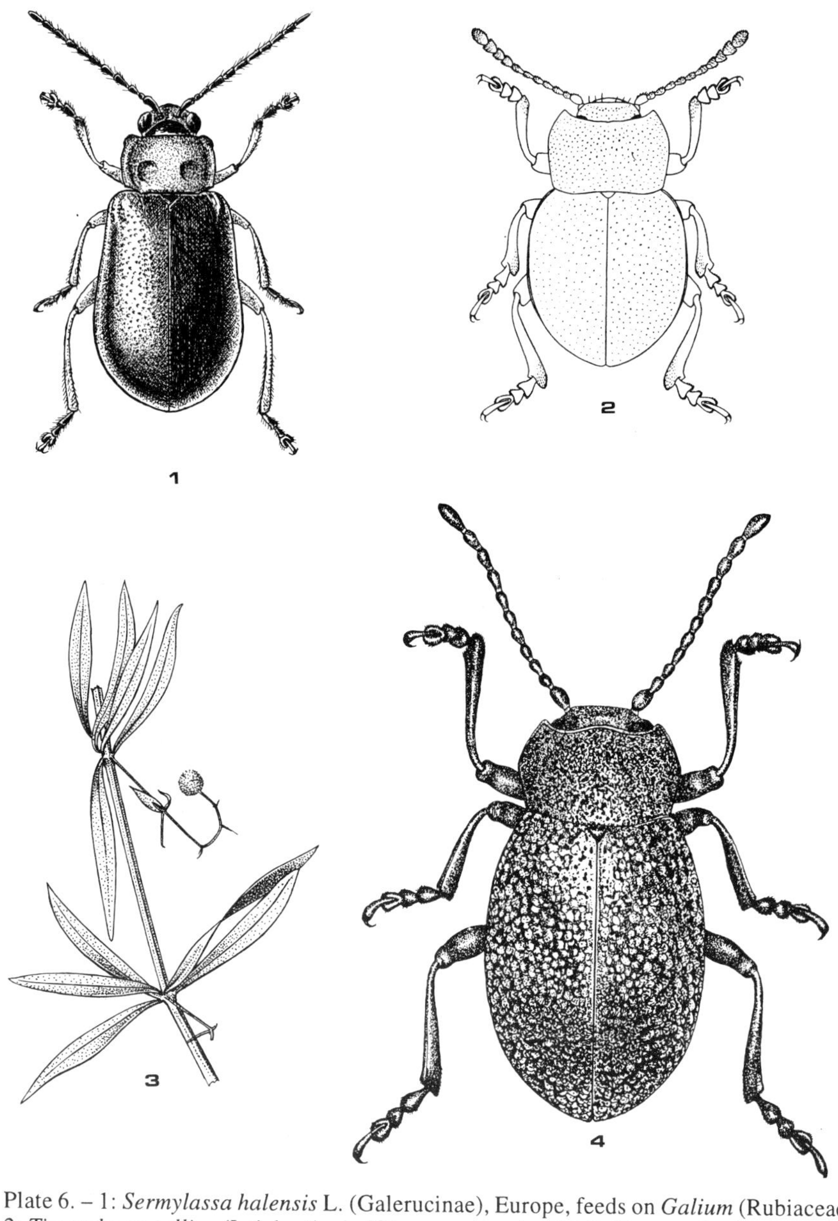

Plate 6. – 1: *Sermylassa halensis* L. (Galerucinae), Europe, feeds on *Galium* (Rubiaceae); 2: *Timarcha metallica* (Laicharting), (Chrysomelinae), Middle Europe, feeds on *Galium*; 3: *Rubia peregrina* (Rubiaceae), food-plant of many European *Timarcha* species, together with *Galium, Asperula, Crucianella, Sherardia*, etc. (Rubiaceae); 4: *Timarcha intricata* Haldeman, female, western USA, feeds on *Rubus parviflorus* (Rosaceae) (1 after Laboissière, others after P. Jolivet).

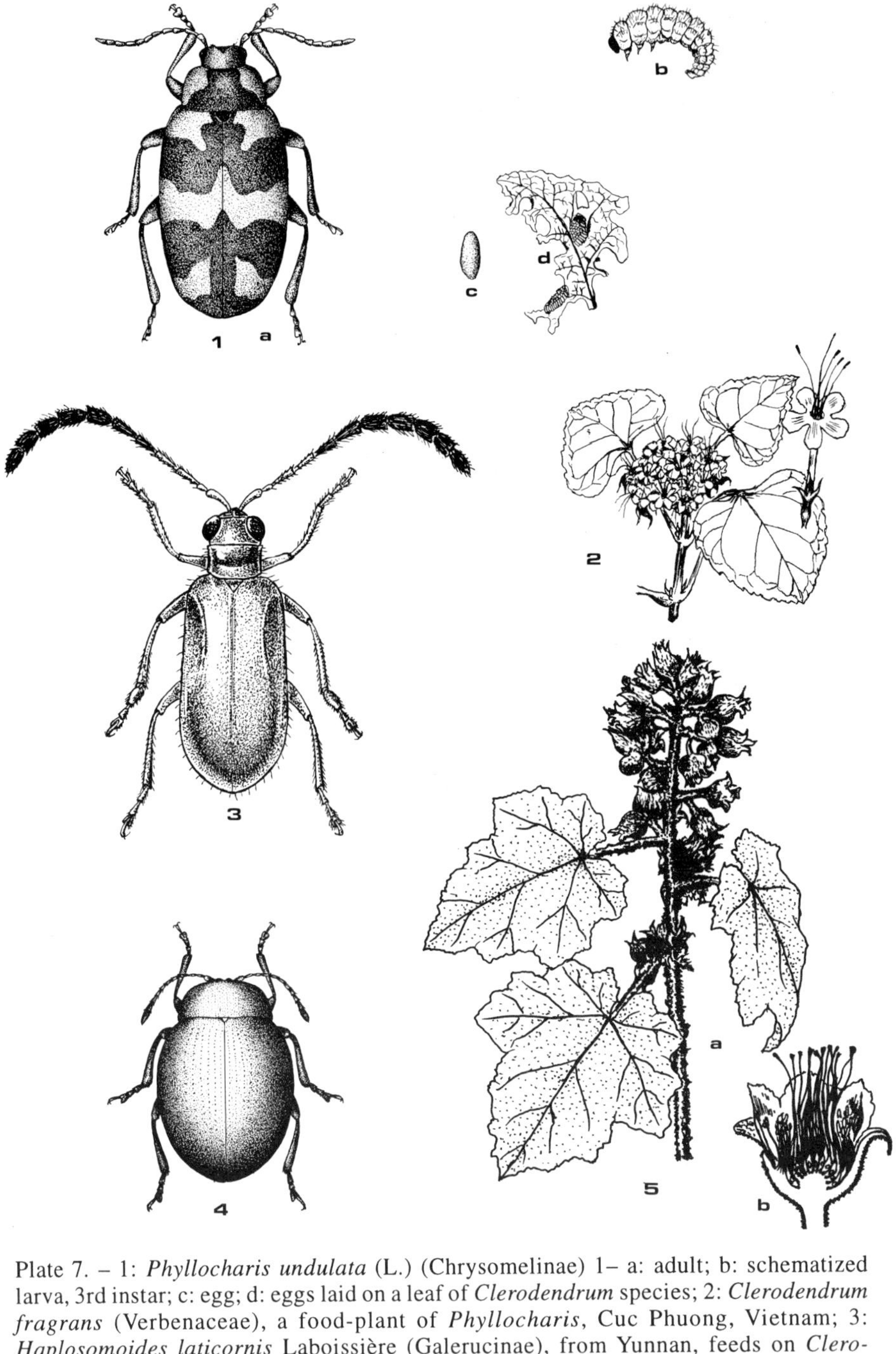

Plate 7. – 1: *Phyllocharis undulata* (L.) (Chrysomelinae) 1– a: adult; b: schematized larva, 3rd instar; c: egg; d: eggs laid on a leaf of *Clerodendrum* species; 2: *Clerodendrum fragrans* (Verbenaceae), a food-plant of *Phyllocharis*, Cuc Phuong, Vietnam; 3: *Haplosomoides laticornis* Laboissière (Galerucinae), from Yunnan, feeds on *Clerodendrum;* 4: *Phaedon fulvescens* Weise (Chrysomelinae), female from Tam Dao, Vietnam, feeds on *Rubus moluccanus* (Rosaceae); 5: *Rubus moluccanus* (Rosaceae), from Vietnam (2 after Laboissière; 1, 3-5 after P. Jolivet).

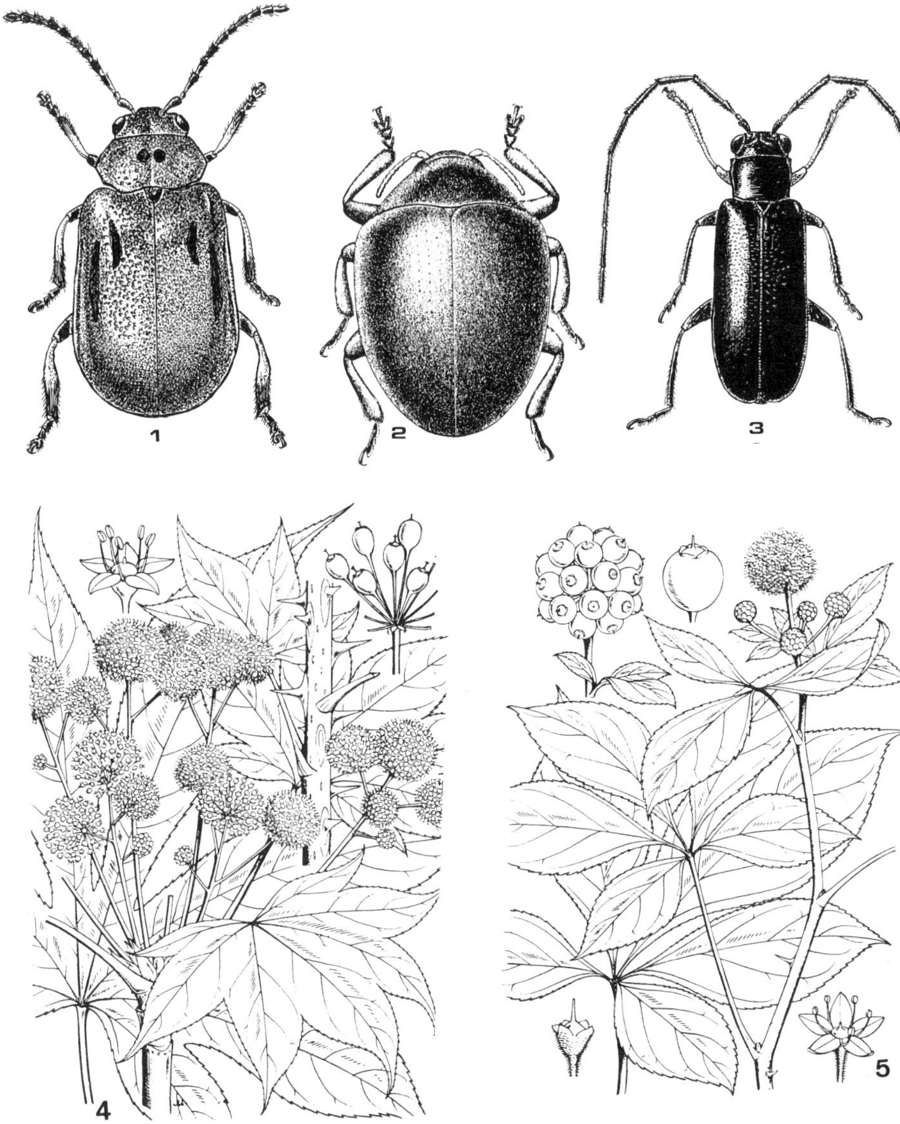

Plate 8. – 1: *Lochmaea crataegi* Forster (Galerucinae) feeds on *Crataegus, Pyrus, Prunus, Malus* (Rosaceae) in the Holarctic region; 2: *Oomorphoides metallicus* Gressitt and Kimoto (Lamprosomatinae), from West China, feeds on *Kalopanax, Acanthopanax* etc. (Araliaceae); 3: *Luperus alpinus* Desbrochers (Galerucinae), from France, feeds on *Salix* (Salicaceae) and probably on other plants, e.g. *Quercus* (Fagaceae) and *Crataegus* (Rosaceae); 4: *Kalopanax pictus* (Araliaceae), food of Lamprosomatinae in the Far East; 5: *Acanthopanax sessiliflorus* (Araliaceae), food of Lamprosomatinae in the Far East (1, 3 after Laboissière; 2, 4, 5 after P. Jolivet).

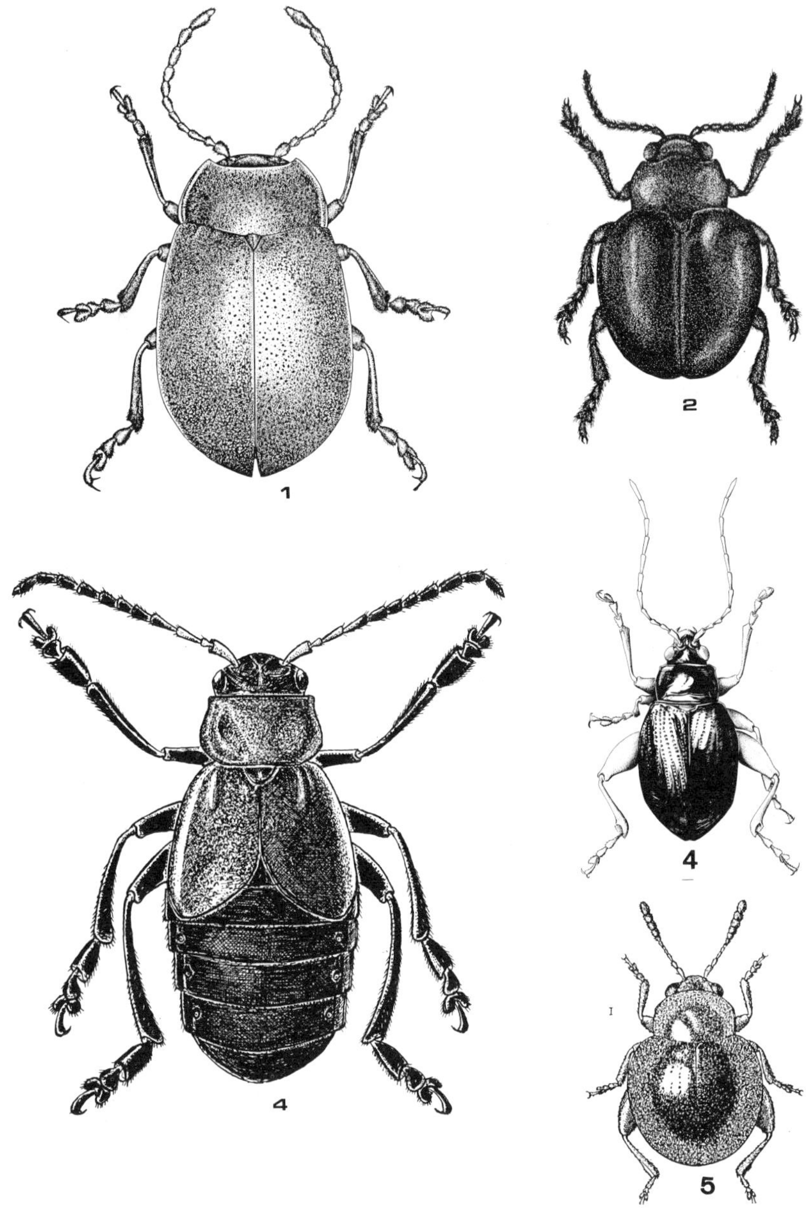

Plate 9. – 1: *Chrysolina staphylea* (L.) (Chrysomelinae), feeds on Lamiaceae, Plantaginaceae and Ranunculaceae, Holarctic but lives in northern or alpine areas; 2: *Nisotra gemella* Erichson (Alticinae) from Thailand, feeds on *Hibiscus* and other Malvaceae; 3: *Xuthea nepalensis* Scherer (Alticinae), from Nepal. The genus *Xuthea* feeds on *Gerardinia, Fleurya* (Urticaceae); 4: *Arima marginata* (Fabricius) (Galerucinae), France, apterous and brachyelytrous, feeds mostly on Lamiaceae and Asteraceae; 5: *Kamala apicipennis* Jacoby (Alticinae), moss-frequenting species, from the Sri Lankan highlands, probably feeds on *Clausena* (Rutacae) (3 after Scherer, 1983; 5 after Maulik, 1926; 1, 2, 4 after P. Jolivet).

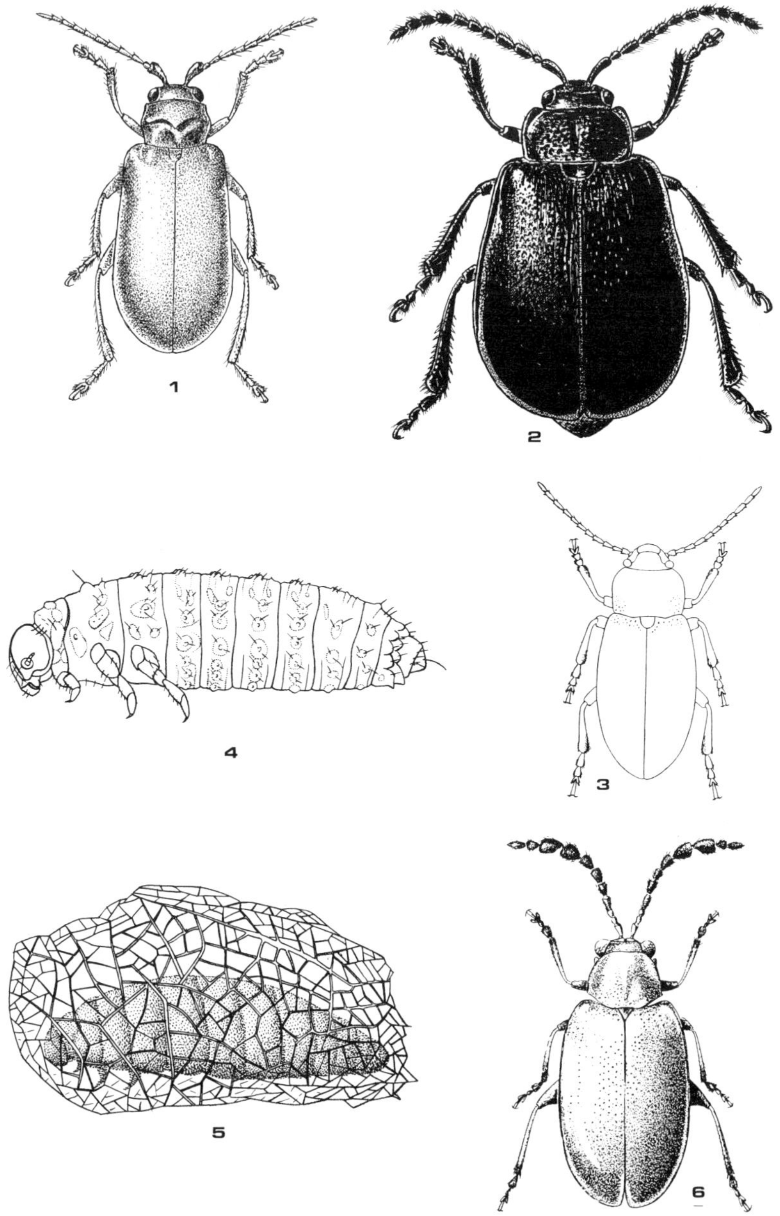

Plate 10. – 1: *Aulacophora foveicollis* Lucas (Galerucinae) from the Mediterranean region, feeds on Cucurbitaceae (leaves, flowers and pollen); 2: *Galeruca tanaceti* L. (Galerucinae), from France, feeds on Asteraceae, Brassicaceae, Caryophyllaceae, etc., polyphagous on low-growing plants; 3: *Microtheca bechynei* Jolivet (Chrysomelinae), from Lake Titicaca, Puno, 4000 m altitude, feeds on wild and cultivated Brassicaceae; 4: Larva of *Microtheca ochroloma* Stål (Chrysomelinae), 4th instar near nymphosis, from Argentina, feeds on Brassicaceae; 5: *Microtheca ochroloma* Stål, pupa and cocoon; 6: *Gabonia gabriela* Scherer (Alticinae), from Kenya, male attracted to withered plants of the genus *Heliotropium* (Boraginaceae) where it feeds upon them, female food is unknown (1, 2 after Laboissière; 6 after Boppré & Scherer, 1981; 3-5 after P. Jolivet).

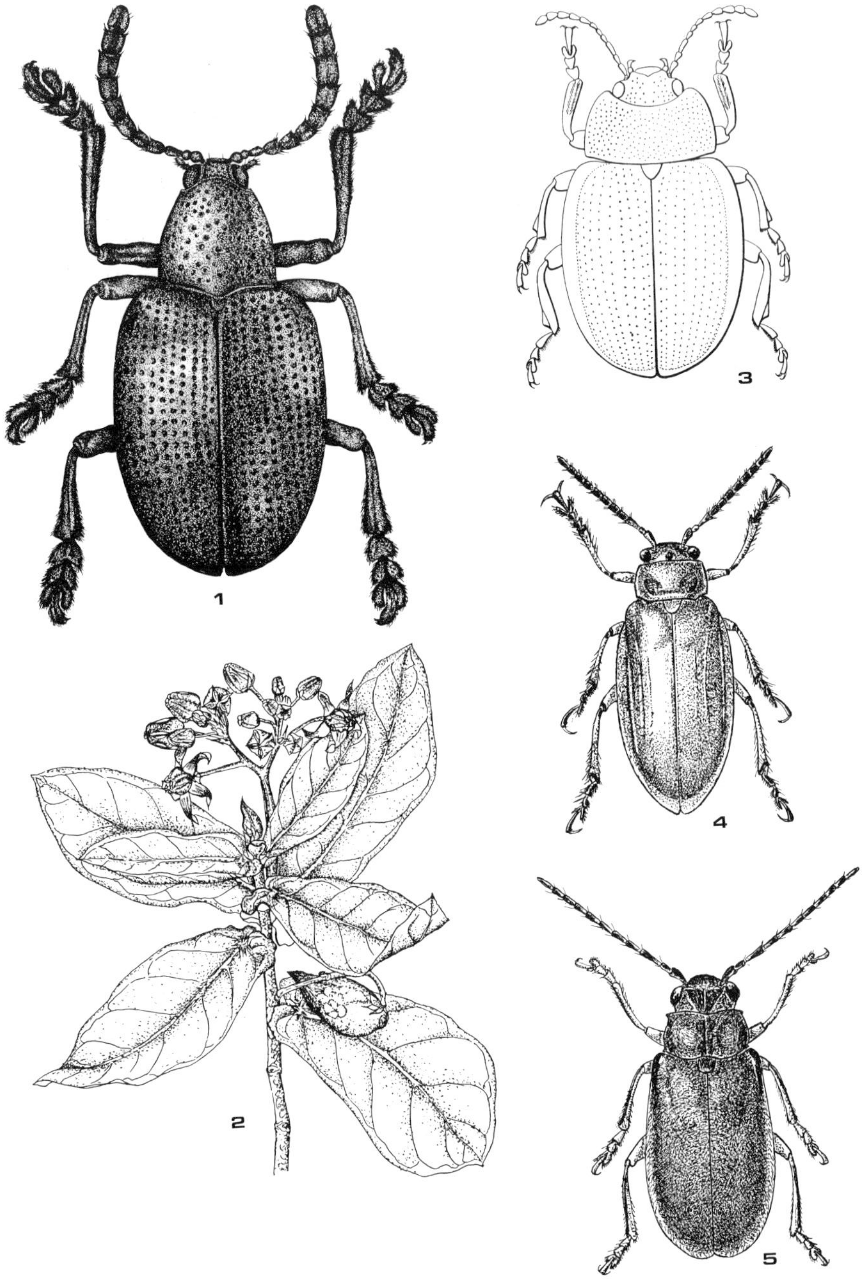

Plate 11. – 1: *Platycorynus peregrinus* (Herbst) (Eumolpinae), from Thailand, feeds on *Calotropis gigantea* (Asclepiadaceae); 2: *Calotropis gigantea* (Asclepiadaceae), from Thailand; 3: *Phaedonia inclusa* Stål (Chrysomelinae), from Java, feeds on various Fabaceae including soya beans (*Glycine max*); 4: *Diorhabda elongata* Brulle (Galercucinae) from north Africa, feeds on *Tamarix* spp. (Tamaricaceae); 5: *Pyrrhalta viburni* Paykull (Galerucinae) from France, feeds on *Viburnum* (Caprifoliaceae) (4, 5 after Laboissière; 1-3 after P. Jolivet).

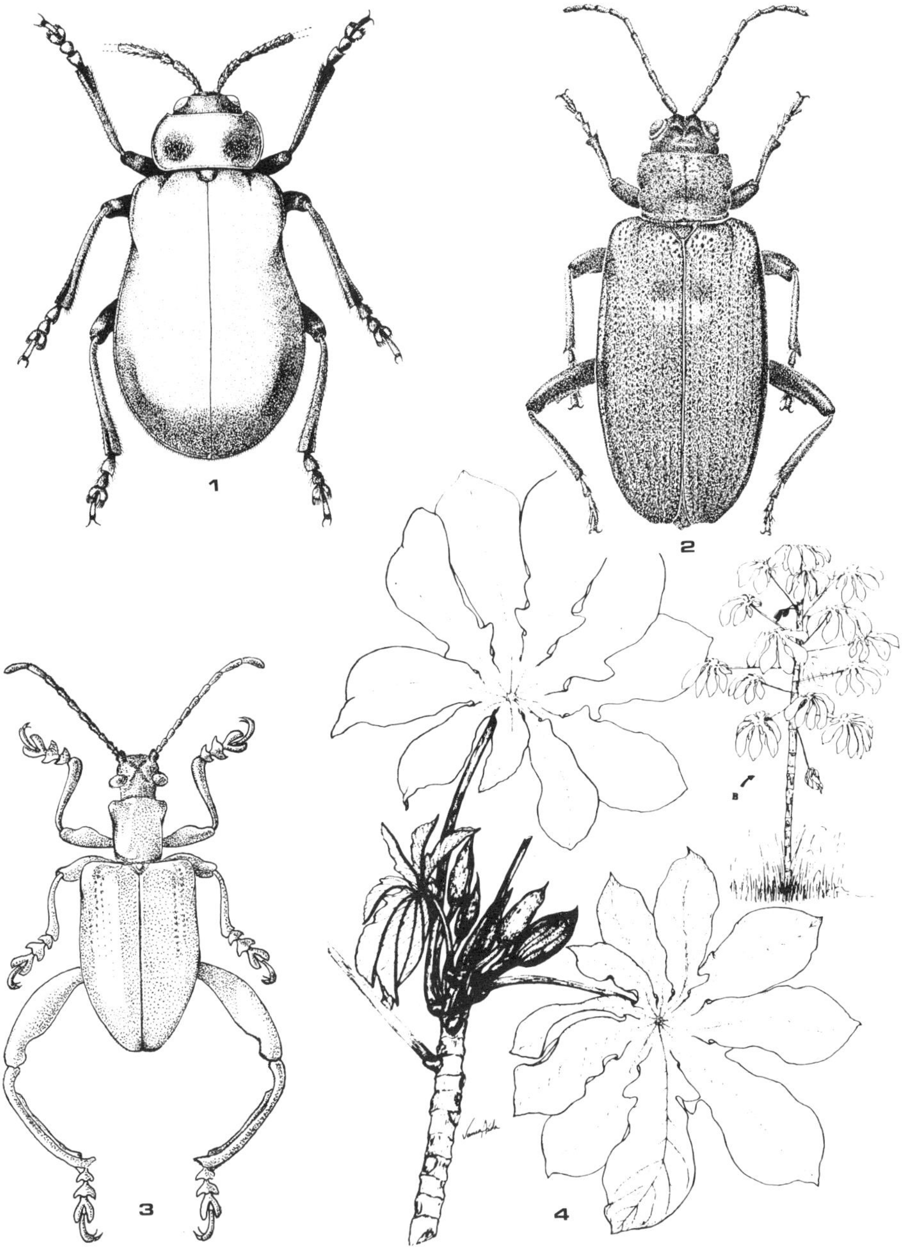

Plate 12. – 1: *Coelomera helenae* Jolivet (Galerucinae), female, from Minas Gerais, Brazil, feeds on *Cecropia hololeuca* (Cecropiaceae), a myrmecophobic plant species; 2: *Doniaciasta garambana* Jolivet (Donaciinae), from Garamba Park, Zaire, feeds on *Potamogeton* (Potamogetonaceae) and *Nymphaea* (Nymphaeaceae) and other aquatic plants without any real specificity; 3: *Sagra* sp. near *Sagra femorata* Drury (Sagrinae), from Thailand, feeds on stems of Leguminosae, mainly Fabaceae, and is gallicolous in the larval stage; 4: *Cecropia lyratiloba* var. *nana* (Cecropiaceae), from Rio, Brazil, food of *Coelomera ruficornis* Baly (Galerucinae) (4 from Carauta and Andrade, 1980; 1-3 from P. Jolivet).

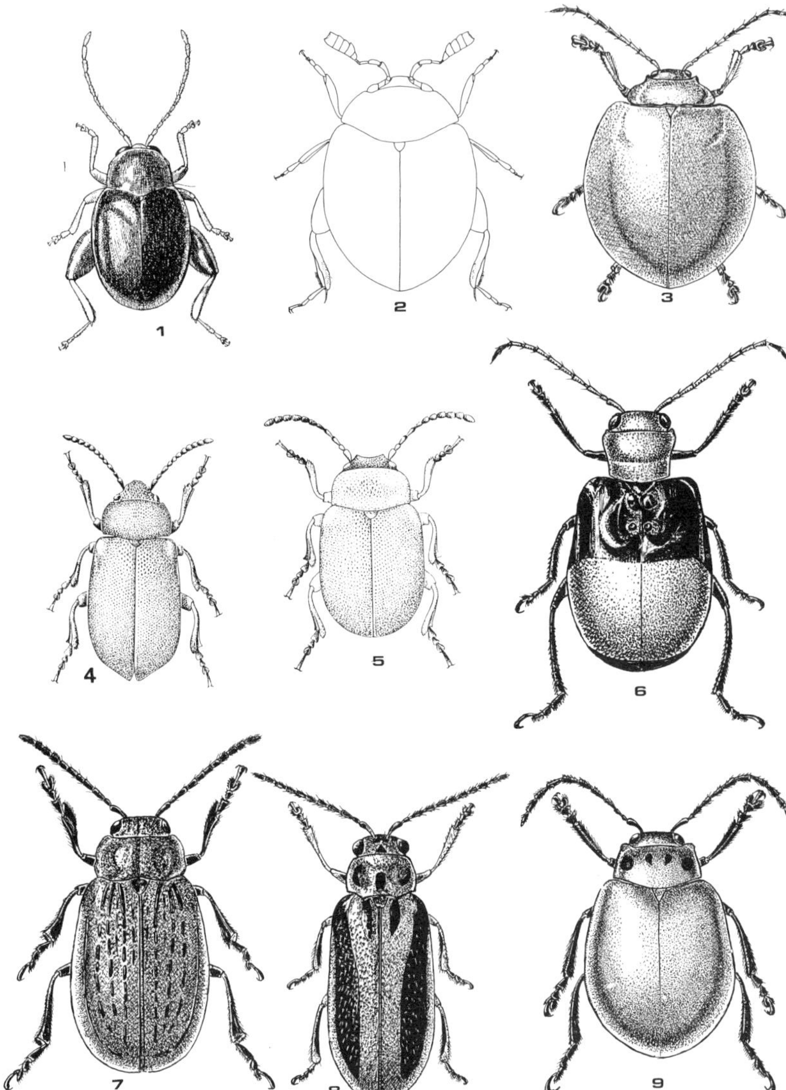

Plate 13. – 1: *Longitarsus nigripennis* Motschulsky (Alticinae) from India. The larva hollows out the pepper berries and the adult shaves the pepper leaves (*Piper nigrum,* Piperaceae) (after Maulik, 1926); 2: *Clavicornaltica besucheti* Scherer (Alticinae) from Sri Lanka. The larva feeds on roots and is totally subterranean, the adult is only partly endogenous at least for some species (after Scherer, 1974); 3: *Oides epipleuralis* Laboissière (Galerucinae) from Taiwan, feeds on *Vitis adstricta* Hance (Vitidaceae) (after Laboissière); 4: *Gastrophysa viridula* Degeer (Chrysomelinae) feeds mostly on *Rumex* (Polygonaceae) (original); 5: *Gastrophysa polygoni* Linne (Chrysomelinae) feeds mostly on *Polygonum* (Polygonaceae) (original); 6: *Pseudocophora apicalis* Laboissière (Galerucinae) from Sumatra, feeds on Cucurbitaceae (after Laboissière); 7: *Galeruca interrupta* Illiger (Galerucinae) from France, feeds on *Artemisia* (Compositae) and many other plants. Polyphagous (after Laboissière); 8: *Pyrrhalta luteola* Müller (Galerucinae) from France, feeds on *Ulmus* (Ulmaceae) and several other related trees (after Laboissière); 9: *Morphosphaera coomani* Laboissière from north Vietnam (Galerucinae), feeds on *Ficus* (Moraceae) (after Laboissière).

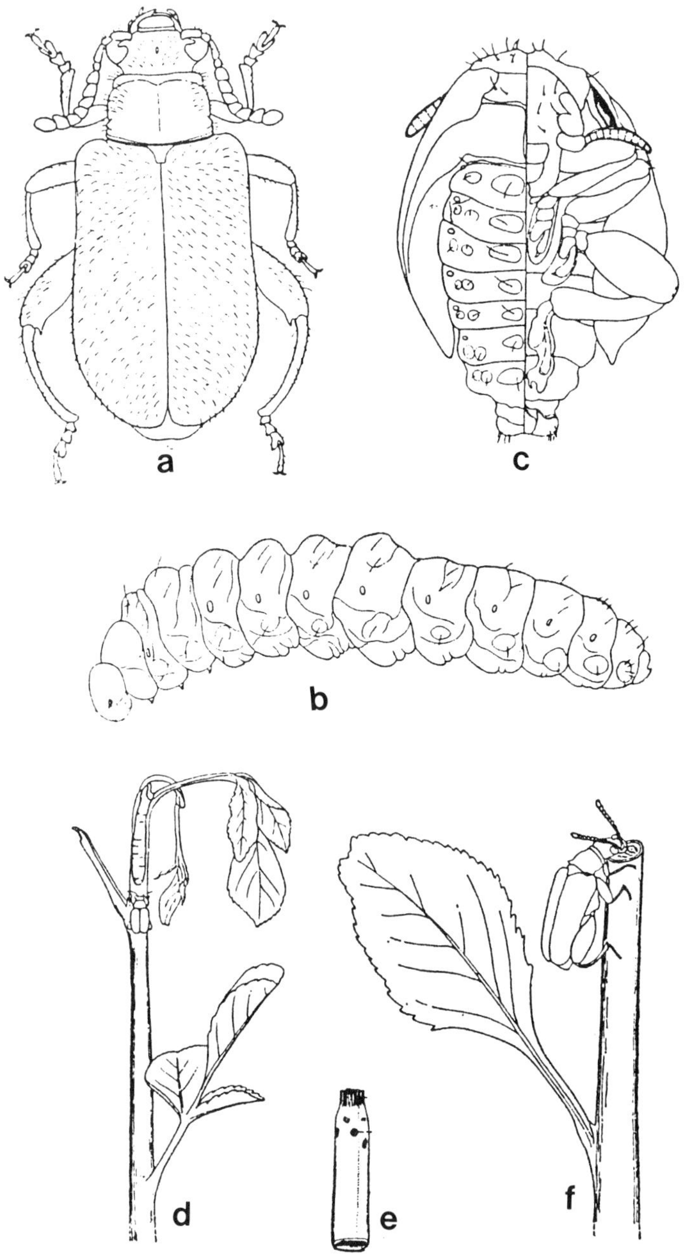

Plate 14. Biology of *Temnaspis nankinea* (Pic) (Megalopodinae) on ash tree, *Fraxinus chinensis* (Oleaceae); a: adult; b: larva; c: pupa (dorsal/ventral); d: incision of the stem; e: cut stem, with egg inside and below circular incision; f: adult of *Temnaspis* drinking sap from the scar before laying an egg inside (after P. Yu, 1977).

Plate 15. Two Chrysomelinae for which the host-plants are unknown; a: *Cyrtonastes weisei* Reitter; b: *Cyrtonastes aeneomicans* Fairmaire-species probably feed during the night on small Asteraceae (after P. Jolivet).

Higher classification system of the family Chrysomelidae (Suzuki, 1993)

Superfamily Chrysomeloidea
- (Family Cerambycidae)
- (Family Bruchidae)
- Family Chrysomelidae
  - Subfamily I. Orsodacninae
    - Tribe 1. Orsodacnini
  - Subfamily II. Megalopodinae
    - Tribe 2. Zeugophorini
    - Tribe 3. Palophagini
    - Tribe 4. Megalopodini
  - Subfamily III. Aulacoscelinae
    - Tribe 5. Aulacoscelini
  - Subfamily IV. Sagrinae
    - Tribe 6. Sagrini
    - Tribe 7. Donaciini
    - Tribe 8. Criocerini
  - Subfamily V. Synetinae
    - Tribe 9. Synetini
  - Subfamily VI. Chrysomelinae
    - Tribe 10. Chrysomelini
  - Subfamily VII. Galerucinae
    - Tribe 11. Galerucini
    - Tribe 12. Alticini
  - Subfamily VIII. Clytrinae
    - Tribe 13. Clytrini
    - Tribe 14. Cryptocephalini
    - Tribe 15. Chlamisini
    - Tribe 16. Lamprosomatini
  - Subfamily IX. Cassidinae
    - Tribe 17. Hispini
    - Tribe 18. Cassidini
  - Subfamily X. Eumolpinae
    - Tribe 19. Megascelini
    - Tribe 20. Eumolpini

Plate 16. Higher classification system of the family Chrysomelidae .

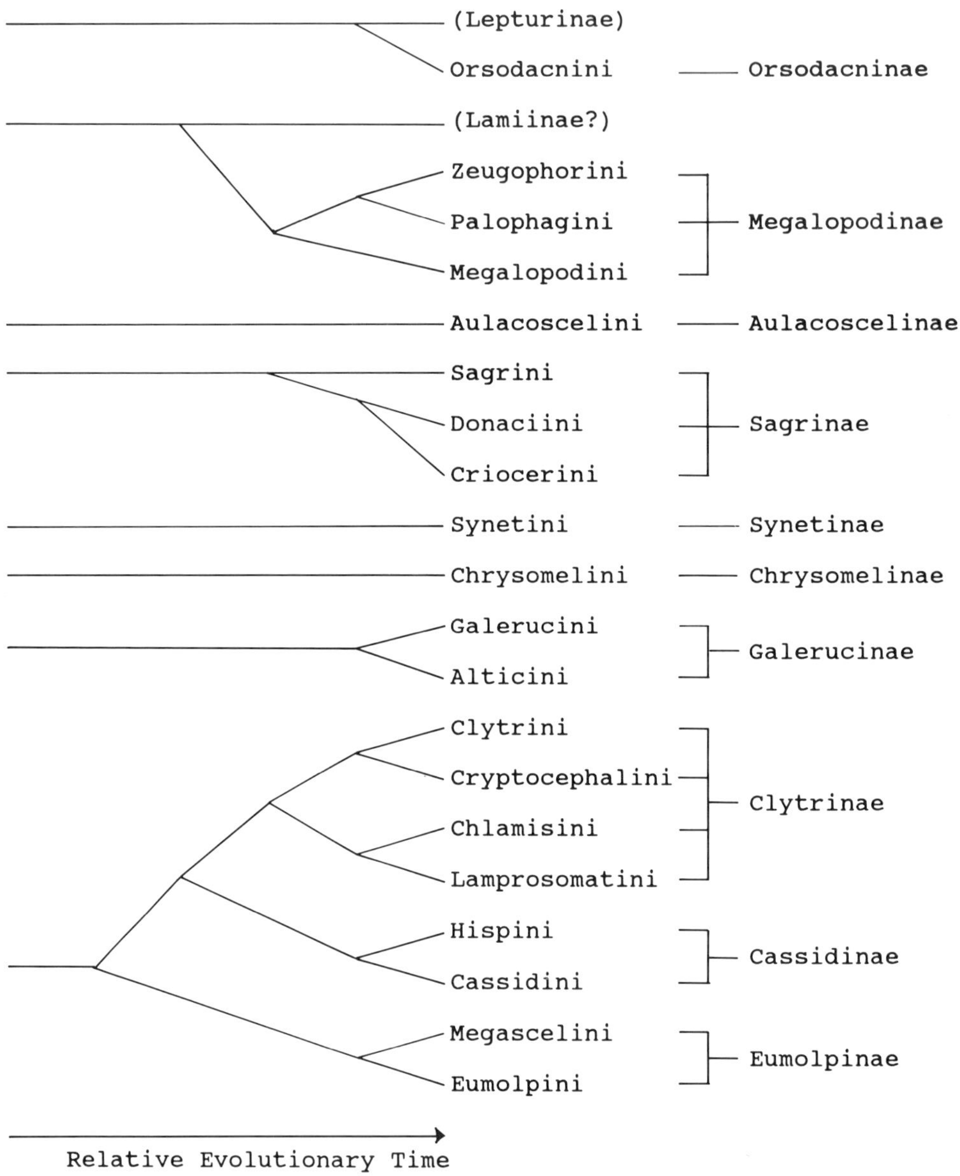

Plate 17. Cladogram of the Chrysomelidae supposed mainly from results of comparative morphology of the internal reproductive system and the hindwing venation. This is a revised version of the cladogram shown in Suzuki (1988). Readers should note the following: 1. the "Palophaginae" was placed near the Zeugophorinae, and 2. the Aulacoscelinae has been moved to between "Megalopodinae" and "Sagrinae". (After Suzuki, 1993).

INDEX TO THE CHRYSOMELIDAE

INDEX TO THE PLANTS

Families are arranged alphabetically and genera are arranged alphabetically within each family.